STUDENT SOLUTIONS MANUAL

to accompany

INTRODUCTION TO

ORDINARY DIFFERENTIAL EQUATION

THIRD EDITION

SHEPLEY L. ROSS
University of New Hampshire

JOHN WILEY & SONS   NEW YORK
CHICHESTER BRISBANE TORONTO SINGAPORE

PREFACE

This manual is a supplement to the author's text, INTRODUCTION TO ORDINARY DIFFERENTIAL EQUATIONS, Third Edition. It contains the answers to the even-numbered exercises and detailed solutions to approximately one third of all exercises, both even- and odd- numbered.

The following abbreviations have been used:

D.E.	differential equation
G.S.	general solution
I.F.	integrating factor
I.C.	initial condition
I.V.P.	initial value problem

The author expresses his thanks to Solange Abbott for typing, to David Douty for providing independent problem solutions for checking, and to his wife for help in proofreading and other tasks.

Shepley L. Ross

TABLE OF CONTENTS

ANSWERS TO EVEN NUMBERED PROBLEMS

Section 1.1, Page 5

2. ordinary; third; linear.

4. ordinary; first; nonlinear.

6. partial; fourth; linear.

8. ordinary; second; nonlinear.

10. ordinary; second; nonlinear.

Section 1.3, Page 20

2. (a) $y = (x^2 + 2)e^{-x}$; (b) $y = (x^2 + 3e^{-1})e^{-x}$.

Section 2.1, Page 35

2. $xy^2 + 3x - 4y = c$.

4. not exact.

6. $(\theta^2 + 1) \sin r = c$.

8. not exact.

10. not exact.

12. $x^3y^2 - y^3x + x^2 + y + 1 = 0$.

14. $e^xy + xy^2 + 2e^x = 8$.

16. $2x^{1/3}y^{-1/3} + 4x^{4/3}y^{1/3} = 9$.

18. (a) $A = 3$; $x^3y + 2xy^2 = c$.

    (b) $A = -2$; $x^{-2}y - x^{-1}y = c$.

20. (a) $xy^4 + 2x^3y^2 + c$; (b) $e^xy^2 + e^{3x}y^3 + c$.

22. (b) $n = -2$; (c) $x + x^2 y^{-1} = c$.

24. $4 \arc\tan \frac{x}{y} + (x^2 + y^2)^2 = c$.

Section 2.2, Page 45

2. $x(x + 2)(y + 2)^2 = c$.

4. $\sin x - \cos y = c$.

6. $(\sin u + 1)(e^v + 1) = c$.

8. $y = x \ln|cx|$.

10. $v^2 = u^2(\ln v^2 + c)$.

12. $t^3 - 3t^2 s - 3ts^2 - 2s^3 = c$.

14. $x\left(\sqrt{\dfrac{x^2 - y^2}{x^2}} + 1\right) = c$.

16. $4x - 2 \sin 2x + \tan y = \frac{\pi}{3}$.

18. $x^2 + y^2 = 5x^3$.

20. $2(3x^2 + 3xy + y^2)^2 = 9x^5$.

22. (a) $x^2 + 4xy - y^2 = c$; (b) $3x^2 - 2xy - y^2 = c$.

26. (a) $y = x \ln|cx|$; (b) $y^2 + xy = cx^3$.

Section 2.3, Page 55

2. $x^2 y + \frac{1}{x} = c$.

4. $y = 2 + ce^{-2x^2}$.

6. $(u^2 + 1)^2 v = \dfrac{3u^4}{4} + \dfrac{3u^2}{2} + c$.

8. $3(x + 2)y = x - 1 + c(x - 1)^{-2}$.

10. $xy = 1 + ce^{-y^2/2}$.

12. $2r = (\theta + \sin\theta \cos\theta + c)\cos\theta$.

14. $(1 + \sin^2 x)y = \sin x + c.$

16. $x^3 y^3 (2x^3 + c) = 1.$

18. $x^2 = 2 + ct^{-1}e^{-t}.$

20. $3y = 1 + 5e^{-x^3}.$

22. $y = -2x^2 - 3.$

24. $x = \frac{1}{5}(2e^t - \sin 2t - 2 \cos 2t).$

26. $\dfrac{1}{\sqrt{xy}} = -\frac{1}{2}x + 1.$

28. $\begin{cases} y = 5 + e^{-x}, & 0 \leq x < 10; \\ y = 1 + (4e^{10} + 1)e^{-x}, & x \geq 10. \end{cases}$

30. $\begin{cases} (x + 2)y = x^2 + 8, & 0 \leq x < 2; \\ (x + 2)y = 4x + 4, & x \geq 2 \end{cases}$

36. (b) $y = ce^{-x} + \displaystyle\sum_{k=1}^{5} \dfrac{\sin kx - k \cos kx}{1 + k^2}$

40. $\dfrac{e^{-\frac{x^2}{2}}}{y - x} = \displaystyle\int e^{-\frac{x^2}{2}}\, dx + c.$

Miscellaneous Review Exercises, Page 58

2. $x^2 y^3 - xy = c.$

4. $y = x^2/4 + c/x^2.$

6. $(e^{2x} - 2)y^2 = c.$

8. $\left(\dfrac{y + x}{y + 2x}\right)^2 = \left|\dfrac{c}{x}\right|.$

10. $y = e^{-x}[-1 + c(x + 1)].$

12. $y^2 = 1/(1 + cx^2).$

4

14. $(y - 2x)(y + x) = cx.$

16. $(x+1) \sqrt{y^2 + 4} = 4(x-1).$

18. $x^3 + x^2y^2 + 2y^3 = 21.$

20. $5(2x + y)^2 = 16(x + 2y).$

22. $y = \begin{cases} 1 - e^{-x}, & 0 \leq x < 2, \\ (e^2 - 1)e^{-x}, & x > 2. \end{cases}$

24. $y = \dfrac{\sqrt{2}\, x}{\sqrt{x^4 + 1}}$.

Section 2.4, Page 66

2. $x^2 \cos y + x \sin y = c.$

4. $x^2 + xy^{-1} + y^2 = c.$

6. $x^{2/3} y^{5/3}(x^2 - y) = c.$

8. $x - 2y + \ln|3x - y - 2| = c.$

10. $(3x + y + \frac{13}{3})^3 (x - y + 1)^2 = c.$

12. $\ln[3(x - 1)^2 + (y + 3)^2] + \dfrac{2}{\sqrt{3}} \text{ arc tan } \dfrac{y + 3}{\sqrt{3}(x-1)} = \ln 4 + \dfrac{\pi}{3\sqrt{3}}$

14. $\ln|2x + y - 1| + \dfrac{x - 2}{2x + y - 1} = 1.$

Section 3.1, Page 75

2. $y^2 + 2x^2 = c^2.$

4. $y^2(\ln|y| - \frac{1}{2}) = -x^2 + c.$

6. $x^2 + 2y^2 = c^2.$

8. $x^2y + \frac{x^4}{4} + c.$

10. $y = cx^{4/3}.$

12. $K = \frac{1}{4}.$

16. $\ln|2\sqrt{3}x^2 - xy + \sqrt{3}y^2| - \frac{6}{\sqrt{23}}$ arc tan $\frac{2\sqrt{3}y-x}{\sqrt{23}x} = c.$

Section 3.2, Page 85

2. (a) $v = 9$ ft./sec.; $x = 539.16$ ft.;

   (b) $v = 9$ ft/sec.

4. (a) $v = \frac{25}{2}(1 - e^{-\frac{t}{250}})$; (b) 12.5 ft./sec.;

   (c) 402 sec.

6. (a) 33.97 cm/sec.; (b) 88.84 cm/sec.

8. (a) 7.16 ft./sec.; (b) 4.95 sec.

10. (a) $v = 100$ tan (arc tan $10 - 0.32t$); (b) 4.60 sec.

12. 3.06 ft./sec.

14. 17.5 ft./sec.

16. 0.25

18. $v = [2gR^2/x + v_0^2 - 2gR]^{1/2}.$

Section 3.3, Page 96

2. (a) 34.5% ; (b) $t = 43.4$ min.

4. 19.8 % ; (b) 9 hrs., 58 min.

6. 40,833.

8. (a) $x = (4219)(10^6) e^{0.02(t - 1978)}$.

(b) 2410 million.

(c) 6551 million.

(d) 886 million.

(e) 48,405 million.

10. (a)

$$x = \frac{kx_0}{\lambda x_0 + (k - \lambda x_0)e^{-k(t - t_0)}}$$

12. 94,742

14. (a) 34.66 % ; (b) 0.85 years.

16. (a) 7.81 lbs.; concentration is 0.0539 lb./gal.

(b) 0.0217 lb./gal.

18. (a) 466.12 lb.; (b) 199.99 lb.

20. 4,119.65 gm.

22. 292.96 $(ft.)^3$/min.

24. 18.91 min.

Section 4.1B, Page 112

2. $y = 0$ for all real x.

4. (b) and (c). Theorem 4.2.

8. (b) $y = c_1 e^x + c_2 x e^x$. (b) $y = e^x + 3xe^x$;

$-\infty < x < \infty$.

10. (b) $y = c_1x^2 + \dfrac{c_2}{x^2}$.

    (c) $y = \frac{1}{4}x^2 + \dfrac{8}{x^2}$; Theorem 4.1; $0 < x < \infty$.

12. $y = c_1e^{-x} + c_2e^{3x} + c_3e^{4x}$.

Section 4.1D, Page 123

2. $y = c_1(x + 1) + c_2(x + 1)^3$.

4. $y = c_1x + c_2(x - 1)e^x$.

6. $y = xe^x$; $y = c_1x^2 + c_2xe^x$.

8. (b) $y = c_1e^x + c_2e^{2x}$.

    (d) $y = c_1e^x + c_2e^{2x} + 2x^2 + 6x + 7$.

Section 4.2, Page 134

2. $y = c_1e^{3x} + c_2e^{-x}$

4. $y = c_1e^{5x} + c_2e^{-\frac{1}{3}x}$.

6. $y = c_1e^{-x} + c_2e^{3x} + c_3e^{4x}$.

8. $y = (c_1 + c_2x)e^{-\frac{1}{2}x}$.

10. $y = e^{-3x}(c_1\sin 4x + c_2\cos 4x)$.

12. $y = c_1\sin \frac{x}{2} + c_2 \cos \frac{x}{2}$.

14. $y = c_1 e^{-2x} + (c_2 + c_3 x)e^{\frac{x}{2}}.$

16. $y = c_1 e^{-3x} + e^{-\frac{1}{2}x}(c_2 \sin \frac{\sqrt{7}}{2}x + c_3 \cos \frac{\sqrt{7}}{2}x).$

18. $y = (c_1 + c_2 x)\sin 2x + (c_3 + c_4 x)\cos 2x.$

20. $y = c_1 e^x + c_2 e^{2x} + (c_3 + c_4 x)e^{-x}.$

22. $y = (c_1 + c_2 x)e^{-2x} + e^{-x}(c_3 \sin \sqrt{2}x + c_4 \cos \sqrt{2}\ x).$

24. $y = c_1 + c_2 x + c_3 x^2 + c_4 x^3 + c_5 x^4.$

26. $y = -6e^{-2x} + 2e^{-5x}.$

28. $y = 2e^{-2x}.$

30. $y = 3xe^{3x/2} + 4e^{3x/2}.$

32. $y = (3 - 2x)\ e^{\frac{1}{3}x}.$

34. $y = e^{-3x}(\frac{2}{7}\sin 7x - \cos 7x).$

36. $y = e^{-x}(4 \sin 2x + 2 \cos 2x).$

38. $y = e^{-\frac{1}{2}x}(-\sin 3x + 2 \cos 3x).$

40. $y = e^{2x} - \sin 2x + \cos 2x.$

42. $y = e^x + e^{2x}(2 \sin x - \cos x).$

44. $y = (c_1 + c_2 x + c_3 x^2 + c_4 x^3 + c_5 x^4 + c_6 x^5)e^{2x}$

$\quad + e^{3x} [(c_7 + c_8 x + c_9 x^2) \sin 4x + (c_{10} + c_{11} x$

$\quad + c_{12} x^2) \cos 4x]$.

46. $y = c_1 e^{-2x} + c_2 e^{-3x} + e^x (c_3 \sin 2x + c_4 \cos 2x)$.

Section 4.3, Page 151

2. $y = c_1 e^{4x} + c_2 e^{-2x} - \frac{1}{2} e^{2x} - 3 e^{-3x}$.

4. $y = e^{-x}(c_1 \sin x + c_2 \cos x) - \frac{7}{13} \sin 4x - \frac{4}{13} \cos 4x$.

6. $y = c_1 e^{4x} + c_2 e^{-x} + 2 e^{2x} - 4x + 3$.

8. $y = e^{-x}(c_1 \sin 3x + c_2 \cos 3x) + x e^{-2x/2} + e^{-2x/10}$.

10. $y = c_1 e^{2x} + e^{-2x}(c_2 \sin x + c_3 \cos x) - 2x e^{-2x} - \frac{1}{2} e^{-2x}$.

12. $y = c_1 e^{-x} + c_2 e^{\frac{1}{2}x} + c_3 e^{\frac{3}{2}x} + x^3 + 5x^2 + 22x + 42$.

14. $y = c_1 e^{x} + c_2 e^{-2x} + x e^{x} - 2x e^{-2x} + 2x^2 + 2x + 3$.

16. $y = c_1 e^{x} + c_2 e^{-x} + c_3 e^{2x} + 3x e^{2x} - e^{3x}$.

18. $y = c_1 + c_2 x + c_3 e^{x} + c_4 e^{2x} + \frac{1}{2} e^{-x} + \frac{3}{2} x e^{2x} - \frac{1}{2} x^3 - \frac{9}{4} x^2$

20. $y = (c_1 + c_2 x)e^{x} + c_3 e^{2x} - \frac{1}{4} x^4 e^{x} - x^3 e^{x} + \frac{1}{2} x^2 e^{x}$.

22. $y = c_1 \sin 2x + c_2 \cos 2x + 3x^2 - \frac{3}{2} - 2x^2 \sin 2x$

    $- x \cos 2x.$

24. $y = c_1 e^{2x} + c_2 e^{3x} + c_3 \sin x + c_4 \cos x + \frac{1}{13}\sin 2x$

    $+ \frac{5}{13} \cos 2x + \frac{1}{4}x \sin x - \frac{1}{4}x \cos x.$

26. $y = 3e^{-x} + 2e^x + 4x - 5.$

28. $y = e^{-2x} - 2xe^{-3x} - e^{-3x}.$

30. $y = (3x - 5)e^{-3x} + 3e^{-6x}.$

32. $y = 2e^{5x}(2 \sin 2x - \cos 2x + 1).$

34. $y = e^{-2x} + 4e^{3x} - 2e^{2x} - xe^{3x}.$

36. $y = \dfrac{9e^x}{8} - \dfrac{e^{-x}}{8} + \dfrac{x^3 e^x}{2} - \dfrac{3x^2 e^x}{4} + \dfrac{3xe^x}{4}.$

38. $y = 5 \sin 2x + 6 \cos 2x - 2x \cos 2x.$

40. $y = (c_1 + c_2 x)e^x + c_3 e^{4x} - 2x^2 - 9x - 15 + 3e^{2x}.$

42. $y_p = Ae^{3x} + Be^{-3x} + Ce^{3x} \sin 3x + De^{3x} \cos 3x.$

44. $y_p = Ax^4 e^x + Bx^3 e^x + Cx^2 e^x + Dxe^x + Ee^x$

    $+ Fx^3 e^{2x} + Gx^2 e^{2x} + Hxe^{2x} + Ie^{2x}$

    $+ Jx^4 e^{3x} + Kx^3 e^{3x} + Lx^2 e^{3x}.$

46. $y_p = Ax^3 e^x + Bx^2 e^x + Cxe^x + Dx^2 e^{2x} + Exe^{2x}$

    $+ Fx^3 + Gx^2 + Hx.$

48. $y_p = Ax^4 e^{-x} + Bx^3 e^{-x} + Cx^2 e^{-x} + Dxe^{-\frac{1}{2}x} \sin \frac{\sqrt{3}}{2}x$

$+ Exe^{-\frac{1}{2}x} \cos \frac{\sqrt{3}}{2}x$.

50. $y_p = Ax^7 + Bx^6 + Cx^5 + Dx^4 + Ex^2 e^{-x} + Fxe^{-x} + Ge^{-x}$.

$+ Hxe^{-x} \sin 2x + Ixe^{-x} \cos 2x$.

52. $y_p = Ax^2 e^{\sqrt{2}x} \sin \sqrt{2}x + Bx^2 e^{\sqrt{2}x} \cos \sqrt{2}x$

$+ Cxe^{\sqrt{2}x} \sin \sqrt{2}x + Dxe^{\sqrt{2}x} \cos \sqrt{2}x$

$+ Exe^{-\sqrt{2}x} \sin \sqrt{2}x + Fxe^{-\sqrt{2}x} \cos \sqrt{2}x$.

54. $y_p = Ax \sin x \sin 2x + Bx \sin x \cos 2x$

$+ Cx \cos x \sin 2x + Dx \cos x \cos 2x$.

Section 4.4, Page 162

2. $y = c_1 \sin x$

$+ c_2 \cos x + (\sin x)[\ln|\sec x + \tan x|] - 2$.

4. $y = c_1 \sin x + c_2 \cos x + \sin x \tan x - \frac{\sec x}{2}$.

OR $y = c_1 \sin x + c_2 \cos x + (\sin x \tan x)/2$.

6. $y = c_1 \sin x + c_2 \cos x$

$- \sin x \ln|\cos x| + x \cos x$.

8. $y = e^x(c_1 \sin 2x + c_2 \cos 2x)$

$- \frac{1}{4} e^x \cos 2x[\ln|\sec 2x + \tan 2x|]$.

10. $y = c_1 e^x + c_2 x e^x - \dfrac{5x^3 e^x}{36} + \dfrac{x^3 e^x \ln|x|}{6}$.

12. $y = c_1 \sin x + c_2 \cos x + \dfrac{\tan x}{2}$

$+ \dfrac{3}{2} \cos x [\ln|\sec x + \tan x|]$.

14. $y = c_1 e^{-x} + c_2 e^{-2x} + e^{-x} \arctan e^x - \dfrac{1}{2} e^{-2x} \ln(1 + e^{2x})$.

16. $y = c_1 e^x + c_2 x e^x + \dfrac{e^x \sin^{-1} x}{4}$

$+ \dfrac{x^2 e^x \sin^{-1} x}{2} + \dfrac{3x e^x \sqrt{1 - x^2}}{4}$.

18. $y = (c_1 + c_2 x)e^x - e^x \left[\displaystyle\int \dfrac{x^2 \ln x}{e^x} dx \right]$

$+ x e^x \left[\displaystyle\int \dfrac{x \ln x}{e^x} dx \right]$.

20. $y = c_1(x + 1) + c_2(x + 1)^2 + \dfrac{1}{2}$.

22. $y = c_1 x + c_2 x e^x - x^2 - x$.

24. $y = c_1 x + c_2 (x + 1)^{-1} + x^2 - \dfrac{(2x^3 + 3x^2)(x + 1)^{-1}}{6}$.

26. $y = c_1 e^x + c_2 e^{-x} + c_3 e^{3x} - \dfrac{x^3 e^x}{12} - \dfrac{x e^x}{8}$.

Section 4.5, Page 168

2. $y = c_1 x^2 + \dfrac{c_2}{x^2}$.

4. $y = c_1 x^2 + c_2 x^2 \ln x.$

6. $y = x^2[c_1 \sin(\ln x^3) + c_2 \cos(\ln x^3)].$

8. $y = c_1 \sin(\ln x^3) + c_2 \cos(\ln x^3).$

10. $y = x^3[c_1 \sin(\ln x) + c_2 \cos(\ln x)].$

12. $y = \dfrac{c_1}{x} + \dfrac{c_2}{x^2} + c_3 x^4.$

14. $y = c_1 x^2 + c_2 x^3 + 2x - 1.$

16. $y = \dfrac{c_1}{x} + \dfrac{c_2}{x^2} + 2 \ln x - 3.$

18. $y = c_1 \sin \ln x + c_2 \cos \ln x$

$$+ \sin(\ln x) \int \frac{\cos \ln x}{1 + x} \, dx$$

$$- \cos(\ln x) \int \frac{\sin \ln x}{1 + x} \, dx.$$

20. $y = \dfrac{3}{x^2} + 2x^5.$

22. $y = -1/x + 2/x^3.$

24. $y = 5x/3 - 2x^2 + 3x^3 - 23x^4/24.$

26. $y = 4x^2 - 2x^3.$

28. $y = c_1(x + 2)^3 + c_2[\dfrac{1}{x + 2}].$

Section 5.2, Page 186

2. (a) $x = \dfrac{1}{4} \sin 8t + \dfrac{1}{3} \cos 8t.$

(b) $x = -\frac{1}{4} \sin 8t + \frac{1}{3} \cos 8t$.

(c) $x = \frac{1}{4} \sin 8t - \frac{1}{3} \cos 8t$.

4. (a) $\frac{1}{2}$ (ft.); $2\sqrt{3}$ ft./sec.

(b) $\frac{\pi}{3} + \frac{2n\pi}{4}$ sec.; $2\sqrt{3}$ ft./sec.

6. $k = 4$; $A = \frac{\sqrt{71}}{4}$.

8. (a) $\theta = c_1 \sin \sqrt{\frac{g}{\ell}} \, t + c_2 \cos \sqrt{\frac{g}{\ell}} \, t$; $A = \theta_0$;

period $= 2\pi \sqrt{\frac{\ell}{g}}$.

(b) $\frac{d\theta}{dt} = \pm \sqrt{\frac{2g}{\ell}} \sqrt{\cos \theta - \cos \theta_0}$.

Section 5.3, Page 196

2. $x = \frac{\sqrt{3}}{3} e^{-6t} \sin 2\sqrt{3} \, t$.

4. $x = -\frac{1}{12} e^{-16t} + \frac{1}{3} e^{-4t}$.

8. (a) $a = 5$.

(b) $x = \frac{1}{2} e^{-8t} + 5te^{-8t}$.

(c) $\frac{5}{8} e^{-\frac{1}{5}}$.

Section 5.4, Page 205

2. (a) $x = e^{-4t}(-\frac{21}{74} \sin 8t + \frac{11}{37} \cos 8t)$

$+ \frac{8}{37} \sin 16t - \frac{11}{37} \cos 16t$.

4. (a) $x = e^{-8t}(\frac{11}{20} \sin 8t + \frac{9}{10} \cos 8t) + \frac{7}{10} \sin 4t$

$- \frac{2}{5} \cos 4t$.

(b) amplitude: $\frac{\sqrt{65}}{10}$.

6. $x = ce^{-t} \cos (\sqrt{k-1}\, t + \phi) + x_{p_1} + x_{p_2} + x_{p_3}$,

where $x_{p_1} = - \dfrac{2}{(k-1)^2 + 4} \cos t + \dfrac{(k-1)}{(k-1)^2 + 4} \sin t$,

$x_{p_2} = \dfrac{-1}{(k-4)^2 + 16} \cos 2t + \dfrac{(k-4)}{(k-4)^2 + 16} \sin 2t$,

$x_{p_3} = - \dfrac{2}{3[(k-9)^2 + 324]} \cos 3t$

$+ \dfrac{k-9}{(k-9)^2 + 324} \sin 3t$.

10. (b) $x = \frac{\sqrt{37}}{8} e^{-2t}[\cos(4t - \phi)] + \frac{\sqrt{5}}{4} [\cos(2t - \theta)]$,

where $\phi = \cos^{-1} \frac{6}{\sqrt{37}}$ and $\theta = \cos^{-1} (- \frac{1}{\sqrt{5}})$

or $x = e^{-2t}[\frac{3}{4} \cos 4t + \frac{1}{8} \sin 4t] + \frac{1}{2} \sin 2t$

$- \frac{1}{4} \cos 2t.$

Section 5.5, Page 211

2. $a = 5 \sqrt{2 - \frac{\pi^2}{32}}$.

16

2. $i = \frac{15}{17}(\cos 200t + 4 \sin 200t - e^{-50t})$.

4. $q = \left(\frac{1}{200} + \frac{t}{2} \right) e^{-100t} - \frac{1}{200} \cos 100t$.

6. $i = e^{-200t}(1.0311 \sin 979.8t + 0.1031 \cos 979.8t)$

     $- 0.1031 e^{-100t}$.

8. (d) $\omega = 50$; ampl.: $\dfrac{100}{\sqrt{(-187.5)^2 + (20)^2}} \approx \dfrac{100}{190}$.

Section 6.1, Page 233

2. $y = c_0(1 + 2x^2 - 2x^4 + \cdots) + c_1(x - \frac{2}{3}x^3 + \frac{2}{3}x^5 + \cdots)$.

4. $y = c_0(1 + 2x^2 + \frac{1}{4}x^4 + \cdots) + c_1(x + \frac{1}{2}x^3 - \frac{1}{20}x^5 + \cdots)$.

6. $y = c_0(1 + x^2 - \frac{1}{2}x^3 + \cdots) + c_1(x + \frac{1}{2}x^3 - \frac{1}{4}x^4 + \cdots)$.

8. $y = c_0(1 + \frac{1}{3}x^3 + \frac{1}{3}x^4 + \cdots) + c_1(x + x^2 + \frac{1}{2}x^3 + \cdots)$.

10. $y = c_0(1 - \frac{1}{6}x^2 + \frac{1}{18}x^3 - \cdots) + c_1(x - \frac{1}{3}x^2 + \frac{1}{36}x^4 + \cdots)$.

12. $y = x + \frac{1}{6}x^3 - \frac{1}{120}x^5 + \cdots$.

14. $y = -1 + 5x - \frac{1}{6}x^2 + \cdots$.

16. $y = c_0[1 + \frac{1}{2}(x-1)^2 - \frac{5}{6}(x-1)^3 + \cdots]$

     $+ c_1[(x-1) - \frac{3}{2}(x-1)^2 + \frac{13}{6}(x-1)^3 - \cdots]$.

18. (a) $y = c_0[1 - \frac{n(n+1)}{2!}x^2 + \frac{n(n-2)(n+1)(n+3)}{4!}x^4 - \cdots]$

$+ c_1[x - \frac{(n-1)(n+2)}{3!}x^3$

$+ \frac{(n-1)(n-3)(n+2)(n+4)}{5!}x^5 - \cdots]$.

Section 6.2, Page 252

2. $x = 0$ and $x = -1$ are regular singular pts.

4. $x = 0$ is an irregular singular pt.;
   $x = -3$ and $x = 2$ are regular singular pts.

6. $y = c_1 x^{3/2}(1 - \frac{1}{9}x^2 + \frac{1}{234}x^4 - \cdots)$

$+ c_2 x^{-1}(1 + x^2 - \frac{1}{6}x^4 + \cdots)$.

8. $y = c_1 x^{5/3}(1 - \frac{3}{10}x^2 + \frac{9}{320}x^4 + \cdots)$

$+ c_2 x^{1/3}(1 - \frac{3}{2}x^2 + \frac{9}{32}x^4 + \cdots)$.

10. $y = c_1 x^{1/2}(1 - \frac{2}{3}x + \frac{2}{15}x^2 - \cdots)$

$+ c_2(1 - \frac{2}{3}x + \frac{4}{21}x^2 - \cdots)$.

12. $y = c_1 x^{-1}(1 - \frac{x^2}{2!} + \frac{x^4}{4!} - \cdots)$

$+ c_2 x^{-1}(x - \frac{x^3}{3!} + \frac{x^5}{5!} - \cdots)$

$= \frac{1}{x}(c_1 \cos x + c_2 \sin x)$.

14. $y = C_1 x(1 - \frac{1}{15}x^3 + \frac{1}{180}x^6 - \cdots)$

$\quad + C_2 x^{-1}(1 + \frac{1}{3}x^3 - \frac{1}{36}x^6 + \cdots)$.

16. $y = C_1 x^2(1 - \frac{1}{2}x + \frac{3}{20}x^2 - \cdots) + C_2 x^{-1}(1 - \frac{1}{2}x)$.

18. $y = c_1 x^{1/2} + c_2 x^{3/2}$.

20. $y = C_1 x^3(1 - \frac{1}{4}x^2 + \frac{5}{128}x^4 - \cdots)$

$\quad + C_2[x^{-1}(1 - \frac{1}{4}x + \frac{1}{192}x^5 - \cdots) + \frac{1}{16}y_1(x) \ln|x|]$,

where $y_1(x)$ denotes the solution of which C_1 is the coefficient.

22. $y = C_1 x^{3/2}[1 + \sum_{n=1}^{\infty} \frac{(-1)^n[3 \cdot 5 \cdot 7 \cdots (2n+1)]}{2^{n-1} \, n!(n+2)!} x^n]$

$\quad + C_2[x^{-1/2}(-\frac{1}{2} + \frac{1}{4}x - \frac{5}{64}x^2 + \cdots)$

$\quad - \frac{1}{16}y_1(x) \ln|x|]$,

where $y_1(x)$ denotes the solution of which C_1 is the coefficient.

24. $y = C_1[1 + \sum_{n=1}^{\infty} \frac{(-1)^n \, x^n}{2^{n-1} \, n!(n+2)!}]$

$\quad + C_2[x^{-2}(-\frac{1}{2} - \frac{1}{4}x + \frac{29}{576}x^2 + \cdots) + \frac{1}{16}y_1(x) \ln|x|]$,

where $y_1(x)$ denotes the solution of which C_1 is the coefficient.

26. $y = C_1 x^3 [1 + \sum\limits_{n=1}^{\infty} \dfrac{(-1)^n x^{2n}}{2^{2n-1} n!(n+2)!}\ ]$

$\quad + C_2 [x^{-1}(-\frac{1}{4} - \frac{1}{16}x^2 + \frac{29}{4608}x^4 + \cdots)$

$\quad + \frac{1}{64}y_1(x)\ \ln|x|]$,

where $y_1(x)$ denotes the solution of which C_1 is
the coefficient.

Section 7.1, Page 278

2. $\begin{cases} x = ce^{-t} - 2, \\ y = -2ce^{-t} - t^2 - 2t + 4 \end{cases}$

4. $\begin{cases} x = 3e^t \\ y = -2e^t - \frac{1}{2}e^{2t} \end{cases}$

6. $\begin{cases} x = c_1 e^{-t} + t - 1 - \frac{1}{2}e^t \\ y = c_2 e^t - \frac{5}{2}c_1 e^{-t} - \frac{1}{2}te^t - 4t + 1 \end{cases}$

8. $\begin{cases} x = c_1 e^{\sqrt{3}t} + c_2 e^{-\sqrt{3}t} + 3t - 3 \\ y = \frac{\sqrt{3}}{3}c_1 e^{\sqrt{3}t} - \frac{\sqrt{3}}{3}c_2 e^{-\sqrt{3}t} - 2t + \frac{4}{3} \end{cases}$

10. $\begin{cases} x = c_1 e^{3t} + c_2 e^t - \frac{1}{6}t - \frac{13}{18} \\ y = -c_1 e^{3t} + \frac{1}{3}c_2 e^t + \frac{1}{6}t - \frac{5}{18} \end{cases}$

12. $\begin{cases} x = e^{\frac{1}{2}t}[c_1 \sin\frac{\sqrt{23}}{2}t + c_2 \cos\frac{\sqrt{23}}{2}t] + \frac{2}{3}t^2 - \frac{7}{9}t - \frac{41}{27} \\[2mm] y = e^{\frac{1}{2}t}[(-\frac{1}{12}c_1 + \frac{\sqrt{23}}{12}c_2) \sin\frac{\sqrt{23}}{2}t + (-\frac{\sqrt{23}}{12}c_1 - \frac{1}{12}c_2) \end{cases}$

$$\cos\frac{\sqrt{23}}{2}t] + \frac{1}{3}t^2 - \frac{5}{9}t - \frac{1}{27}$$

14. $\begin{cases} x = c_1 \cos t + c_2 \sin t - t - 3, \\[2mm] y = -(c_1 + c_2) \cos t + (c_1 - c_2) \sin t - 1 \end{cases}$

16. $\begin{cases} x = c_1 e^{2t} + \frac{1}{2}t^2 + \frac{3}{2}t + \frac{3}{4} \\[2mm] y = -3c_1 e^{2t} + c_2 e^t - \frac{1}{2}t^2 + \frac{3}{2}t + \frac{15}{4} \end{cases}$

18. $\begin{cases} x = c_1 + c_2 e^t + c_3 te^t + \frac{1}{2}e^{2t} \\[2mm] y = -c_1 - c_2 e^t - c_3(t+1)e^t - \frac{1}{2}e^{2t} \end{cases}$

20. $\begin{cases} x = c_1 + c_2 e^{2t} + c_3 e^{-2t} - \frac{2}{3}e^t \\[2mm] y = -4c_1 + 2c_2 e^{2t} - 2c_3 e^{-2t} - \frac{5}{3}e^t \end{cases}$

22. $\begin{cases} x = c_1 e^{-t} + e^{-2t}(c_2 \sin t + c_3 \cos t) - \frac{4}{51}e^{2t} \\[2mm] y = \frac{1}{4}c_1 e^{-t} + e^{-2t}[(\frac{2}{5}c_2 + \frac{1}{5}c_3)\sin t \end{cases}$

$$+ (-\frac{1}{5}c_2 + \frac{2}{5}c_3) \cos t] + \frac{5}{51}e^{2t}$$

24.
$$\begin{cases} \dfrac{dx_1}{dt} = x_2 \\[2mm] \dfrac{dx_2}{dt} = x_3 \\[2mm] \dfrac{dx_3}{dt} = 2x_1 + x_2 - 2x_3 + e^{3t} \end{cases}$$

26.
$$\begin{cases} \dfrac{dx_1}{dt} = x_2 \\[2mm] \dfrac{dx_2}{dt} = x_3 \\[2mm] \dfrac{dx_3}{dt} = x_4 \\[2mm] \dfrac{dx_4}{dt} = -2tx_1 + t^2 x_3 + \cos t \end{cases}$$

Section 7.2, Page 290

2. (b)
$$\begin{cases} x = c_1 + c_2 e^{-\frac{k}{m}t} \\[3mm] y = c_3 + c_4 e^{-\frac{k}{m}t} - \dfrac{mg}{k}t \end{cases}$$

6.
$$\begin{cases} x = 10e^{-t/5} + 20 \\[2mm] y = -20e^{-t/5} + 20 \end{cases}$$

Section 7.3, Page 300

2. (b)
$$\begin{cases} x = 3c_1 e^{7t} + c_2 e^{-t} \\ y = 2c_1 e^{7t} - 2c_2 e^{-t} \end{cases}$$

(c)
$$\begin{cases} x = 3e^{7t} - 3e^{-t} \\ y = 2e^{7t} + 6e^{-t} \end{cases}$$

Section 7.4, Page 310

2.
$$\begin{cases} x = c_1 e^{4t} + c_2 e^{2t} \\ y = c_1 e^{4t} + 3c_2 e^{2t} \end{cases}$$

4.
$$\begin{cases} x = c_1 e^{t} + c_2 e^{-t} \\ y = -3c_1 e^{t} - c_2 e^{-t} \end{cases}$$

6.
$$\begin{cases} x = c_1 e^{5t} + c_2 e^{3t} \\ y = c_1 e^{5t} + 3c_2 e^{3t} \end{cases}$$

8.
$$\begin{cases} x = e^{2t}(c_1 \cos 3t + c_2 \sin 3t) \\ y = e^{2t}(3c_1 \sin 3t - 3c_2 \cos 3t) \end{cases}$$

10.
$$\begin{cases} x = c_1 e^{5t} + c_2 e^{-3t} \\ y = c_1 e^{5t} - 3c_2 e^{-3t} \end{cases}$$

12. $\begin{cases} x = -2c_1 e^{5t} - c_2(2t + 1)e^{5t} \\ y = c_1 e^{5t} + c_2 e^{5t} \end{cases}$

14. $\begin{cases} x = c_1 e^{-t} + c_2(2t + 1)e^{-t} \\ y = c_1 e^{-t} + 2c_2 t e^{-t} \end{cases}$

16. $\begin{cases} x = e^{2t}(c_1 \cos 3t - c_2 \sin 3t) \\ y = e^{2t}(c_1 \cos 3t + c_2 \sin 3t) \end{cases}$

18. $\begin{cases} x = 2e^{3t}(c_1 \cos 2t + c_2 \sin 2t) \\ y = e^{3t}[c_1(\cos 2t + \sin 2t) + c_2(\sin 2t - \cos 2t)] \end{cases}$

20. $\begin{cases} x = 5(c_1 \cos 3t + c_2 \sin 3t) \\ y = -c_1(\cos 3t + 15 \sin 3t) \\ \qquad + c_2(3 \cos 3t - 5 \sin 3t) \end{cases}$

22. $\begin{cases} x = -5e^{4t}(c_1 \cos t + c_2 \sin t) \\ y = e^{4t}[(c_2 - 2c_1) \cos t - (c_1 + 2c_2) \sin t] \end{cases}$

24. $\begin{cases} x = e^{5t} + 5e^{-3t} \\ y = 7e^{5t} - 5e^{-3t} \end{cases}$

26. $\begin{cases} x = e^{5t} - te^{5t} \\ y = 3e^{5t} - 2te^{5t} \end{cases}$

28. $\begin{cases} x = e^{4t}(5 \cos 3t - \frac{10}{3} \sin 3t) \\ y = -e^{4t}(\cos 3t + \frac{11}{3} \sin 3t) \end{cases}$

30. $\begin{cases} x = c_1 t^2 + c_2 t^4 \\ y = c_1 t^2 + 3c_2 t^4 \end{cases}$

Section 7.5A, Page 321

2. (a) $\begin{pmatrix} 3 & 6 \\ 21 & -9 \end{pmatrix}$ (b) $\begin{pmatrix} -4 & 12 & -20 \\ -24 & 8 & 0 \\ 12 & -4 & -8 \end{pmatrix}$

(c) $\begin{pmatrix} -15 & 3 & -6 \\ -12 & 9 & -6 \\ 0 & -9 & 18 \end{pmatrix}$

4. (a) $\begin{pmatrix} 2x_1 + x_2 - 4x_3 \\ 5x_1 - 2x_2 + 3x_3 \\ x_1 - 3x_2 + 2x_3 \end{pmatrix}$ (b) $\begin{pmatrix} -35 \\ 10 \\ -7 \end{pmatrix}$

(c) $\begin{pmatrix} x_1 - 2x_2 + 3x_3 \\ -3x_1 - 4x_2 - 4x_3 \\ -2x_1 + x_2 - 2x_3 \end{pmatrix}$

6. (a) (i.) $\begin{pmatrix} 10t \\ -18t^2 + 2t \\ 4t - 5 \end{pmatrix}$ (ii.) $\begin{pmatrix} 5t^3/3 \\ -3t^4/2 + t^3/3 \\ 2t^3/3 - 5t^2/2 \end{pmatrix}$

(b) (i.) $\begin{pmatrix} 3e^{3t} \\ (6t + 11)e^{3t} \\ (3t^2 + 2t)e^{3t} \end{pmatrix}$

(ii.) $\begin{pmatrix} (e^{3t} - 1)/3 \\ [(6t + 7)e^{3t} - 7]/9 \\ [(9t^2 - 6t + 2)e^{3t} - 2]/27 \end{pmatrix}$

(c) (i.) $\begin{pmatrix} 3 \cos 3t \\ -3 \sin 3t \\ 3t \cos 3t + \sin 3t \\ -3t \sin 3t + \cos 3t \end{pmatrix}$

(ii.) $\begin{pmatrix} (1 - \cos 3t)/3 \\ \sin 3t/3 \\ \sin 3t/9 - t \cos 3t/3 \\ \cos 3t/9 + t \sin 3t/3 - 1/9 \end{pmatrix}$

Section 7.5B, Page 331

2. $AB = \begin{pmatrix} -9 & 50 \\ 2 & 17 \end{pmatrix}$; $BA = \begin{pmatrix} 27 & 26 \\ -10 & -19 \end{pmatrix}$.

4. $AB = \begin{pmatrix} 6 & 20 & 11 & 21 \\ 5 & 11 & 12 & 26 \end{pmatrix}$; BA not defined.

6. $AB = \begin{pmatrix} -5 & -6 & -3 \\ -4 & -6 & 9 \\ -11 & -14 & 1 \end{pmatrix}$; $BA = \begin{pmatrix} -9 & -9 & -9 \\ -13 & -8 & -3 \\ 23 & 15 & 7 \end{pmatrix}$.

8. $AB = \begin{pmatrix} 7 & 14 & 21 \\ 5 & -4 & 6 \\ 5 & 17 & 20 \end{pmatrix}$; $BA = \begin{pmatrix} 11 & 7 & -3 \\ 29 & 19 & -8 \\ -3 & -4 & -7 \end{pmatrix}$.

10. $AB = \begin{pmatrix} 12 & 8 & -16 \\ 21 & 13 & 19 \\ -6 & 6 & -22 \end{pmatrix}$; $BA = \begin{pmatrix} 20 & 8 & 20 \\ 2 & 7 & -4 \\ 12 & -2 & -24 \end{pmatrix}$.

12. $\begin{pmatrix} 6 & 6 & 18 \\ 0 & 2 & 4 \\ -16 & -10 & -8 \end{pmatrix}$.

14. $\begin{pmatrix} 4 & -5/2 \\ 1 & -1/2 \end{pmatrix}$.

16. $\begin{pmatrix} 5/3 & 2 \\ 2/3 & 1 \end{pmatrix}$.

18. $\begin{pmatrix} -1/3 & 7/6 & -1/2 \\ -1 & 3/2 & -1/2 \\ 2/3 & -5/6 & 1/2 \end{pmatrix}$.

20. $\begin{pmatrix} 3/5 & -2/5 & 1 \\ -2/5 & 3/5 & -1 \\ -6/5 & 4/5 & -1 \end{pmatrix}$.

22. $\begin{pmatrix} 9 & -3 & 1 \\ -4 & 3/2 & -1/2 \\ 1 & -1/2 & 1/2 \end{pmatrix}$.

24. $\begin{pmatrix} 5/2 & 3 & -3/2 \\ -1/12 & 1/6 & 1/12 \\ 25/12 & 17/6 & -13/12 \end{pmatrix}$.

Section 7.5D, Page 344

2. Characteristic values: 5 and -3;

   Respective corresponding characteristic vectors:

   $$\begin{pmatrix} k \\ k \end{pmatrix} \quad \text{and} \quad \begin{pmatrix} k \\ -3k \end{pmatrix} \quad ,$$

   where in each vector k is an arbitrary nonzero real number.

4. Characteristic values: 5 and -5;

   Respective corresponding characteristic vectors:

   $$\begin{pmatrix} k \\ k \end{pmatrix} \quad \text{and} \quad \begin{pmatrix} 7k \\ -3k \end{pmatrix} \quad ,$$

   where in each vector k is an arbitrary nonzero real number.

6. Characteristic values: 7 and -2;

   Respective corresponding characteristic vectors:

   $$\begin{pmatrix} 5k \\ -4k \end{pmatrix} \quad \text{and} \quad \begin{pmatrix} k \\ k \end{pmatrix} \quad ,$$

   where in each vector k is an arbitrary nonzero real number.

8. Characteristic values: 2, 3, and 5;

Respective corresponding characteristic vectors:

$$\begin{pmatrix} 10k \\ -7k \\ -3k \end{pmatrix}, \quad \begin{pmatrix} k \\ -k \\ -k \end{pmatrix}, \quad \text{and} \quad \begin{pmatrix} k \\ -k \\ -3k \end{pmatrix},$$

where in each case k is an arbitrary nonzero real number.

10. Characteristic values: 1, 2, and -1;

Respective corresponding characteristic vectors:

$$\begin{pmatrix} k \\ 0 \\ -k \end{pmatrix}, \quad \begin{pmatrix} k \\ k \\ k \end{pmatrix}, \quad \text{and} \quad \begin{pmatrix} k \\ -2k \\ k \end{pmatrix},$$

where in each case k is an arbitrary nonzero real number.

12. Characteristic values: 1, 3, and 4;

Respective corresponding characteristic vectors:

$$\begin{pmatrix} k \\ 0 \\ k \end{pmatrix}, \quad \begin{pmatrix} 0 \\ k \\ 2k \end{pmatrix}, \quad \text{and} \quad \begin{pmatrix} 2k \\ -k \\ k \end{pmatrix},$$

where in each vector k is an arbitrary nonzero real number.

14. Characteristic values: 1, -1, and 4;

Respective corresponding characteristic vectors:

$$\begin{pmatrix} 2k \\ k \\ 0 \end{pmatrix}, \quad \begin{pmatrix} 0 \\ 3k \\ k \end{pmatrix}, \quad \text{and} \quad \begin{pmatrix} k \\ -2k \\ -k \end{pmatrix},$$

where in each vector k is an arbitrary nonzero real number.

Section 7.6B, Page 367

2. yes.

4. yes.

6. yes.

8. (b) $\begin{cases} x_1 = 3e^{5t} - e^{3t} \\ x_2 = 3e^{5t} - 3e^{3t} \end{cases}$

10. (b) $\begin{cases} x_1 = 2e^t - e^{5t} \\ x_2 = -4e^t - 2e^{5t} \end{cases}$

12. (b) $\begin{cases} x_1 = 3e^t - 2e^{3t} \\ x_2 = 3e^t - 2e^{3t} - e^{-2t} \\ x_3 = 9e^t - 4e^{3t} - e^{-2t} \end{cases}$

Section 7.6C, Page 378

2. (a) $\begin{pmatrix} 3e^t & e^{-t} \\ -e^t & -e^{-t} \end{pmatrix}$, (b) $\begin{pmatrix} 17e^{3t}/8 \\ -5e^{3t}/8 \end{pmatrix}$.

4. (a) $\begin{pmatrix} e^t & (t+1)e^t \\ 2e^t & (2t+1)e^t \end{pmatrix}$, (b) $\begin{pmatrix} 6e^{2t} + 2 \\ 8e^{2t} + 6 \end{pmatrix}$.

6. (a) $\begin{pmatrix} 5\cos 3t & 5\sin 3t \\ 3\sin 3t + \cos 3t & \sin 3t - 3\cos 3t \end{pmatrix}$,

(b) $\begin{pmatrix} 3 \sin 2t + \cos 2t \\ \sin 2t \end{pmatrix}$.

8. (a) $\begin{pmatrix} 2e^{2t} - e^{3t} + e^{4t} \\ 6e^{2t} - 4e^{3t} + 2e^{4t} \end{pmatrix}$.

Section 7.7, Page 390

2. $\begin{cases} x_1 = 10c_1 e^{2t} + c_2 e^{3t} + c_3 e^{5t} \\ x_2 = -7c_1 e^{2t} - c_2 e^{3t} - c_3 e^{5t} \\ x_3 = -3c_1 e^{2t} - c_2 e^{3t} - 3c_3 e^{5t} \end{cases}$

4. $\begin{cases} x_1 = c_1 e^{t} + c_3 e^{4t} \\ x_2 = 2c_2 e^{3t} - c_3 e^{4t} \\ x_3 = c_2 e^{3t} - c_3 e^{4t} \end{cases}$

6. $\begin{cases} x_1 = c_1 e^{t} + c_2 e^{2t} + c_3 e^{-t} \\ x_2 = c_2 e^{2t} - 2c_3 e^{-t} \\ x_3 = -c_1 e^{t} + c_2 e^{2t} + c_3 e^{-t} \end{cases}$

8. $\begin{cases} x_1 = c_1 e^{2t} + c_2 e^{t} \\ x_2 = 2c_1 e^{2t} - c_2 e^{t} + c_3 e^{t} \\ x_3 = c_1 e^{2t} + 2c_3 e^{t} \end{cases}$

10. $\begin{cases} x_1 = c_1 e^t + c_2 e^{3t} \\ x_2 = -c_1 e^t + 2c_3 e^{3t} \\ x_3 = 3c_1 e^t + c_3 e^{3t} \end{cases}$

12. $\begin{cases} x_1 = c_1 e^{28t} + c_2 e^t \\ x_2 = 2c_1 e^{28t} + c_3 e^t \\ x_3 = c_1 e^{28t} - c_3 e^t \end{cases}$

14. $\begin{cases} x_1 = 5c_1 e^{-t} + 5c_2 e^{2t} + c_3 e^{3t} \\ x_2 = 3c_1 e^{-t} + 5c_2 e^{2t} + c_3 e^{3t} \\ x_3 = -2c_1 e^{-t} - c_2 e^{2t} \end{cases}$

Section 8.2, Page 405

2. $y = 4 + 32x + 256x^2 + \dfrac{6145}{3} x^3 + \cdots$

4. $y = 3 + 27x + \dfrac{729}{2} x^2 + \dfrac{32805}{6} x^3 + \cdots$

6. $y = x + \dfrac{1}{3} x^3 + \dfrac{1}{15} x^5 + \cdots$

8. $y = 4 + 17(x-1) + 69(x-1)^2 + \dfrac{842}{3}(x-1)^3 + \cdots$

10. $y = 1 + 3(x-1) + 5(x-1)^2 + 8(x-1)^3 + \cdots$

12. $y = \dfrac{\pi}{2} + 2(x-1) + \dfrac{3}{2}(x-1)^2 - \dfrac{1}{3}(x-1)^3 + \cdots$

32

Section 8.3, Page 409

2. $\phi_1(x) = 1 + x + \dfrac{x^2}{2}$, $\phi_2(x) = 1 + x + x^2 + \dfrac{x^3}{6}$,

$\phi_3(x) = 1 + x + x^2 + \dfrac{x^3}{3} + \dfrac{x^4}{24}$

4. $\phi_1(x) = x,$ $\phi_2(x) = x + \dfrac{x^4}{4}$,

$\phi_3(x) = x + \dfrac{x^4}{4} + \dfrac{x^7}{14} + \dfrac{x^{10}}{160}$

6. $\phi_1(x) = 1 - \cos x,$

$\phi_2(x) = 1 + \dfrac{3}{2}x - \cos x - 2 \sin x + \dfrac{1}{4} \sin 2x$

8. $\phi_1(x) = x$, $\phi_2(x) = x + x^6$,

$\phi_3(x) = x + x^6 + \dfrac{24}{11}x^{11} + \dfrac{9}{4}x^{16} + \dfrac{8}{7}x^{21} + \dfrac{6}{25}x^{25}$

Section 8.4B, Page 414

2. (a) 2.200, 2.430, 2.693, 2.992

(b) 2.208, 2.447, 2.720, 3.032

(c) $y = -x - 1 + 3e^{-x}$; 2.216, 2.464, 2.750, 3.075

Section 8.4C, Page 418

2. (a) 2.724, 2.494, 2.305

(b) 2.723, 2.492, 2.303

(c) $y = 2x - 2 + 5e^{-x}$; 2.725, 2.495, 2.305

Section 8.4D, Page 423

2. (a) 2.7242, 2.4937, 2.3041

 (b) $y = 2x - 2 + 5e^{-x}$; 2.7242, 2.4937, 2.3041

Section 8.4E, Page 425

2. 2.1516

Section 9.1A, Page 434

2. $\dfrac{1}{s^2 - 1}$.

4. $\dfrac{2}{s}(2 - e^{-3s})$.

6. $(e^{-s} - e^{-2s})(\dfrac{1}{s} + \dfrac{1}{s^2})$.

Section 9.1B, Page 439

2. $\dfrac{2abs^2}{[s^2 + (a - b)^2][s^2 + (a + b)^2]}$.

4. $\dfrac{s^3 + 7a^2s}{(s^2 + a^2)(s^2 + 9a^2)}$; $\dfrac{as^2 + 3a^3}{(s^2 + a^2)(s^2 + 9a^2)}$.

6. $24/s^5$.

8. $\dfrac{3s + 11}{s^2 + 4s - 8}$.   12. $\dfrac{24bs(s^2 - b^2)}{(s^2 + b^2)^4}$.

10. $\dfrac{2b^2}{(s - a)[(s - a)^2 + 4b^2]}$.   14. $24/(s - a)^5$.

34

2. $\dfrac{-3e^{-10s}}{s}$.

4. $\dfrac{2}{s}(1 - e^{-5s})$.

6. $\dfrac{6}{s}(e^{-9s} - e^{-3s})$.

8. $\dfrac{3}{s}(3 - e^{-5s} - e^{-10s} - e^{-15s})$.

10. $\dfrac{1}{s}(4 - 4e^{-5s} + 3e^{-10s})$.

12. $3e^{-4s}(\dfrac{1}{s^2} + \dfrac{4}{s})$.

14. $\dfrac{2}{s^2}(1 - e^{-5s})$.

16. $\dfrac{e^{-2(s + 1)}}{s + 1}$.

18. $\dfrac{6}{s} - \dfrac{2e^{-s}}{s^2} + \dfrac{2e^{-3s}}{s^2}$

Section 9.2A, Page 453

2. $3 \cos 2t$.

4. $5e^{-2t}(1 - 2t)$.

6. $e^{-4t}(3 \sin 2t + \cos 2t)$.

8. $\dfrac{1}{2} - \dfrac{1}{2} \cos \sqrt{2}t + \dfrac{1}{\sqrt{2}} \sin \sqrt{2}t$.

10. $-\dfrac{13}{4} + \dfrac{5}{2}t + 4e^{-t} - \dfrac{3}{4}e^{-2t}$.

12. $\frac{7}{16}t^2e^{-t/2}$.

14. $te^{t/2}(1 + 3t/4)$.

16. $-\frac{1}{2}e^{-t/2} + \frac{3}{4}e^{3t/4}$.

18. $(18e^{2t} + 17e^{-5t})/7$.

20. $f(t) = \begin{cases} 0, & 0 < t < 2 , \\ -e^{-4(t-2)} + 2e^{2(t-2)}, & t > 2 \end{cases}$.

22. $f(t) = \begin{cases} 0, & 0 < t < 3 , \\ e^{-2(t-3)}[2\cos 3(t-3) + \frac{5}{3}\sin 3(t-3)], & t > 3 \end{cases}$.

24. $f(t) = \begin{cases} 0, & 0 < t < 3 , \\ (t-3)^2/2, & 3 < t < 8 , \\ 5t - 55/2, & t > 8 \end{cases}$.

26. $f(t) = \begin{cases} 2(\sin 3t)/3, & 0 < t < 3 , \\ 2(\sin 3t)/3 - \sin 3(t - 3), & t > 3 \end{cases}$.

28. $f(t) = \begin{cases} \cos 2t - 1, & 0 < t < 2, \\ \cos 2t - \cos 2(t-2), & t > 2 \end{cases}$.

Section 9.2B, Page 457

2. $(e^t - e^{-4t})/5$.

4. $[3 - e^{-2t}(2\sin 3t + 3\cos 3t)]/39$.

6. $(2 \sin t - \cos t + e^{-2t})/5$.

Section 9.3, Page 469

2. $y = \sin t - \cos t$.

4. $y = (15e^{3t} + 13e^{-4t})/7$.

6. $y = e^{-t}(3 \sin 2t + 2 \cos 2t)$.

8. $y = -e^{-t} + 2te^{-t} + 2e^{-2t} + te^{-2t}$.

10. $y = 3e^{3t} - 2e^{5t} + 4e^{2t} + 3te^{2t}$.

12. $y = -4e^{t} + 14e^{3t} + 6te^{4t} - 11e^{4t}$.

14. $y = \begin{cases} 1 - 3e^{-2t} + 2e^{-3t} & , \ 0 < t < 2 \ , \\ 3(e^4 - 1)e^{-2t} + 2(1 - e^6)e^{-3t} & , \ t > 2 \ . \end{cases}$

16. $y = \begin{cases} \frac{3}{8} + \frac{3}{4}e^{-2t} - \frac{1}{8}e^{-4t} \ , \ 0 < t < 2\pi \ , \\ \frac{3}{4}e^{-2t}(1 + e^{4\pi}) - \frac{1}{8}e^{-4t}(1 + 3e^{8\pi}), \ t > 2\pi \ . \end{cases}$

18. $y = \begin{cases} 2 \sin t + 2 \cos t + t, \ 0 < t < \pi \ , \\ \sin t + 2 \cos t + \pi, \ t > \pi \ . \end{cases}$

Section 9.4, Page 473

2. $\begin{cases} x = 2 + 2e^{t} - e^{2t} \ , \\ y = e^{t} - e^{2t} \ . \end{cases}$

4. $\begin{cases} x = -3t - e^{-t} \ , \\ y = 2t - 1 + e^{-t} \ . \end{cases}$

6. $\begin{cases} x = e^{2t} + 2 \cos 2t - \sin 2t, \\ y = 2e^{2t} + 5 \sin 2t \ . \end{cases}$

8. $\begin{cases} x = 1 - t + 3e^{4t} - e^{-2t} \ , \\ y = t - 3e^{4t} - e^{-2t} \ . \end{cases}$

10. $\begin{cases} x = 2e^{t} - e^{2t} - 1 \ , \\ y = e^{t} - 2 \ . \end{cases}$

CHAPTER 1

Section 1.2, Page 11

1. (d) We must show that $f(x) = (1 + x^2)^{-1}$
satisfies the D.E. $(1 + x^2)y'' + 4xy' + 2y = 0$.
Differentiating $f(x)$, we find $f'(x) = -(1 + x^2)^{-2}(2x)$
and $f''(x) = (6x^2 - 2)(1 + x^2)^{-3}$. We now substitute
$f(x)$ for y, $f'(x)$ for y', and $f''(x)$ for y'' in the
stated D.E. We obtain
$(1+x^2)(6x^2-2)(1+x^2)^{-3} - 4x(1+x^2)^{-2}(2x) + 2(1+x^2)^{-1} = 0$.
This reduces to
$$(6x^2-2-8x^2)(1+x^2)^{-2} + 2(1+x^2)^{-1} = 0 \quad,$$
and hence to
$$[(-2x^2-2) + (2+2x^2)](1+x^2)^{-2} = 0,$$
that is, $0(1 + x^2)^{-2} = 0$ or $0 = 0$. Hence the given
D.E. is satisfied by $f(x) = (1 + x^2)^{-1}$.

2. (a) We must show that the relation $x^3+3xy^2 = 1$
defines at least one real function which is an explicit
solution of the given D.E. on $0 < x < 1$. Solving the
given relation for y, we obtain

38

$$y^2 = \frac{1-x^3}{3x} \quad , \quad y = \pm[\frac{1-x^3}{3x}]^{1/2} \qquad (1)$$

We choose the plus sign and consider the function f defined by

$$f(x) = [\frac{1-x^3}{3x}]^{1/2} \quad , \quad 0 < x < 1 \quad .$$

We note that $(1 - x^3)/3x > 0$ for $0 < x < 1$, and hence $f(x)$ is indeed defined on $0 < x < 1$. Now differentiating and simplifying, we find

$$f'(x) = -\frac{1}{6}[\frac{1-x^3}{3x}]^{-1/2} [\frac{2x^3+1}{x^2}] \quad .$$

Substituting $f(x)$ for y and $f'(x)$ for y' in the given D.E., we find

$$2x[\frac{1-x^3}{3x}]^{1/2} \left\{ -\frac{1}{6}[\frac{1-x^3}{3x}]^{-1/2}[\frac{2x^3+1}{x^2}] \right\} + x^2 + \frac{1-x^3}{3x} = 0.$$

This simplifies to $-\frac{2x^3+1}{3x} + x^2 + \frac{1-x^3}{3x} = 0$ and thence to $\frac{0}{3x} = 0$. Thus $f(x)$ is an explicit solution of the D.E. on $0 < x < 1$, and so the given relation $x^3 + 3xy^2 = 1$ is an implicit solution on $0 < x < 1$. We note that choosing the minus sign in (1) would also have led to an explicit solution of the D.E. on $0 < x < 1$.

    5. (a) To determine the values of m for which

$f(x) = e^{mx}$ satisfies the given D.E., we differentiate
$f(x)$ the required number of times and substitute into
the D.E. We have $f(x) = e^{mx}$, $f'(x) = me^{mx}$,
$f''(x) = m^2 e^{mx}$, $f'''(x) = m^3 e^{mx}$. Then substituting
$f(x)$ for y, $f'(x)$ for y', $f''(x)$ for y'', and $f'''(x)$
for y''' , we find $m^3 e^{mx} - 3m^2 e^{mx} - 4m e^{mx} + 12 e^{mx} = 0$
or $e^{mx}(m^3 - 3m^2 - 4m + 12) = 0$.
Then since $e^{mx} \neq 0$ for all m and x, we must have
$m^3 - 3m^2 - 4m + 12 = 0$. That is, if $f(x) = e^{mx}$ is a
solution of the given D.E., then the constant m must
satisfy the cubic equation $m^3 - 3m^2 - 4m + 12 = 0$. By
inspection we observe that $m = 2$ is a root, and so
$(m - 2)$ is a factor of the left member. We now use
synthetic division to find the other factor. We have

$$
\begin{array}{r|rrrr}
2 & 1 & -3 & -4 & 12 \\
  &   & 2 & -2 & -12 \\
\hline
  & 1 & -1 & -6 & 0 \\
\end{array}
$$

From this we see that the reduced quadratic factor is
$m^2 - m - 6$. Hence the cubic equation may be written
$(m-2)(m^2 - m - 6) = 0$ and so $(m-2)(m-3)(m+2) = 0$.
Thus we see that its roots are $m = 2, 3, -2$. These
then are the values of m for which $f(x) = e^{mx}$ is a
solution of the given D.E.

6. (b) From $f(x) = 3e^{2x} - 2xe^{2x} - \cos 2x$, we
find $f'(x) = 4e^{2x} - 4xe^{2x} + 2 \sin 2x$,
$f''(x) = 4e^{2x} - 8xe^{2x} + 4 \cos 2x$. We substitute $f(x)$
for y, $f'(x)$ for y', $f''(x)$ for y'' in the given D.E.,
obtaining

$$(4e^{2x} - 8xe^{2x} + 4 \cos 2x)$$
$$-4(4e^{2x} - 4xe^{2x} + 2 \sin 2x)$$
$$+4(3e^{2x} - 2xe^{2x} - \cos 2x) = -8 \sin 2x.$$

Collecting like terms in the left member, we find

$$(4 - 16 + 12)e^{2x} + (-8 + 16 - 8)xe^{2x}$$
$$+ (4 - 4) \cos 2x - 8 \sin 2x = -8 \sin 2x$$

or $-8 \sin 2x = -8 \sin 2x$.

Thus $f(x)$ satisfies the D.E. Now note that
$f(0) = 3e^{0} - 2(0)e^{0} - \cos 0 = 3 - 0 - 1 = 2$ and
$f'(0) = 4e^{0} - 4(0)e^{0} + 2 \sin 0 = 4 - 0 + 0 = 4$.
Hence $f(x)$ also satisfies the stated conditions.

Section 1.3, Page 20

2. (a) We apply the I.C. $y(0) = 2$ to the given
family of solutions. That is, we let $x = 0$, $y = 2$ in
$y = (x^2 + c)e^{-x}$. We obtain $2 = (0 + c)e^{0}$ and hence
$c = 2$. We thus obtain the particular solution
$y = (x^2 + 2)e^{-x}$ satisfying the stated I.V.P.

3. (a) We first apply the I.C. $y(0) = 5$ to the

given family of solutions. That is, we let $x = 0$,

$y = 5$ in $y = c_1 e^{4x} + c_2 e^{-3x}$. We obtain $c_1 + c_2 = 5$.

We next differentiate the given family to obtain

$y' = 4c_1 e^{4x} - 3c_2 e^{-3x}$. We apply the I.C. $y'(0) = 6$ to

this derived relattion. That is, we let $x = 0$, $y' = 6$

in $y' = 4c_1 e^{4x} - 3c_2 e^{-3x}$. We obtain $4c_1 - 3c_2 = 6$.

The two equations

$$\begin{cases} c_1 + c_2 = 5, \\ 4c_1 - 3c_2 = 6 \end{cases} \quad \text{determine } c_1 \text{ and } c_2 \text{ uniquely.}$$

Solving this system, we find $c_1 = 3$, $c_2 = 2$.

Substituting these values back into $y = c_1 e^{4x} + c_2 e^{-3x}$

we obtain the particular solution

$y = 3e^{4x} + 2e^{-3x}$ satisfying the stated I.V.P.

    5. We are given that every solution of the stated

D.E. may be written in the form

$$y = c_1 x + c_2 x^2 + c_3 x^3 \tag{1}$$

for some choice of the constants c_1, c_2, c_3. We must

determine these constants so that (1) will satisfy the

three stated conditions. We differentiate (1) twice to

obtain

$$y' = c_1 + 2c_2 x + 3c_3 x^2 \tag{2}$$

and

$$y'' = 2c_2 + 6c_3 x \tag{3}$$

We now apply the condition $y(2) = 0$ to (1), letting
$x = 2$, $y = 0$ in (1). We obtain $2c_1 + 4c_2 + 8c_3 = 0$.
Similarly, we apply the condition $y'(2) = 2$ to (2),
thereby obtaining $c_1 + 4c_2 + 12c_3 = 2$. Finally, we
apply the condition $y''(2) = 6$ to (3), obtaining
$2c_2 + 12c_3 = 6$. Thus we have the three equations

$$\begin{cases} 2c_1 + 4c_2 + 8c_3 = 0, \\ c_1 + 4c_2 + 12c_3 = 2, \\ 2c_2 + 12c_3 = 6, \end{cases}$$

in the three unknowns. These can be solved in various
ways. One easy way is to eliminate c_1 from the first
two equations, obtaining the equivalent of $c_2 + 4c_3 = 1$.
Combining this last with $c_2 + 6c_3 = 3$, which is equivalent to the third equation of the system, we readily
find $c_2 = -3$, $c_3 = 1$. Then from the second equation
one finds $c_1 = 2$. Thus $c_1 = 2$, $c_2 = -3$, $c_3 = 1$.
Substituting these values back into (1), we find the
solution of the stated I.V.P. is

$$y = 2x - 3x^2 + x^3 \quad .$$

CHAPTER 2

Section 2.1, Page 35

In these solutions we denote derivatives by primes and partial derivatives by subscripts. The solutions of Exercises 3 and 7 follow the pattern of Example 2.5 on page 30.

3. Here $M(x,y) = 2xy + 1$, $N(x,y) = x^2 + 4y$. From these we find $M_y(x,y) = 2x = N_x(x,y)$, so the D.E. is exact. We seek $F(x,y)$ such that

$$F_x(x,y) = M(x,y) = 2xy + 1 \text{ and } F_y(x,y) = N(x,y) = x^2 + 4y$$

From the first of these, we find

$$F(x,y) = \int M(x,y)\partial x + \phi(y) = \int (2xy + 1)\partial x + \phi(y)$$
$$= x^2 y + x + \phi(y). \text{ From this,}$$
$$F_y(x,y) = x^2 + \phi'(y).$$

But we must have $F_y(x,y) = N(x,y) = x^2 + 4y$. Therefore

$$x^2 + \phi'(y) = x^2 + 4y$$

or $\dfrac{d\phi(y)}{dy} = 4y$. Then $\phi(y) = 2y^2 + c_0$. Thus

$F(x,y) = x^2 y + x + 2y^2 + c_0$. The one-param. family of

solutions $F(x,y) = c_1$ is $\quad x^2 y + x + 2y^2 = c$, where

$c = c_1 - c_0$.

44

Alternatively, by the method of grouping, we first write the D.E. in the form $2xy\ dx + x^2\ dy + dx + 4y\ dy = 0$. We recognize this as $d(x^2y) + d(x) + d(2y^2) = d(c)$ or $d(x^2y + x + 2y^2) = d(c)$. Hence we obtain the solution $x^2y + x + 2y^2 = c$.

4. Here $M(x,y) = 3x^2y + 2$, $N(x,y) = -(x^3 + y)$. From these we find $M_y(x,y) = 3x^2 \neq -3x^2 = N_x(x,y)$. Since $M_y(x,y) \neq N_x(x,y)$, the D.E. is not exact.

7. Here $M(x,y) = y\sec^2 x + \sec x \tan x$, $N(x,y) = \tan x + 2y$. From these we find $M_y(x,y) = \sec^2 x = N_x(x,y)$, so the D.E. is exact. We seek $F(x,y)$ such that $F_x(x,y) = M(x,y) = y\sec^2 x + \sec x \tan x$ and $F_y(x,y) = N(x,y) = \tan x + 2y$. From the first of these, we find $F(x,y) = \int M(x,y)\partial x + \phi(y)$

$= \int (y\sec^2 x + \sec x \tan x)\partial x + \phi(y) =$ $y\tan x + \sec x + \phi(y)$. From this, $F_y(x,y) = \tan x + \phi'(y)$. But we must have $F_y(x,y) = N(x,y) = \tan x + 2y$. Therefore, $\tan x + \phi'(y) = \tan x + 2y$ or $\phi'(y) = 2y$. Then $\phi(y) = y^2 + c_0$. Thus $F(x,y) = y\tan x + \sec x + y^2 + c_0$. The one-parameter family of solutions $F(x,y) = c_1$ is $y\tan x + \sec x + y^2 = c$, where $c = c_1 - c_0$.

Alternatively, by the method of grouping, we first write the D.E. in the form $(y \sec^2 x \, dx + \tan x \, dy) + \sec x \tan x \, dx + 2y \, dy = 0$. We recognize this as $d(y \tan x) + d(\sec x) + d(y^2) = d(c)$, and hence obtain the solution $y \tan x + \sec x + y^2 = c$.

The solutions of Exercises 13 and 16 follow the pattern of Example 2.6 on page 31.

13. Here $M(x,y) = 2y \sin x \cos x + y^2 \sin x$, $N(x,y) = \sin^2 x - 2y \cos x$. From these we find $M_y(x,y) = 2 \sin x \cos x + 2y \sin x = N_x(x,y)$, so D.E. is exact. We first seek $F(x,y)$ such that $F_x(x,y) = M(x,y) = 2y \sin x \cos x + y^2 \sin x$ and $F_y(x,y) = N(x,y) = \sin^2 x - 2y \cos x$. From the first of these, we have

$$F(x,y) = \int M(x,y) \, \partial x + \phi(y) =$$
$$\int (2y \sin x \cos x + y^2 \sin x) \partial x + \phi(y) =$$

$y \sin^2 x - y^2 \cos x + \phi(y)$. From this, $F_y(x,y) = \sin^2 x - 2y \cos x + \phi'(y)$. But we must have $F_y(x,y) = N(x,y) = \sin^2 x - 2y \cos x$. Therefore, $\sin^2 x - 2y \cos x + \phi'(y) = \sin^2 x - 2y \cos x$ or $\phi'(y) = 0$. Then $\phi(y) = c_0$. Thus $F(x,y) = y \sin^2 x - y^2 \cos x + c_0$. The one-parameter family of solutions $F(x,y) = c_1$ is

$$y \sin^2 x - y^2 \cos x = c, \qquad\qquad (*)$$

where $c = c_1 - c_0$.

Applying the I.C. $y(0) = 3$, we let $x = 0$, $y = 3$ in
($*$), obtaining $3 \sin^2 0 - 9 \cos 0 = c$, from which
$c = -9$. Thus the particular solution of the stated I.V.
problem is $y \sin^2 x - y^2 \cos x = -9$ or

$$y^2 \cos x - y \sin^2 x = 9. \qquad (**)$$

Alternatively, the one-parameter family of solutions
($*$) could also be found by the method of grouping. To
do so, we first write the D.E. in the form
$(2y \sin x \cos x \, dx + \sin^2 x \, dy) +$
$(y^2 \sin x \, dx - 2y \cos x \, dy) = 0$. We recognize this as
$d(y \sin^2 x) + d(-y^2 \cos x) = d(c)$ and hence again obtain
the solution ($*$) in the form $y \sin^2 x - y^2 \cos x = c$.
The I.C. again yields the particular solution ($**$).

16. Here $M(x,y) = x^{-2/3} y^{-1/3} + 8x^{1/3} y^{1/3}$,
$N(x,y) = 2x^{4/3} y^{-2/3} - x^{1/3} y^{-4/3}$. From these we find
$M_y(x,y) = -\frac{1}{3} x^{-2/3} y^{-4/3} + \frac{8}{3} x^{1/3} y^{-2/3} = N_x(x,y)$, so
the D.E. is exact. We first seek $F(x,y)$ such that
$F_x(x,y) = M(x,y) = x^{-2/3} y^{-1/3} + 8x^{1/3} y^{1/3}$ and $F_y(x,y) =$
$N(x,y) = 2x^{4/3} y^{-2/3} - x^{1/3} y^{-4/3}$. From the first of
these, we have $F(x,y) = \int M(x,y) \partial x + \phi(y) =$
$\int (x^{-2/3} y^{-1/3} + 8x^{1/3} y^{1/3}) \partial x + \phi(y) =$
$3x^{1/3} y^{-1/3} + 6x^{4/3} y^{1/3} + \phi(y)$. From this, $F_y(x,y) =$
$-x^{1/3} y^{-4/3} + 2x^{4/3} y^{-2/3} + \phi'(y)$. But we must have

$F_y(x,y) = N(x,y) = 2x^{4/3}y^{-2/3} - x^{1/3}y^{-4/3}$. Thus

$-x^{1/3}y^{-4/3} + 2x^{4/3}y^{-2/3} + \phi'(y) = 2x^{4/3}y^{-2/3} - x^{1/3}y^{-4/3}$

or $\phi'(y) = 0$. Then $\phi(y) = c_0$. Thus $F(x,y) =$

$3x^{1/3}y^{-1/3} + 6x^{4/3}y^{1/3}$. The one-parameter family of

solutions $F(x,y) = c_1$ is $3x^{1/3}y^{-1/3} + 6x^{4/3}y^{1/3} = c_2$,

where $c_2 = c_1 - c_0$. We can simplify this slightly by

dividing through by 3 and replacing $c_2/3$ by c, thus

obtaining $\qquad x^{1/3}y^{-1/3} + 2x^{4/3}y^{1/3} = c.$ $\qquad$ (*)

Applying the I.C. $y(1) = 8$, we let $x = 1$, $y = 8$ in

(*) to obtain $\frac{1}{2} + 4 = c$, from which $c = 9/2$. Thus the

particular solution of the stated I.V. problem is

$x^{1/3}y^{-1/3} + 2x^{4/3}y^{1/3} = 9/2$ or

$$2x^{1/3}y^{-1/3} + 4x^{4/3}y^{1/3} = 9. \qquad (**)$$

Alternatively, the one-parameter family of solutions (*) could also be found by the method of grouping.
To do so, we first write the D.E. in the form

$(x^{-2/3}y^{-1/3} dx - x^{1/3}y^{-4/3} dy) +$

$(8x^{1/3}y^{1/3} dx + 2x^{4/3}y^{-2/3} dy) = 0$. We recognize

this as $d(3x^{1/3}y^{-1/3}) + d(6x^{4/3}y^{1/3}) = d(c_2)$ and hence

obtain the solution $3x^{1/3}y^{-1/3} + 6x^{4/3}y^{1/3} = c_2$. Once

again, this quickly reduces to (*), and the I.C. again

yields the particular solution (**).

21. Here $M(x,y) = 4x + 3y^2$, $N(x,y) = 2xy$.

(a) Since $M_y(x,y) = 6y \neq 2y = N_x(x,y)$, the D.E. is not exact.

(b) We multiply the given equation through by x^n to obtain $(4x^{n+1} + 3x^n y^2)dx + 2x^{n+1}y \, dy = 0$. For this equation, we have $M(x,y) = 4x^{n+1} + 3x^n y^2$, $N(x,y) = 2x^{n+1}y$. For this equation to be exact, we must have $M_y(x,y) = 6x^n y = 2(n+1) x^n y = N_x(x,y)$, and hence $6 = 2(n+1)$, from which $n = 2$. Thus an I.F. of the form x^n is x^2.

(c) We multiply the given equation through by the I.F. x^2, obtaining $(4x^3 + 3x^2 y^2)dx + 2x^3 y \, dy = 0$. Here $M(x,y) = 4x^3 + 3x^2 y^2$, $N(x,y) = 2x^3 y$. Since $M_y(x,y) = 6x^2 y = N_x(x,y)$, this D.E. is indeed exact. We seek $F(x,y)$ such that $F_x(x,y) = M(x,y) = 4x^3 + 3x^2 y^2$ and $F_y(x,y) = N(x,y) = 2x^3 y$. From the first of these, $F(x,y) = \int M(x,y) \partial x + \phi(y) = \int (4x^3 + 3x^2 y^2)\partial x + \phi(y) = x^4 + x^3 y^2 + \phi(y)$. From this $F_y(x,y) = 2x^3 y + \phi'(y)$. But we must have $F_y(x,y) = N(x,y) = 2x^3 y$, so $\phi'(y) = 0$. Then $\phi(y) = c_0$. Thus $F(x,y) = x^4 + x^3 y^2 + c_0$. The one-parameter family of solutions $F(x,y) = c_1$ is $x^4 + x^3 y^2 = c$, where $c = c_1 - c_0$. Alternatively, by the __method of grouping__, we first write the D.E. $(4x^3 + 3x^2 y^2)dx + 2x^3 y \, dy = 0$ in the form $4x^3 \, dx + (3x^2 y^2 \, dx + 2x^3 y \, dy) = 0$. We recognize this as $d(x^4) + d(x^3 y^2) = d(c)$, and hence we obtain the solution $x^4 + x^3y^2 = c$.

23.   (a) Here $M(x,y) = y + x f(x^2 + y^2)$ and $N(x,y) = y f(x^2 + y^2) - x$. Since $M_y(x,y) = 1 + 2xy f'(x^2 + y^2) \neq 2xy f'(x^2 + y^2) - 1 = N_x(x,y)$, the given D.E. is not exact.

(b) We multiply the given D.E. through by $1/(x^2 + y^2)$ to obtain

$$\left[\frac{y}{x^2+y^2} + \frac{xf(x^2+y^2)}{x^2 + y^2}\right] dx + \left[\frac{yf(x^2+y^2)}{x^2 + y^2} - \frac{x}{x^2+y^2}\right] dy = 0.$$

For this equation, we have

$$M(x,y) = \frac{y}{x^2+y^2} + \frac{xf(x^2+y^2)}{x^2 + y^2} \text{ and } N(x,y) = \frac{yf(x^2+y^2)}{x^2 + y^2} - \frac{x}{x^2+y^2}$$

From these we find

$$M_y(x,y) = \frac{2xy(x^2+y^2)f'(x^2+y^2)-2xyf(x^2+y^2)+x^2-y^2}{(x^2 + y^2)^2} = N_x(x,y)$$

So the D.E. of part (b) is exact, and hence $1/(x^2+y^2)$ is an I.F. of the given D.E.

24.   Applying Exercise 23(a) with $f(x^2+y^2) = (x^2 + y^2)^2$, we see that the given D.E. is not exact. By Exercise 23(b), we know that $1/(x^2+y^2)$ is an I.F. of the given D.E.   Hence we multiply the given D.E. through by $1/(x^2 + y^2)$ to obtain the equivalent D.E.

$$[\frac{-y}{x^2+y^2} + x(x^2+y^2)] \, dx + [y(x^2+y^2) - \frac{x}{x^2+y^2}]dy = 0,$$

which is therefore exact. Here $M(x,y) =$
$y/(x^2+y^2) + x(x^2+y^2)$, $N(x,y) = y(x^2+y^2) - x/(x^2+y^2)$,
$M_y(x,y) = (x^2-y^2)/(x^2+y^2)^2 + 2xy = N_x(x,y)$. We seek
$F(x,y)$ such that $F_x(x,y) = M(x,y) = y/(x^2+y^2) + x(x^2+y^2)$
and $F_y(x,y) = N(x,y) = y(x^2+y^2) - x/(x^2+y^2)$. From the
first of these, we find $F(x,y) = \int M(x,y)\partial x + \phi(y)$
$= \int [y/(x^2+y^2) + x(x^2+y^2)]\partial x + \phi(y)$
$=$ arc tan $(x/y) + x^4/4 + x^2y^2/2 + \phi(y)$. From this,
$F_y(x,y) = -x/(x^2+y^2) + x^2y + \phi'(y)$. But we must have
$F_y(x,y) = N(x,y) = x^2y + y^3 - x/(x^2+y^2)$. Therefore,
$\phi'(y) = y^3$ or $\frac{d\phi(y)}{dy} = y^3$. Then

$\phi(y) = y^4/4 + c_0$. Thus $F(x,y) =$ arc tan $(x/y) + x^4/4 +$
$x^2y^2/2 + y^4/4 + c_0$, or more simply, $F(x,y) =$
arc tan $(x/y) + (x^2+y^2)^2/4 + c_0$. The one-parameter
family of solutions $F(x,y) = c$ is arc tan$(x/y) +$
$(x^2 + y^2)^2/4 = c$, where $c = c_1 - c_0$.

Section 2.2, Page 45

The equations in Exercises 1 - 7 and 15 - 17 are
separable, and those in Exercises 8 - 14 and 18-20 are
homogeneous.

3. The D.E. is separable. We first separate

variables to obtain $2r\ dr/(r^4 + 1) + ds/(s^2 + 1) = 0$.
Next we integrate. By a well-known formula, $\int ds/(s^2+1)$
$=$ arc tan s. We next apply the same formula with $s = r^2$,
$ds = 2r\ dr$. Thus we obtain $\int 2r\ dr/(r^4 + 1) =$
arc tan r^2. Hence we find the one-parameter family of
solutions in the form

$$\text{arc tan } r^2 + \text{arc tan } s = \text{arc tan } c, \quad (*)$$

(where we write arc tan c for the arbitrary constant,
since each term on the left is an arc tan). We could
leave the solutions in this form, but they are unwieldy.
We take the tangent of each side of ($*$), applying the
formula $\tan(A+B) = \dfrac{\tan A + \tan B}{1-\tan A \tan B}$ with $A =$ arc tan r^2
and $B =$ arc tan s, to the left member. We obtain

$$\frac{\tan(\text{arc tan } r^2) + \tan(\text{arc tan } s)}{1-\tan(\text{arc tan } r^2)\tan(\text{arc tan } s)} = \tan(\text{arc tan } c),$$

which reduces to

$$\frac{r^2 + s}{1 - r^2 s} = c \text{ or } r^2 + s = c(1 - r^2 s) .$$

7. This equation is separable. We first separate
variables to obtain

$$(x+4)\ dx/(x^2 + 3x + 2) + y\ dy/(y^2 + 1) = 0. \text{ Next}$$
we integrate: $\int \dfrac{y\ dy}{y^2 + 1} = \dfrac{1}{2} \ln(y^2 + 1)$. To integrate

the dx term, we use partial fractions. We set

$$\frac{x + 4}{x^2 + 3x + 2} = \frac{x+ 4}{(x+1)(x+2)} = \frac{A}{x + 1} + \frac{B}{x + 2} ,$$

$x + 4 = A(x + 2) + B(x + 1)$.   Then $x = -1$ gives $A = 3$

and $x = -2$ gives $B = -2$.   Thus we find $\int \dfrac{(x+4)\ dx}{x^2 + 3x + 2}$

$= 3 \int \dfrac{dx}{x+1} - 2 \int \dfrac{dx}{x+2} = 3 \ln|x+1| - 2 \ln|x+2| =$

$\ln \dfrac{|x+1|^3}{(x+2)^2}$.   Hence we obtain solutions in the form

$\ln \dfrac{|x+1|^3}{(x+2)^2} + \frac{1}{2} \ln(y^2 + 1) = \ln c_1$ (where we write $\ln c_1$

for the arbitrary constant, since each term on the left

is a $\ln$ term).   We multiply by 2 and simplify to obtain

$\ln \dfrac{(x+1)^6}{(x+2)^4} + \ln(y^2 + 1) = \ln c_1^2$ or

$\ln\dfrac{(x+1)^6(y^2+1)}{(x+2)^4} = \ln c_1^2$.   From this, we have

$(x+1)^6 (y^2+1) = c(x+2)^4$.

9.   We first write the D.E. in the form

$\dfrac{dy}{dx} = \dfrac{2xy + 3y^2}{2xy + x^2}$ and thence $\dfrac{dy}{dx} = \dfrac{2(y/x) + 3(y/x)^2}{2(y/x) + 1}$.   In

this form we recognize the D.E. is homogeneous.   We let

$y = vx$.   Then $\dfrac{dy}{dx} = v + x \dfrac{dv}{dx}$ and $v = \dfrac{y}{x}$.   We make these

substitutions to obtain $v + x \dfrac{dv}{dx} = \dfrac{2v + 3v^2}{2v + 1}$ or

$x \dfrac{dv}{dx} = \dfrac{2v + 3v^2}{2v + 1} - v = \dfrac{2v + 3v^2 - 2v^2 - v}{2v + 1}$ or

$x \dfrac{dv}{dx} = \dfrac{v^2 + v}{2v + 1}$.   We now separate variables to obtain

$\dfrac{(2v + 1) \, dv}{v^2 + v} = \dfrac{dx}{x}$. We integrate to obtain $\ln|v^2 + v|$

$= \ln|x| + \ln|c|$, or $\ln|v^2 + v| = |cx|$. From this, we

have $|v^2 + v| = |cx|$. We now resubstitute $v = y/x$ to

obtain $\left| \dfrac{y^2}{x^2} + \dfrac{y}{x} \right| = |cx|$. We simplify to obtain

$|y^2 + xy| = |cx|x^2$, from which we find $y^2 + xy = cx^3$.

14. This D.E. is homogeneous. Recognizing this, we could let $y = vx$ and substitute in order to separate the variables. However, the resulting separable equation is not readily tractable. This being the case, we assume $x > y > 0$ and divide the entire equation through by $\sqrt{y}$. Then we solve for dx/dy, putting the D.E. in th form

$$\frac{dx}{dy} = \frac{\sqrt{x/y + 1} - \sqrt{x/y - 1}}{\sqrt{x/y + 1} + \sqrt{x/y - 1}} \quad .$$

We now let $x = uy$ (see Exercise 24 concerning this). Then $\dfrac{dx}{dy} = u + y \dfrac{du}{dy}$ and $x/y = u$. Substituting in the D. it takes the form

$$u + y \frac{du}{dy} = \frac{\sqrt{u + 1} - \sqrt{u - 1}}{\sqrt{u + 1} + \sqrt{u - 1}} \quad .$$

We simplify the right member by multiplying both its

numerator and denominator by $\sqrt{u + 1} - \sqrt{u - 1}$ and then

simplifying. As a result of this, the D.E. takes the

form $u + y \dfrac{du}{dy} = u - \sqrt{u^2 - 1}$ which readily simplifies to

the separable equation

$$\frac{du}{\sqrt{u^2 - 1}} = -\frac{dy}{y}$$. Integrating (tables are

useful!), we obtain $\ln|u + \sqrt{u^2 - 1}| = -\ln|y| + \ln|c|$ or

$\ln|u + \sqrt{u^2 - 1}| = \ln|\frac{c}{y}|$. Since $x > y > 0$ and $u = x/y$,

we can write this more simply as $\ln(u + \sqrt{u^2 - 1}) =$

$\ln(c/y)$, from which we at once have $u + \sqrt{u^2 - 1} = c/y$.

Now substituting $u = x/y$, we have $x/y + \sqrt{(x/y)^2 - 1}$

$= c/y$ which readily simplifies to $x + \sqrt{x^2 - y^2} = c$.

17. The D.E. is separable. We separate variables
to obtain

$$(3x + 8)\ dx/(x^2 + 5x + 6) - 4y\ dy/(y^2 + 4) = 0$$

To integrate the dx term, we use partial fractions.
We write

$$\frac{3x + 8}{x^2 + 5x + 6} = \frac{3x + 8}{(x+2)(x+3)} = \frac{A}{x+2} + \frac{B}{x+3}\ ,$$

and so $3x + 8 = A(x + 3) + B(x + 2)$. Then $x = -2$ gives
$A = 2$; and $x = -3$ gives $B = 1$. Thus we find

$$\int \frac{3x + 8}{x^2 + 5x + 6}\ dx = 2 \int \frac{dx}{x + 2} + \int \frac{dx}{x + 3}$$

$= 2 \ln|x + 2| + \ln|x + 3| = \ln(x + 2)^2 (x + 3)$. Using

this, we obtain solutions in the form

$$\ln(x+2)^2 (x+3) - 2 \ln(y^2 + 4) = \ln|c| \qquad \text{or}$$

$\ln \dfrac{(x+2)^2 (x+3)}{(y^2 + 4)^2} = \ln|c|$. From this, $|x + 3| =$

$x + 3 \geq 0$ and $|c| = c \geq 0$, we have $(x+2)^2 (x+3) =$

$c(y^2 + 4)^2$. We now apply the I.C. $y(1) = 2$ to this,

obtaining $36 = 64c$, $c = \dfrac{9}{16}$. Thus we find the

particular solution

$$16(x + 2)^2 (x + 3) = 9(y^2 + 4).$$

20. We first write the D.E. in the form

$$\frac{dy}{dx} = \frac{3x^2 + 9xy + 5y^2}{6x^2 + 4xy}$$

or

$$\frac{dy}{dx} = \frac{3 + 9(y/x) + 5(y/x)^2}{6 + 4(y/x)} \quad .$$

We recognize that this D.E. is homogeneous. We let $y = vx$, and then $\dfrac{dy}{dx} = v + x\dfrac{dv}{dx}$ and $v = \dfrac{y}{x}$. Making these

substitutions, the D.E. becomes

$$v + x\frac{dv}{dx} = \frac{3 + 9v + 5v^2}{6 + 4v}$$

or

$$x\frac{dv}{dx} = \frac{v^2 + 3v + 3}{4v + 6} \quad .$$

We now separate variables to obtain

$$\frac{4v + 6}{v^2 + 3v + 3} \, dv = \frac{dx}{x} \quad .$$

Integrating, we find $2 \ln|v^2 + 3v + 3| = \ln|x| + \ln|c|$
or $\ln(v^2 + 3v + 3)^2 = \ln|cx|$. From this,
$(v^2 + 3v + 3)^2 = |cx|$. We resubstitute $v = y/x$ to
obtain
$$\left[\frac{y^2 + 3xy + 3x^2}{x^2} \right]^2 = |cx|.$$

We simplify, taking $|x| = x > 0$, thereby obtaining
$$(y^2 + 3xy + 3x^2)^2 = cx^5.$$

Applying the I.C. $y(2) = -6$, we let $x = 2$, $y = -6$ and
find $144 = 32c$ or $c = 9/2$. We thus obtain the particular solution $2(y^2 + 3xy + 3x^2)^2 = 9x^5$.

25. Since the D.E. is homogeneous, it can be
expressed in the form $\frac{dy}{dx} = g(\frac{y}{x})$. Let $x = r \cos \theta$,
$y = r \sin \theta$. Then
$$\frac{dy}{dx} = \frac{r \cos \theta \, d\theta + \sin \theta \, dr}{-r \sin \theta \, d\theta + \cos \theta \, dr} = \frac{r \cos \theta + \sin \theta \, dr/d\theta}{-r \sin \theta + \cos \theta \, dr/d\theta}$$

Substitute into $\frac{dy}{dx} = g(\frac{y}{x})$. The D.E. reduces
successively to
$$\frac{r \cos \theta + \sin \theta \, dr/d\theta}{-r \sin \theta + \cos \theta \, dr/d\theta} = g(\tan \theta),$$

$r \cos \theta + \sin \theta \frac{dr}{d\theta} = g(\tan \theta)[-r \sin \theta + \cos \theta \frac{dr}{d\theta}]$,

$[\sin \theta - \cos \theta \, g(\tan \theta)] \frac{dr}{d\theta} = -[g(\tan \theta) \sin \theta + \cos \theta]r$,

$\frac{dr}{r} = \frac{\sin \theta \, g(\tan \theta) + \cos \theta}{\cos \theta \, g(\tan \theta) - \sin \theta} \, d\theta$, which is separable in r and Θ.

26. (a). The D.E. of Exercise 8 is $(x+y)dx - x\,dy = 0$. This can be written $\frac{dy}{dx} = 1 + \frac{y}{x}$, and so is homogeneous. Using the method of Exercise 25, we let $x = r\cos\Theta$, $y = r\sin\Theta$. Then $\frac{dy}{dx} = \frac{r\cos\Theta\,d\Theta + \sin\Theta\,dr}{-r\sin\Theta\,d\Theta + \cos\Theta\,dr}$, We substitute this into $\frac{dy}{dx} = 1 + \frac{y}{x}$. The D.E. reduces successively to

$$\frac{r\cos\Theta\,d\Theta + \sin\Theta\,dr}{-r\sin\Theta\,d\Theta + \cos\Theta\,dr} = 1 + \tan\Theta,$$

$$r\cos\Theta\,d\Theta + \sin\Theta\,dr = (1 + \tan\Theta)(-r\sin\Theta\,d\Theta + \cos\Theta\,dr),$$

$$[\sin\Theta - (1 + \tan\Theta)\cos\Theta]\,dr = -r[\cos\Theta + (1 + \tan\Theta)\sin\Theta]\,d\Theta,$$

$$\frac{dr}{r} = \frac{\cos\Theta + (1 + \tan\Theta)\sin\Theta}{(1 + \tan\Theta)\cos\Theta - \sin\Theta}\,d\Theta.$$

$$= \left[\frac{\cos\Theta + \sin\Theta + \tan\Theta\sin\Theta}{\cos\Theta}\right]\,d\Theta$$

$$= (1 + \tan\Theta + \tan^2\Theta)\,d\Theta,$$

$$= (\sec^2\Theta + \tan\Theta)\,d\Theta,$$

or finally

$$\frac{dr}{r} = (\sec^2\Theta + \tan\Theta)\,d\Theta.$$

This is separable. Integrating, assuming $r > 0$, $\cos\Theta > 0$, we obtain $\ln r = \tan\Theta - \ln\cos\Theta + \ln c$, where $c > 0$. Now resubstitute, according to $x = r\cos\Theta$,

$y = r \sin \Theta$; that is, let $r = \sqrt{x^2 + y^2}$, $\tan \Theta = y/x$.
We obtain successively

$$\ln \sqrt{x^2 + y^2} = y/x - \ln(x/\sqrt{x^2 + y^2}) - \ln c,$$

$$\ln\sqrt{x^2 + y^2} + \ln x/\sqrt{x^2 + y^2} + \ln c = y/x,$$

$\ln(cx) = y/x$, or finally $y = x \ln(cx)$.

Section 2.3, Page 55

The equations of Exercises 1 through 14 are linear.
In solving, first express the equation in the standard
form of equation (2.26) and then follow the procedure
of Example 2.14.

1. The D.E. is already in the standard form (2.26)
with $P(x) = 3/x$, $Q(x) = 6x^2$. An I.F. is

$$e^{\int P(x)dx} = e^{\int (3/x)dx} = e^{3 \ln|x|} = e^{\ln|x|^3} = |x|^3$$

$= \pm x^3$ (+ if $x \geq 0$, $-$ if $x < 0$). We multiply
the D.E. through by this I.F. to obtain

$$x^3 \frac{dy}{dx} + 3x^2 y = 6x^5$$

or $\frac{d}{dx} (x^3 y) = 6x^5$. Integrating, we obtain $x^3 y = x^6 + c$
or $y = x^3 + cx^{-3}$.

6. This equation is linear in v. We divide through
by $u^2 + 1$ to put it in the standard form

$$\frac{dv}{du} + \frac{4u}{u^2 + 1} v = \frac{3u}{u^2 + 1} \qquad\qquad , \text{ with } P(u) = 4u/(u^2+1)$$

and $Q(u) = 3u(u^2 + 1)$. An I.F. is $e^{\int P(u)du} =$

$\exp \int \dfrac{4u}{u^2 + 1}\, du = e^{2\,\ln(u^2 + 1)} = (u^2 + 1)^2$. We

multiply the standard form equation through by this to

obtain $(u^2 + 1)^2 \dfrac{dv}{du} + 4u(u^2 + 1)v = 3u(u^2 + 1)$ or

$\dfrac{d}{du}[(u^2 + 1)^2 v] = 3u^3 + 3u$. Integrating, we obtain

$(u^2 + 1)^2\, v = \dfrac{3u^4}{4} + \dfrac{3u^2}{2} + c$.

8. We first divide through by $x^2 + x - 2$ to put the

equation in the standard form $\dfrac{dy}{dx} + \dfrac{3(x+1)}{(x+2)(x-1)}\, y = \dfrac{1}{x+2}$,

where $P(x) = 3(x+1)/(x+2)(x-1)$ and $Q(x) = 1/(x+2)$. An

I.F. is
$$e^{\int P(x)dx} = \exp \int \left[\dfrac{3(x+1)}{(x+2)(x-1)} \right] dx =$$

$e^{\ln|x+2|\, +\, 2\,\ln|x-1|} = |x+2|\,(x-1)^2 = \pm\,(x+2)(x-1)^2$,

(+ if $x \geq -2$, - if $x \leq -2$), where partial fractions

have been used to perform the integration. In either

case ($x \geq -2$ or $x \leq -2$), upon multiplying the standard

form equation through by the I.F., we obtain

$(x+2)(x-1)^2 \dfrac{dy}{dx} + 3(x+1)(x-1)y = (x-1)^2$, that is,

$\dfrac{d}{dx}[(x+2)(x-1)^2] = (x-1)^2$. Integrating, we find

$(x+2)(x-1)^2 y = (x-1)^3/3 + c/3$ or $3(x+2)y = x - 1 + c(x-1)^{-2}$.

10. This D.E. is linear in x (like Example 2.16).
We divide through by y and dy to put it in the standard
form $\dfrac{dx}{dy} + \dfrac{xy^2 + x - y}{y} = 0,$

or $\dfrac{dx}{dy} + (y + \dfrac{1}{y})x = 1$, where $P(y) = y+1/y$, $Q(y) = 1$.
An I.F. is
$$e^{\int P(y)dy} = e^{\int (y+1/y)dy} = e^{y^2/2 + \ln|y|} = |y|e^{y^2/2}$$

$= \pm\, ye^{y^2/2}$(+ if $y \geq 0$; - if $y < 0$). Multiplying the
standard form equation through by this, we obtain

$$ye^{y^2/2}\ \dfrac{dx}{dy} + (y^2+1)e^{y^2/2}\ x = ye^{y^2/2}$$

or $\dfrac{d}{dy}(ye^{y^2/2}x) = ye^{y^2/2}$. Integrating, we find

$ye^{y^2/2}\ x = e^{y^2/2} + c$ or $xy = 1 + ce^{-y^2/2}$.

15. This a Bernoulli D.E., where n = 2. We multi-
ply through by y^{-2} to obtain $y^{-2}\dfrac{dy}{dx} - \dfrac{1}{x}\,y^{-1} = -\dfrac{1}{x}$.
Let $v = y^{1-n} = y^{-1}$; then $\dfrac{dv}{dx} = -y^{-2}\dfrac{dy}{dx}$. The preceeding
D.E. readily transforms into the linear equation
$\dfrac{dv}{dx} + \dfrac{1}{x}\,v = \dfrac{1}{x}$. An I.F. is
$e^{\int \frac{dx}{x}} = e^{\ln|x|} = |x| = \pm\, x$. Multiplying through by
this, we find $x\dfrac{dv}{dx} + v = 1$ or $\dfrac{d}{dx}(xv) = 1$. Integrating
we find $xv = x + c$, from which $v = 1 + cx^{-1}$. But

$v = 1/y$. Thus we obtain the solution in the form
$1/y = 1 + cx^{-1}$.

18.   This is a Bernoulli D.E. in the dependent variable
x and independent variable t, where n = -1. We multiply
through by x to obtain $x \frac{dx}{dt} + \frac{t+1}{2t} x^2 = \frac{t+1}{t}$. Let
$v = x^{1-m} = x^2$; then $\frac{dv}{dt} = 2x \frac{dx}{dt}$. The preceeding D.E.
readily transforms into the linear equation $\frac{dv}{dt} + \frac{t+1}{t}$ v
$= 2(\frac{t+1}{t})$. An I.F. is $e \int \frac{t+1}{t} dt = e^{t + \ln|t|} = |t|e^t =$
$\pm t e^t$. Multiplying through by this, we find
$t e^t \frac{dv}{dt} + (t+1)e^t v = 2(t+1) e^t$ or $\frac{d}{dt}[t e^t v] = 2(t+1)e^t$.
Integrating, we obtain $t e^t v = 2 t e^t + c$ or
$v = 2 + c t^{-1} e^{-t}$. But $v = x^2$. Thus we obtain the
solution in the form $x^2 = 2 + c t^{-1} e^{-t}$.
- - - - - - - - - -
The equations of Exercises 19 through 24 are linear.

21.   We divide thru by $(e^x + 1)$ and dx to put in the
form $e^x[y(e^x + 1)^{-1} - 3(e^x + 1)] + \frac{dy}{dx} = 0$ and thence in
the standard form

$$\frac{dy}{dx} + \frac{e^x}{e^x + 1} y = 3e^x(e^x + 1) ,$$

where $P(x) = e^x/(e^x + 1)$, $Q(x) = 3e^x(e^x + 1)$. An I.F.
is $e \int P(x)dx = \exp \int \frac{e^x}{e^x + 1} dx = e^{\ln(e^x + 1)} = e^x + 1$.

We multiply the standard form equation through by this, obtaining $(e^x + 1) \frac{dy}{dx} + e^x y = 3e^x(e^x + 1)^2$

or $\frac{d}{dx}[(e^x + 1)y] = 3e^x(e^x + 1)^2$. Integrating, we obtain $(e^x + 1) y = (e^x + 1)^3 + c$ or

$$y = (e^x + 1)^2 + c(e^x + 1)^{-1} \qquad (\ast)$$

We apply the I.C. $y(0) = 4$: Let $x = 0$, $y = 4$ in ($\ast$) to obtain $4 = 4 + c/2$. Thus $c = 0$, and the particular solution of the stated I.V.P. is $y = (e^x+1)^2$.

24. This equation is linear in the dependent variable x with t as the independent variable, and is already in the standard form, with $P(t) = -1$, $Q(t) = \sin 2t$. An I.F. is
$e^{\int P(t)dt} = e^{\int(-1)dt} = e^{-t}$. We multiply the equation through by this to obtain $e^{-t} \frac{dx}{dt} - e^{-t} x = e^{-t} \sin 2t$ or $\frac{d}{dt}(e^{-t} x) = e^{-t} \sin 2t$. We next integrate, using integration by parts twice, or an integral table, on the right. We obtain $e^{-t} x = e^{-t} (-\sin 2t - 2 \cos 2t)/5 + c$ or

$x = -(\sin 2t + 2 \cos 2t)/5 + ce^t. \qquad (\ast)$

We apply the I.C. $x(0) = 0$: Let $t = x = 0$ in ($\ast$). We obtain $0 = -\frac{1}{5}(0 + 2) + c$ from which $c = 2/5$. Thus the general solution of the stated I.V.P. is

$x = (2e^t - \sin 2t - 2\cos 2t)/5.$

25.   This is a Bernoulli D.E. with n = -3. We first
multiply through by y^3 to obtain $y^3 \frac{dy}{dx} + \frac{y^4}{2x} = x.$
Now let $v = y^{1-m} = y^4.$ Then $\frac{dv}{dx} = 4\,y^3 \frac{dy}{dx}$ and the
preceeding D.E. transforms into $\frac{1}{4} \frac{dv}{dx} + \frac{v}{2x} = x,$ which
is linear in v. In the standard form this linear
equation is $\frac{dv}{dx} + \frac{2}{x}v = 4x,$ with $P(x) = 2/x,$ $Q(x) = 4x.$
An I.F. is

$$e^{\int P(x)dx} = e^{\int (2/x)dx} = e^{2\ln|x|} = x^2.$$

We multiply the standard form through by this to obtain
$x^2 \frac{dv}{dx} + 2\,x\,v = 4x^3$ or $\frac{d}{dx}(x^2\,v) = 4x^3.$ We integrate
to obtain $x^2\,v = x^4 + c.$ But $v = y^4.$ Hence we obtain
the one-parameter family of solutions of the given
Bernoulli Equation in the form

$$x^2 y^4 = x^4 + c \ . \qquad\qquad (*)$$

We apply the I.C. y(1) = 2. Let x = 1, y = 2 in
(*) to obtain 16 = 1 + c, so c = 15. Thus the particular solution of the stated I.V.P. is $x^2 y^4 = x^4 + 15.$

27.   Here we actually have two I.V. problems:

(I) $\begin{cases} \text{For } 0 \le x < 1, \\ \frac{dy}{dx} + y = 2, \\ y(0) = 0, \end{cases}$ (II) $\begin{cases} \text{For } x \ge 1, \\ \frac{dy}{dx} + y = 0, \\ y(1) = a, \end{cases}$

where a is the value of $\lim_{x \to 1^-} \phi(x)$ and ϕ denotes the

solution of (I). This is prescribed so that the
solution of the entire problem will be continuous at
x = 1.

We first solve (I). The D.E. of this problem is
linear in standard form with P(x) = 1, Q(x) = 2. An
I.F. is $e^{\int P(x)dx} = e^{\int (1)dx} = e^x$. We multiply the
D.E. of (I) through by this to obtain $e^x \frac{dy}{dx} + e^x y =$
$2e^x$ or $\frac{d}{dx}[e^x y] = 2e^x$. We integrate to obtain

$e^x y = 2e^x + c$ or $y = 2 + ce^{-x}$. We apply the I.C. of
(I), y(0) = 0, to this, obtaining 0 = 2 + c or c = -2.
Thus the particular solution of problem (I) is
$y = 2 - 2e^{-x}$. This is valid for $0 \le x < 1$.

Letting ϕ denote the solution just obtained, we
note that $\lim_{x \to 1^-} \phi(x) = \lim_{x \to 1^-} (2 - 2e^{-x}) = 2 - 2e^{-1}$. This
is the a of Problem (II); that is, the I.C. of (II) is
$y(1) = 2 - 2e^{-1}$.

Now solve (II). The D.E. of this problem is linear
in standard form with P(x) = 1, Q(x) = 0. An I.F., as
in problem (I), is e^x. We multiply through by this to
obtain $e^x \frac{dy}{dx} + e^x y = 0$ or $\frac{d}{dx} (e^x y) = 0$. We integrate
to obtain $e^x y = c$ or $y = c e^{-x}$. Now apply the I.C. of
(II), $y(1) = 2 - 2e^{-1}$, to this. Let x = 1, $y = 2 - 2e^{-1}$.

We obtain $2 - 2e^{-1} = c\ e^{-1}$, from which $c = 2e - 2$. Thus the particular solution of problem (II) is $y = (2e - 2)e^{-x}$. This is valid for $x \geq 1$.

We write the solution of the entire problem, showing intervals where each part is valid, as

$$y = \begin{cases} 2(1 - e^{-x}), & 0 \leq x < 1, \\ 2(e - 1)\ e^{-x}, & x \geq 1 \end{cases} \quad .$$

31. (a) The D.E. is linear. In the standard form it is

$$\frac{dy}{dx} + \frac{b}{a}\ y = \frac{k}{a}\ e^{-\lambda x} \quad, \text{ with } P(x) = \frac{b}{a}\ ,\ Q(x) = \frac{k}{a}\ e^{-\lambda x}\ .$$

An I.F. is $e^{\int P(x)dx} = e^{\int (b/a)dx} = e^{(b/a)x}$.

Multiplying the standard form equation through by this, we obtain

$$e^{(b/a)x}\ \frac{dy}{dx} + \frac{b}{a}\ e^{(b/a)x}\ y = \frac{k}{a}\ e^{(b/a - \lambda)x}$$

or $\frac{d}{dx}[e^{(b/a)x}\ y] = \frac{k}{a}\ e^{(b/a - \lambda)x}$. $\qquad\qquad$ (1)

We now consider two cases: (i) $\lambda \neq b/a$; and (ii) $\lambda = b/a$.

In case (i), where $\lambda \neq b/a$, integrating (1) we obtain $e^{(b/a)x}\ y = \frac{k}{a}(\frac{b}{a} - \lambda)^{-1}\ e^{(b/a - \lambda)x} + c$ or

$$y = \frac{ke^{-\lambda x}}{b - a\lambda} + ce^{-bx/a} \quad . \qquad\qquad (2)$$

In case (ii), where λ = b/a, (1) becomes

$\frac{d}{dx}[e^{(b/a)x}$ y] = $\frac{k}{a}$. Integrating, we obtain $e^{(b/a)x}$ y =

kx/a + c or

$$y = \frac{k \; x \; e^{-bx/a}}{a} + ce^{-bx/a} \; . \qquad (3)$$

(b) Suppose λ = 0. Then since b/a > 0, we can have only case (i), $\lambda \neq$ b/a, here. Then with λ = 0, (2) becomes y = k/b + $ce^{-bx/a}$. As x → ∞, $e^{-bx/a}$ → 0 and y → k/b.

Suppose λ > 0. Then either case (i) or case (ii) can occur. In case (i), y is given by (2); and since $e^{-\lambda x}$ → 0 and $e^{-bx/a}$ → 0 as x → ∞, we have y → 0 as x → ∞. In case (ii), y is given by (3); and since $xe^{-bx/a}$ → 0 and $e^{-bx/a}$ → 0 as x → ∞, we again have y → 0 as x → ∞.

33. (a) The function f such that f(x) = 0 for all x ε I has derivative f'(x) = 0 for all x ε I. We substitute this f(x) for y and f'(x) for dy/dx in the D.E. We obtain f'(x) + P(x)f(x) = 0 + P(x)0 = 0 for all x ε I, so f(x) is a solution.

(b) By the Theorem 1.1, there is a unique solution f of the D.E. satisfying the I.C. $f(x_0)$ = 0, where x_0 ε I. By (a), $\phi(x)$ = 0 for all x ε I is a solution of the D.E., and obviously it satisfies the

I.C. Thus by the uniqueness of Theorem 1, we must have
$f(x) = \phi(x)$, that is, $f(x) = 0$ for <u>all</u> $x \in I$.

(c) Since f and g are solutions of the D.E.,
so is their difference $h = f - g$ (by Example 32(a)).
Also $h(x_0) = f(x_0) - g(x_0) = 0$. So h is a solution
such that $h(x_0) = 0$ for some $x_0 \in I$. By part (b),
$h(x) = 0$ for all $x \in I$. But this means $f(x) - g(x) = 0$
and hence $f(x) = g(x)$ for all $x \in I$.

37. (a) This is of the form (2.41) with $f(y) =$
$\sin y$, $P(x) = 1/x$, and $Q(x) = 1$. We let $v = f(y) =$
$\sin y$, from which $dv/dx = (\cos y) \, dy/dx$. Substituting
this into the stated D.E., it becomes $\frac{dv}{dx} + \frac{1}{x} v = 1$,
which is linear in v. An I.F. is
$e^{\int P(x)dx} = e^{\int (1/x)dx} = e^{\ln|x|} = |x|$. Multiplying the
linear equation through by this, we have $x \frac{dv}{dx} + v = x$
or $\frac{d}{dx}(xv) = x$. Integrating, we find $xv = x^2/2 + c_0$, or
$2xv = x^2 + c$, where $c = 2c_0$. Now replacing v by $\sin y$,
we obtain the solution in the form $2x \sin y - x^2 = c$.

40. This is a Riccati Equation of the form (A) of
Exercise 38, with $A(x) = -1$, $B(x) = x$, $C(x) = 1$. The
solution $f(x) = x$ is given. By Exercise 38(b), we make
the transformation $y = f(x) + 1/v = x + 1/v$, from which
$dy/dx = 1 - (1/v^2) \, dv/dx$. Substituting in the given

D.E., it takes the form

$$1 - \frac{1}{v^2}\frac{dv}{dx} = -(x + \frac{1}{v})^2 + x(x + \frac{1}{v}) + 1.$$

This reduces to $-\frac{1}{v^2}\frac{dv}{dx} = -\frac{1}{v^2} - \frac{x}{v}$ or $\frac{dv}{dx} - xv = 1$,

which is linear in v, with $P(x) = -x$, $Q(x) = 1$. An I.F.
is $e^{\int P(x)dx} = e^{\int(-x)dx} = e^{-x^2/2}$. Multiplying the

linear D.E. through by this, we obtain

$$e^{-x^2/2}\frac{dv}{dx} - e^{-x^2/2}xv = e^{-x^2/2}$$

or $\quad\quad \frac{d}{dx}(e^{-x^2/2}v) = e^{-x^2/2}$.

Integrating, we obtain $e^{-x^2/2}v = \int e^{-x^2/2}dx + c.$

Since the integral on the right cannot be expressed as a
finite sum of known elementary functions, we leave it as
indicated. But from the transformation $y = x + 1/v$, we
see that $v = 1/(y - x)$. We replace v accordingly and
thus obtain the solution,

$$\frac{e^{-x^2/2}}{y - x} = \int e^{-x^2/2}dx + c .$$

Section 2.3, Miscellaneous Exercises, Pages 58-59

The solution of each of these review problems is
given, but most of these solutions are presented in
abbreviated form with many details omitted.

1. The D.E. is both separable and linear. Upon

separating variables and integrating, we obtain
$2\ln(x^3 + 1) + \ln|c| = \ln|y|$, from which we find

$y = c(x^3 + 1)^2.$

The solution as a linear equation is almost as eas; The I.F. is $(x^3 + 1)^{-2}.$

2. The D.E. is exact. We seek $F(x,y)$ such that $F_x(x,y) = M(x,y) = 2xy^3 - y$ and $F_y(x,y) = N(x,y) = 3x^2y^2 - x.$ From the first of these, we find $F(x,y) = x^2y^3 - xy + \phi(y).$ From this, $F_y(x,y) = 3x^2y^2 - x + \phi'(y).$ But since $F_y(x,y) = N(x,y) = 3x^2y^2 - x,$ we find $\phi'(y) = 0,$ $\phi(y) = c_0.$ Thus $F(x,y) = x^2y^3 - xy + c_0,$ and the family of solutions is $x^2y^3 - xy = c.$

3. The D.E. is both separable and linear. First consider it as a separable equation. Upon separating variables and using partial fractions (preparatory to integrating), we find $[1/x - 1/(x+1)] \, dx + dy/(y-1) = ($ Integration gives $\ln|x| - \ln|x+1| + \ln|y-1| = \ln c_1$. Upon simplifying and taking antilogs, we find $|x(y-1)|$ $|c_1(x+1)|.$ Assuming $x > 0,$ $y > 1$ and simplifying, we can write this as $xy = c\,x + (c - 1),$ or $xy + 1 = c(x+$ where $c = 1 + c_1.$

Alternately, consider the D.E. as a linear equatioi The I.F. is $\exp\left[\int dx/x(x+1)\right] = \pm\, x/(x+1).$ Multiplyini the "standard form" of the D.E. through by this we havi $[xy/(x+1)]' = (x+1)^{-2}.$ Integrating, we find $xy/(x+1)$ $-(x+1)^{-1} + c.$ From this we find $xy + 1 = c(x+1),$ as before.

    4. The D.E. is linear. The I.F. is x. Multiplying the "standard form" of the equation through by this, we find $x^2 y' + 2xy = x^3$ and hence $(x^2 y)' = x^3$. Integration and simplification at once give $y = x^2/4 + c/x^2$.

    5. Writing the D.E. in the form $\frac{dy}{dx} = \frac{5(y/x) - 3}{1 + (y/x)}$, we recognize that it is homogeneous. We let $y = vx$. Then the D.E. becomes $v + x \frac{dv}{dx} = \frac{5v - 3}{1 + v}$. Simplification yields $(v + 1) \, dv/(v^2 - 4v + 3) = -dx/x$. Then upon applying partial fractions to the left member, we find

$$[- \frac{1}{v-1} + \frac{2}{v-3}] \, dv = - \frac{dx}{x} \, .$$

Integration gives $-\ln|v-1| + 2 \ln|v-3| = -\ln|x| + \ln|c|$. Simplification and taking of antilogarithms results in $(v - 3)^2/|v - 1| = |c/x|$. But $v = y/x$. Upon resubstituting accordingly and simplifying, we find $(3x - y)^2 = |c(y - x)|$.

    6. The D.E. is exact, separable, and linear. Considered as an exact equation, we seek $F(x,y)$ such that $F_x(x,y) = M(x,y) = e^{2x}y^2$ and $F_y(x,y) = N(x,y) = e^{2x}y - 2y$. From the first of these, we find $F(x,y) = e^{2x} y^2/2 + \phi(y)$. From this, $F_y(x,y) = e^{2x}y + \phi'(y)$. But since $F_y(x,y) = N(x,y) = e^{2x}y - 2y$, we find $\phi'(y) = -2y$, $\phi(y) = -y^2 + c_0$. Thus $F(x,y) = e^{2x}y^2/2 - y^2 + c_0$,

and the family of solutions is $e^{2x}y^2 - 2y^2 = c$, [where we have set $F(x,y) = c_1$, multiplied through by 2, and let $c = 2(c_1 - c_0)$]. Assuming $e^{2x} - 2 \geq 0$, this can also be written as $y = c_2(e^{2x} - 2)^{-1/2}$ where $c_2 = \sqrt{c}$.

We also consider the D.E. as a linear equation. We rewrite it as $(e^{2x} - 2) y' + e^{2x} y = 0$ and then put it in the "standard form" $y' + e^{2x} y/(e^{2x} - 2) = 0$. From this we find the I.F. is $(e^{2x} - 2)^{1/2}$, where we assume $e^{2x} - 2 \geq 0$. Multiplying the standard form through by the I.F., we find $(e^{2x} - 2)^{1/2} y' + e^{2x} y (e^{2x} - 2)^{-1/2} = 0$ or $[(e^{2x} - 2)^{1/2} y]' = 0$. Integration and simplification readily yield $y = c(e^{2x} - 2)^{-1/2}$.

7. The D.E. is both separable and linear. We first solve this as a separable equation. Separating variables, we have $4x^3 dx/(x^4 + 1) + dy/(2y - 3) = 0$. Integration yields $\ln(x^4 + 1) + \frac{1}{2} \ln|2y - 3| = \ln|c_1|$. Simplify, assuming $2y - 3 \geq 0$, to obtain $\ln(x^4 + 1)(2y - 3)^{1/2} = \ln|c_1|$. From this we at once have $(x^4 + 1)^2 (2y - 3) = c_1^2$. Then $y = \frac{3}{2} + c(x^4+1)^{-2}$, where $c = c_1^2/2$.

Consider the D.E. as a linear equation. In standard form the D.E. is $y' + 8x^3 y/(x^4 + 1) = 12x^3/(x^4 + 1)$. The I.F. is $(x^4 + 1)^2$. Multiplying the standard form equation through by this we find

$[(x^4 + 1)^2 y]' = (x^4 + 1)(12x^3)$. Integration and division by $(x^4 + 1)^2$ then give $y = \frac{3}{2} + c(x^4 + 1)^{-2}$.

8. Writing the D.E. in the form $\frac{dy}{dx} =$ $- [2 + y/x + (y/x)^2]/2$, we see that the D.E. is homogeneous. Let $y = vx$. Then $\frac{dy}{dx} = v + x\frac{dv}{dx}$, and the D.E. becomes $v + x\frac{dv}{dx} = -1 - v/2 - v^2/2$. Simplification yields $x\, dv/dx = - (v^2 + 3v + 2)/2$, and hence $2dv/(v^2 + 3v + 2) = -dx/x$. Use of partial fractions gives $2[\frac{1}{v+1} - \frac{1}{v+2}]\, dv = - \frac{dx}{x}$. Integration and immediate simplification give $2 \ln \left|\frac{v+1}{v+2}\right| = \ln \left|\frac{c}{x}\right|$. Further simplification gives $[(v + 1)/(v + 2)]^2 = |c/x|$. But $v = y/x$. Resubstituting accordingly and again simplifying, we obtain $[(y+x)/(y+2x)]^2 = |c/x|$.

9. Rewriting the D.E. in the form $(4x^3y^2 - 3x^2y)dx + (2x^4y - x^3)\, dy = 0$, we find that it is exact. We seek $F(x,y)$ such that $F_x(x,y) = M(x,y) = 4x^3y^2 - 3x^2y$ and $F_y(x,y) = N(x,y) = 2x^4y - x^3$. From the first of these, $F(x,y) = x^4y^2 - x^3y + \phi(y)$. From this, $F_y(x,y) = 2x^4y - x^3 + \phi'(y)$. But since $F_y(x,y) = N(x,y) = 2x^4y - x^3$, we must have $\phi'(y) = 0$, $\phi(y) = c_0$. Thus $F(x,y) = x^4y^2 - x^3y + c_0$, and the solution is $x^4y^2 - x^3y = c$.

10. The D.E. is linear. The standard form is $y' + xy/(x+1) = e^{-x}(x+1)$, and the I.F. is $\pm e^{x}(x+1)$. Multiplying the standard form through by this, we obtain the equivalent of $[e^{x}y/(x+1)]' = (x+1)^{-2}$. Integration and multiplication by $(x+1)e^{-x}$ result in $y = e^{-x}[-1 + c(x+1)]$.

11. Writing the D.E. in the form $\dfrac{dy}{dx} = \dfrac{2 - 7(y/x)}{3(y/x) - 8}$, we recognize that it is homogeneous. We let $y = vx$. Then $\dfrac{dy}{dx} = v + x\dfrac{dv}{dx}$. The D.E. becomes $v + x\dfrac{dv}{dx} = \dfrac{2 - 7v}{3v - 8}$. Simplification gives $(3v - 8)dv/(3v^2 - v - 2) = -dx/x$. Partial fractions decomposition puts this in the form $[\dfrac{6}{3v+2} - \dfrac{1}{v-1}]\, dv = -\dfrac{dx}{x}$. Integration gives $2\ln(3v + 2) - \ln|v - 1| = -\ln|x| + \ln|c|$. Simplification and taking of antilogarithms results in $(3v + 2)^2/|v - 1| = |c/x|$. But $v = y/x$. Resubstituting accordingly and simplifying gives

$$\frac{(3y + 2x)^2}{x^2} \cdot \frac{|x|}{|y - x|} = \left|\frac{c}{x}\right| \quad \text{and hence}$$

$(2x + 3y)^2 = |c(y - x)|$.

12. The D.E. is a Bernoulli Equation with $n = 3$. We multiply through by y^{-3} to obtain $x^2 y^{-3}\dfrac{dy}{dx} + xy^{-2} = x$. Let $v = y^{1-n} = y^{-2}$; then $\dfrac{dv}{dx} = -2y^{-3}\dfrac{dy}{dx}$. The preceeding

D.E. readily transforms into the linear equation
$\frac{dv}{dx} - \frac{2}{x} v = - \frac{2}{x}$. An I.F. is

$e^{-\int \frac{2}{x}dx} = 1/x^2$. Multiplying through by this, we obtain
$\frac{d}{dx}[v/x^2] = -2/x^3$. Integrating, we find $v/x^2 = 1/x^2 + c$.
Replacing v by y^{-2} and simplifying, we find $y^2 = \dfrac{1}{1 + cx^2}$.

13. This is a linear equation. The standard form
is $y' + 6x^2y/(x^3 + 1) = 6x^2/(x^3 + 1)$. An I.F. is
$(x^3 + 1)^2$. Multiplying through by this we obtain
$[(x^3 + 1)y]' = (x^3 + 1)(6x^2)$. Integration and division
by $(x^3 + 1)^2$ then give the solutions $y = 1 + c(x^3+1)^{-2}$.

14. Writing the D.E. in the form $\frac{dy}{dx} = \frac{2 + (y/x)^2}{2(y/x) - 1}$,
we recognize that it is homogeneous. We let $y = vx$.
Then $\frac{dy}{dx} = v + x \frac{dv}{dx}$. The D.E. becomes $v + x \frac{dv}{dx}$

$= \frac{2 + v^2}{2v - 1}$. Simplification gives $(2v - 1)dv/(v^2 - v - 2)$
$= -dx/x$. Integration gives $\ln|v^2 - v - 2| = \ln|c/x|$.
Taking antilogarithms and simplifying slightly, we
obtain $v^2 - v - 2 = c/x$. But $v = y/x$. Resubstituting
accordingly and simplifying, we find $y^2 - xy - 2x^2 = cx$
or $(y - 2x)(y+x) = cx$.

15. The D.E. is both homogeneous and Bernoulli.
We first consider it as a homogeneous equation, which
form we recognize by writing it as

$\frac{dy}{dx} = \frac{1 + (y/x)^2}{2(y/x)}$. We let $y = vx$. Then $\frac{dy}{dx} = v + x\frac{dv}{dx}$.
The D.E. becomes $v + x\frac{dv}{dx} = \frac{1+v^2}{2v}$. Simplification
quickly leads to $2vdv/(v^2 - 1) = -dx/x$, and then
integration gives $\ln|v^2 - 1| = \ln|c/x|$. From this, we
readily find $v^2 - 1 = c/x$. Resubstituting $v = y/x$ and
simplifying, we find $y^2 - x^2 = cx$.

Now apply the initial condition $y(1) = 2$. Letting
$x = 1$, $y = 2$, we get $3 = c$, and hence obtain the
solution in the form $y^2 = x^2 + 3x$ or $y = \sqrt{x^2 + 3x}$.

Alternatively, we recognize that the given D.E. is
a Bernoulli equation, with $n = -1$, by writing in the
equivalent form $2x\frac{dy}{dx} - y = x^2y^{-1}$. We rewrite this as
$2xy\frac{dy}{dx} - y^2 = x^2$; let $v = y^2$ and $\frac{dv}{dx} = 2y\frac{dy}{dx}$; and
transform the D.E. into the linear equation $x\frac{dv}{dx} - v = x^2$. The standard form of this is $\frac{dv}{dx} - \frac{1}{x}v = x$, and an
I.F. is x^{-1}. Multiplying the standard form equation
through by this, we obtain $(x^{-1}v)' = 1$. Integration
then gives $x^{-1}v = x + c$. Now resubstituting $v = y^2$
and simplifying slightly, we find $y^2 = x^2 + cx$. Application of the I.C. again gives the particular solution
$y^2 = x^2 + 3x$ or $y = \sqrt{x^2 + 3x}$.

16. The D.E. is both a separable equation and a
Bernoulli Equation. We first solve it as a separable

equation. Separating variables, we get $\dfrac{2\,dx}{1-x^2} + \dfrac{y\,dy}{y^2+4} = 0.$

We can use partial fractions or an integral table to integrate the dx term. Integrating the equation and simplifying slightly, we find

$$\ln\left|\frac{1+x}{1-x}\right| + \ln\,(y^2 + 4)^{1/2} = \ln|c| \; .$$

Further simplification results in the one-parameter family of solutions

$$\left|(1 + x)\,(y^2 + 4)^{1/2}\right| = \left|c\,(1 - x)\right| \; .$$

Now apply the I.C. $y(3) = 0$. Letting $x = 3$, $y = 0$ in the preceeding gives $8 = 2|c|$, so $|c| = 4$. Substituting this value of $|c|$ back into the equation of the family of solutions, we get

$$\left|(1 + x)(y^2 + 4)^{1/2}\right| = 4|1 - x| \; .$$

Since the initial x value is $3 > 1$, we take $x > 1$ and $1 - x < 0$, so $|1 - x| = x - 1$. Thus we obtain the particular solution

$$(x + 1)(y^2 + 4)^{1/2} = 4(x - 1) \; .$$

Alternately, we now solve the given D.E. as a Bernoulli Equation. We first rewrite it as $\dfrac{dy}{dx} + \dfrac{2}{1-x^2}\,y$

$= -\dfrac{8}{1-x^2}\,y^{-1}$, in which we recognize that it is indeed a Bernoulli D.E. with $n = -1$. We thus let $v = y^{1-n} = y^2$, find $dv/dx = 2\,y\,dy/dx$, and transform the D.E. into

$$\frac{dv}{dx} + \frac{4}{1 - x^2}\, v = -\frac{16}{1 - x^2}\ ,$$

which is linear in v, with $P(x) = 4/(1 - x^2)$,
$Q(x) = -16/(1 - x^2)$. An I.F. is

$$e^{\int P(x)dx} = e^{\int 4dx/(1-x^2)} = e^{2\,\ln|(1+x)/(1-x)|} = \left(\frac{1+x}{1-x}\right).$$

Multiplying the linear D.E. through by this, we obtain

$$\frac{d}{dx}[(\frac{1+x}{1-x})^2 v] = -\frac{16(1+x)}{(1-x)^3}\ .$$

To integrate the right member, we note that

$$-\frac{16(1+x)}{(1-x)^3} = -16[-\frac{1}{(1-x)^2} + \frac{2}{(1-x)^3}]\ .$$

Integrating the D.E., we thus find

$$(\frac{1+x}{1-x})^2\, v = -16[-\frac{1}{1-x} + \frac{1}{(1-x)^2}] + c,$$

which readily simplifies into

$$(\frac{1+x}{1-x})^2\, v = \frac{-16x}{(1-x)^2} + c.$$

Application of the I.C. $x = 3$ at $y = 0$ gives $c = 12$.
Substituting this back and multiplying by $(1-x)^2$, we
obtain $(1+x)^2\, y^2 = -16\, x + 12(1-x)$. Adding $4(1+x)^2$ to
both sides of this and simplifying, we find
$(1+x)^2(y^2+4) = 16(x-1)^2$, from which the solution
previously found is readily obtained.

Thus we obtain the solution in the form $x^3 + x^2y^2 + 2y^3$ = 21.

19. This D.E. is both a separable equation and a Bernoulli Equation. We first solve it as a separable D.E., writing it as $4 y \, dy/(y^2 + 1) = dx/x$. Integration then gives $2 \ln(y^2 + 1) = \ln|x| + \ln|c|$. Simplifying we find $\ln(y^2 + 1)^2 = \ln|cx|$ and hence $(y^2 + 1)^2 = |cx|$. Applying the I.C. $y(2) = 1$, we set $x = 2$, $y = 1$ in this to obtain $4 = 2|c|$ and hence $|c| = 2$. Thus we obtain the solution $(y^2 + 1)^2 = 2|x|$. Since the initial x value is $2 > 0$, we take $x > 0$ and write the solution as $(y^2 + 1)^2 = 2x$. Alternately, writing the D.E. as $4x \frac{dy}{dx} - y = y^{-1}$, we recognize that it is a Bernoulli Equation with $n = -1$. Thus we let $v = y^{1-n} = y^2$, $dv/dx = 2y \, dy/dx$, and substitute into the stated equation $4xy \frac{dy}{dx} = y^2 + 1$, obtaining $2x \frac{dv}{dx} = v + 1$ or $\frac{dv}{dx} - \frac{1}{2x} v = \frac{1}{2x}$. This is linear in v, with I.F. $e^{-\int (1/2x)dx} = x^{-1/2}$. Multiplying the standard form linear equation through by this, we find $\frac{d}{dx}(x^{-1/2}v) = x^{-3/2}/2$. Integrating, we obtain $x^{-1/2} v = -x^{-1/2} + c$ or $v = -1 + cx^{1/2}$. But $v = y^2$, so we have $y^2 = -1 + cx^{1/2}$. Applying the I.C. $y(2) = 1$ gives $c = \sqrt{2}$. Thus we obtain the solution $y^2 = -1 + \sqrt{2}x^{1/2}$, from

which we readily obtain the solution $(y^2 + 1)^2 = 2x$
previously found.

20. Writing the D.E. in the form $\frac{dy}{dx} = \frac{2+7(y/x)}{2-2(y/x)}$,
we see that it is homogeneous. Thus we let $y = vx$,
$\frac{dy}{dx} = v + x\frac{dv}{dx}$. The D.E. successively reduces to
$v + x\frac{dv}{dx} = \frac{2+7v}{2-2v}$, $x\frac{dv}{dx} = \frac{2v^2 + 5v + 2}{2 - 2v}$, and

$\frac{(2v-2)\,dv}{2v^2 + 5v + 2} = -\frac{dx}{x}$. This last equation is separable,

with the variables separated. Using partial fractions,
we find $\frac{2v - 2}{2v^2 + 5v + 2} = -\frac{2}{2v + 1} + \frac{2}{v + 2}$. We thus have

$[\frac{-2}{2v+1} + \frac{2}{v+2}]\,dv = -\frac{dx}{x}$. Integrating, we obtain

$-\ln|2v+1| + 2\ln|v+2| = -\ln|x| + \ln|c|$, which reduces to

$$\ln\frac{(v + 2)^2}{|2v + 1|} = \ln\left| \frac{c}{x} \right| .$$

From this we find $\frac{(v+2)^2}{|2v+1|} = \left| \frac{c}{x} \right|$. But $v = y/x$; and
resubstituting and simplifying, we find

$$\frac{(y + 2x)^2}{x^2} \cdot \frac{|x|}{|2y + x|} = \left| \frac{c}{x} \right|$$

or

$(2x + y)^2 = |c(2y + x)|$. Applying the I.C.
$y(1) = 2$, we let $x = 1$, $y = 2$, to obtain $16 = |c|5$,
from which $|c| = 16/5$. Replacing $|c|$ with this value,
and taking $x + 2y \geq 0$, we express the solution as

$$5(2x + y)^2 = 16(x + 2y) \quad .$$

21. This equation is both separable and linear.
Treating it as a separable equation, we have $dy/y =$
$x \, dx/(x^2 + 1)$. Integrating then gives $2 \ln|y| =$
$\ln(x^2 + 1) + \ln|c|$, which quickly simplifies to
$y^2 = c(x^2 + 1)$, with $c > 0$. Applying the I.C.
$y(\sqrt{15}) = 2$, we let $x = \sqrt{15}$, $y = 2$, to obtain $16c = 4$,
$c = 1/4$. Thus we find the solution $y^2 = (x^2 + 1)/4$
or $y = (x^2 + 1)^{1/2}/2$. Now treating the equation as a
linear equation, we have $\frac{dy}{dx} - \left[\frac{x}{x^2 + 1}\right] y = 0$, with
$P(x) = -x/(x^2 + 1)$, $Q(x) = 0$. An I.F. is $e^{\int P(x)dx} =$
$e^{-\int [x/(x^2+1)]dx} = e^{-(1/2)\ln(x^2+1)} = (x^2 + 1)^{-1/2}$.
Multiplying the D.E. through by this, we obtain
$\frac{d}{dx}[(x^2 + 1)^{-1/2} y] = 0$. Integrating, we find
$(x^2 + 1)^{-1/2} y = c$ or $y = c(x^2 + 1)^{1/2}$. Applying the
I.C. $y(\sqrt{15}) = 2$ gives $c = 1/2$. Thus we again obtain
$y = (x^2 + 1)^{1/2}/2$.

22. Here we actually have two I.V. problems

$$\text{(I)} \begin{cases} \text{For } 0 \le x < 2, \\ \frac{dy}{dx} + y = 1, \\ y(0) = 0; \end{cases} \qquad \text{(II)} \begin{cases} \text{For } x > 2, \\ \frac{dy}{dx} + y = 0 \\ y(2) = a; \end{cases}$$

where a is the value of $\lim\limits_{x \to 2-} \phi(x)$ and ϕ denotes the

solution of (I). This is prescribed so that the
solution of the entire problem will be continuous at
$x = 2$.

We first solve (I). The D.E. is linear in standard
form, with $P(x) = 1$, $Q(x) = 1$. An I.F. is $e^{\int P(x)dx} =$
$e^{\int dx} = e^x$. Multiplying the D.E. through by this we
obtain $e^x \frac{dy}{dx} + e^x y = e^x$ or $\frac{d}{dx}(e^x y) = e^x$. Integrating,
we find $e^x y = e^x + c$ or $y = 1 + ce^{-x}$. Applying the I.C.
$y(0) = 0$ gives $0 = 1 + ce^0$, $c = -1$. Hence the solution
of Problem (I) is $y = 1 - e^{-x}$, valid for $0 \leq x < 2$.

Letting ϕ denote the solution just obtained, we note
that $\lim_{x \to 2-} \phi(x) = \lim_{x \to 2-} (1 - e^{-x}) = 1 - e^{-2}$. This is the
a of problem (II); that is, the I.C. of problem (II) is
$y(2) = 1 - e^{-2}$.

We now solve problem (II). The D.E. is linear in
standard form, with $P(x) = 1$, $Q(x) = 0$. An I.F., as in
Problem (I), is e^x. Multiplying the D.E. through by
this and integrating, we obtain $e^x \frac{dy}{dx} + e^x y = 0$ or
$\frac{d}{dx}(e^x y) = 0$. Integrating, we find $e^x y = c$ or $y = ce^{-x}$.
Now apply the I.C. of problem (II), namely, $y(2) =$
$1 - e^{-2}$. We have $1 - e^{-2} = ce^{-2}$, from which we find
$c = e^2 - 1$. Thus the solution of problem (II) is
$y = (e^2 - 1) e^{-x}$, valid for $x > 2$.

We write the solution of the entire problem, showing
intervals where each part is valid, as

$$y = \begin{cases} 1 - e^{-x}, & 0 \le x < 2, \\ (e^2 - 1)\, e^{-x}, & x > 2. \end{cases}$$

23. Here, as in problem 22, we actually have two
I.V. problems:

$$(I) \begin{cases} \text{For } 0 \le x \le 2, \\ (x+2)\dfrac{dy}{dx} + y = 2x, \\ y(0) = 4; \end{cases} \qquad (II) \begin{cases} \text{For } x > 2, \\ (x+2)\dfrac{dy}{dx} + y = 4, \\ y(2) = a; \end{cases}$$

where a is the value of $\lim\limits_{x \to 2-} \phi(x)$ and ϕ denotes the
solution of (I).

We first solve (I). The D.E. of this problem is
linear. In standard form it is $\dfrac{dy}{dx} + (\dfrac{1}{x+2})y = \dfrac{2x}{x+2}$ with
$P(x) = 1/(x+2)$, $Q(x) = 2x/(x+2)$. An I.F. is $e^{\int P(x)dx} =$
$e^{\int [1/(x+2)]dx} = e^{\ln|x+2|} = |x+2|$. Multiplying the
standard form equation through by this, we obtain the
originally stated D.E., $(x+2)\dfrac{dy}{dx} + y = 2x$, or $\dfrac{d}{dx}[(x+2)y]$
$= 2x$. Integrating, we find $(x+2)y = x^2 + c$. Applying
the I.C. of (I), $y(0) = 4$, we find $c = 8$. Thus the
particular solution of Problem (I) is $(x+2)\,y = x^2 + 8$,
valid for $0 \le x < 2$.

We can write this solution as $y = (x^2+8)/(x+2)$.

Denoting this solution by ϕ, we note that $\lim\limits_{x\to 2-} \phi(x) =$

$\lim\limits_{x\to 2-} [(x^2+8)/(x+2)] = 3$. This is the a of problem (II);

that is, the I.C. of problem (II) is $y(2) = 3$.

We now solve (II). The D.E. is linear. In standard

form it is $\frac{dy}{dx} + (\frac{1}{x+2})y = 4$. Just as in (I), an I.F. is

$|x+2|$. Multiplying the standard form equation through

by this, we obtain the originally stated D.E.,

$(x+2)\frac{dy}{dx} + y = 4$ or $\frac{d}{dx}[(x+2)y] = 4$. Integrating, we

find $(x+2)y = 4x + c$. Applying the I.C. of (II), $y(2) =$

3, we find $c = 4$. Thus the particular solution of

problem (I) is $(x+2)y = 4x + 4$, valid for $x > 2$.

We write the solution of the entire problem as

$$\begin{cases} (x+2)y = x^2 + 8, \ 0 \le x \le 2, \\ (x+2)y = 4x + 4, \ x > 2. \end{cases}$$

24. This D.E. is both a Bernoulli D.E. and a

homogeneous D.E. We first solve it as a Bernoulli

Equation, with $n = 3$. We thus multiply through by

y^{-3}, expressing the D.E. in the equivalent form

$x^2 y^{-3}\frac{dy}{dx} + xy^{-2} = x^{-1}$ or $-2y^{-3}\frac{dy}{dx} - \frac{2}{x}y^{-2} = -\frac{2}{x^3}$.

We let $v = y^{1-n} = y^{-2}$, $dv/dx = -2y^{-3}\, dy/dx$. The D.E.

then transforms into the linear D.E. in v, $\frac{dv}{dx} - \frac{2}{x}v =$

$-\frac{2}{x^3}$. This linear equation is in standard form with

$P(x) = -2/x$, $Q(x) = -2/x^3$. An I.F. is

$e \int P(x)dx = e^{-\int (2/x)dx} = e^{-2 \ln|x|} = x^{-2}$. Multiplying the linear equation through by this we get

$x^{-2} \frac{dv}{dx} - 2x^{-3} v = -2x^{-5}$ or $\frac{d}{dx} (x^{-2} v) = -2x^{-5}$.

Integrating, we find $x^{-2} v = x^{-4}/2 + c$ or

$v = x^{-2}/2 + c x^2$. But $v = y^{-2}$, thus we obtain

$y^{-2} = x^{-2}/2 + cx^2$. Applying the I.C. $y(1) = 1$ to this,

we find $c = 1/2$. Thus we obtain the solution $y^{-2} = x^{-2}/2 + x^2/2$, which can be written as $y^2 = 2x^2/(x^4 + 1)$,

or
$$y = \frac{\sqrt{2} x}{(x^4 + 1)^{1/2}} .$$

Now we express the D.E. in the form $\frac{dy}{dx} = - \frac{y}{x} + (\frac{y}{x})^3$

and solve it as a homogeneous equation. We let $y = vx$,

$\frac{dy}{dx} = v + x \frac{dv}{dx}$ and obtain the transformed separable

equation $v + x \frac{dv}{dx} = -v + v^3$, which reduces to

$\frac{dv}{v(v^2 - 2)} = \frac{dx}{x}$. We rewrite the left member, using

partial fractions, and thus have

$$(- \frac{1}{v} + \frac{v}{v^2 - 2}) dv = 2 \frac{dx}{x} .$$

Integrating gives $-\ln|v| + (\ln|v^2 - 2|)/2 = 2\ln|x|$

$+ \ln|c_1|$. This simplifies to

$$- \ln v^2 + \ln|v^2 - 2| = \ln x^4 + \ln c ,$$

$\ln(|v^2 - 2|/v^2) = \ln cx^4$. From this, $|v^2 - 2| = cv^2x^4$. Resubstituting $v = y/x$, we find $|y^2 - 2x^2| = cy^2x^4$. Applying the I.C. $y(1) = 1$, we find $c = 1$.

17. This D.E., with $M(x,y) = e^{2x}y^2 - 2x$ and
$N(x,y) = e^{2x}y$, is exact; for $M_y(x,y) = 2e^{2x}y = N_y(x,y)$.
We seek $F(x,y)$ such that $F_x(x,y) = M(x,y) = e^{2x}y^2 - 2x$
and $F_y(x,y) = N(x,y) = e^{2x}y$. From the first of these,
$F(x,y) = e^{2x} y^2/2 - x^2 + \phi(y)$. From this, $F_y(x,y) =$
$e^{2x}y + \phi'(y)$. But this must equal $N(x,y) = e^{2x}y$. Thus
we must have $\phi'(y) = 0$ and hence $\phi(y) = c_0$. Thus
$F(x,y) = e^{2x} y^2/2 - x^2 + c_0$; and from this the solutions
$F(x,y) = c_1$ take the form $e^{2x} y^2/2 - x^2 = c$, where
$c = c_1 - c_0$. Now apply the given I.C. $y(0) = 2$ to this.
Thus letting $x = 0$, $y = 2$, we find $c = 2$. Thus we
obtain the solution $e^{2x} y^2/2 - x^2 = 2$ or $y^2 e^{2x} = 2x^2+4$
or, finally, $y = e^{-x}(2x^2 + 4)^{1/2}$.

18. This D.E., with $M(x,y) = 3x^2 + 2xy^2$ and
$N(x,y) = 2x^2y + 6y^2$, is exact; for $M_y(x,y) = 4xy = $
$N_x(x,y)$. We seek $F(x,y)$ such that $F_x(x,y) = M(x,y) =$
$3x^2 + 2xy^2$ and $F_y(x,y) = N(x,y) = 2x^2y + 6y^2$. From the
first of these, $F(x,y) = x^3 + x^2y^2 + \phi(y)$. From this,
$F_y(x,y) = 2x^2y + \phi'(y)$. But this must equal $N(x,y) =$
$2x^2y + 6y^2$. Thus we must have $\phi'(y) = 6y^2$ and hence
$\phi(y) = 2y^3 + c_0$. Thus $F(x,y) = x^3 + x^2y^2 + 2y^3 + c_0$;
and from this the solutions $F(x,y) = c_1$ take the form
$x^3 + x^2y^2 + 2y^3 = c$, where $c = c_1 - c_0$. Applying the
I.C. $y(1) = 2$, we let $x = 1$, $y = 2$, to obtain $c = 21$.

Thus we find the solution $|y^2 - 2x^2| = y^2x^4$. Since $y^2 - 2x^2$ is initially -1, we take $|y^2 - 2x^2| = 2x^2-y^2$, and write the solution as $2x^2-y^2 = x^4y^2$ or $(x^4 + 1)y^2 = 2x^2$ or $y^2 = 2x^2/(x^4 + 1)$), as we obtained before.

Section 2.4, Page 66

1. This equation is of the form (2.42) of Theorem 2.6 with $M(x,y) = 5xy + 4y^2 + 1$, $N(x,y) = x^2 + 2xy$. To apply that theorem, we first find

$$\frac{M_y(x,y) - N_x(x,y)}{N(x,y)} = \frac{5x + 8y - 2x - 2y}{x^2 + 2xy} = \frac{3(x + 2y)}{x(x + 2y)} = \frac{3}{x}.$$

This depends on x only, so

$$e^{\int(3/x)dx} = e^{3 \ln|x|} = e^{\ln|x|^3} = |x|^3$$

is an I.F. of the given equation. Multiplying the given equation through by this and simplifying, we find $(5x^4y + 4x^3y^2 + x^3)dx + (x^5 + 2x^4y)dy = 0$. For this equation $M(x,y) = 5x^4y + 4x^3y^2 + x^3$, $N(x,y) = x^5 + 2x^4y$, and $M_y(x,y) = 5x^4 + 8x^3y = N_x(x,y)$. Thus this equation is exact. We seek $F(x,y)$ such that $F_x(x,y) = M(x,y)$ and $F_y(x,y) = N(x,y)$. From the first of these, $F(x,y) = \int M(x,y)\partial x = \int (5x^4y + 4x^3y^2 + x^3)\partial x = x^5y + x^4y^2 + x^4/4 + \phi(y)$. From this, $F_y(x,y) =$

$x^5 + 2x^4 y + \phi'(y)$. But we must have $F_y(x,y) = N(x,y) =$
$x^5 + 2x^4 y$. Thus $\phi'(y) = 0$, $\phi(y) = c_0$. Hence $F(x,y) =$
$x^5 y + x^4 y^2 + x^4/4 + c_0$. The family of solutions is then
$F(x,y) = c_1$, and multiplying through by 4 we express
this as $4x^5 y + 4x^4 y^2 + x^4 = c$, where $c = 4(c_1 - c_0)$.

4. This equation is of the form (2.42) of Theorem
2.6 with $M(x,y) = 2xy^2 + y$, $N(x,y) = 2y^3 - x$. To
apply that theorem, we first find

$$\frac{M_y(x,y) - N_x(x,y)}{N(x,y)} = \frac{2(2xy + 1)}{2y^3 - x} \; ; \text{ but since this}$$

depends on y as well as x, we cannot proceed using this.
We next find

$$\frac{N_x(x,y) - M_y(x,y)}{M(x,y)} = \frac{-2(1 + 2xy)}{y(1 + 2xy)} = -\frac{2}{y} \; . \text{ This depends}$$

on y only, so

$$e^{-\int (2/y)dy} = e^{-2 \ln|y|} = e^{\ln|y|^{-2}} = |y|^{-2} = 1/y^2$$

is an I.F. of the given equation. Multiplying the
given equation through by this and simplifying, we find

$$(2x + y^{-1}) \, dx + (2y - xy^{-2})dy = 0.$$

For this equation $M(x,y) = 2x + y^{-1}$, $N(x,y) = 2y - xy^{-2}$,
$M_y(x,y) = -y^{-2} = N_x(x,y)$. Thus this equation is exact.
We seek $F(x,y)$ such that $F_x(x,y) = M(x,y)$ and
$F_y(x,y) = N(x,y)$. From the first of these,

$$F(x,y) = \int M(x,y)\partial x = \int (2x + y^{-1})\partial x = x^2 + xy^{-1} + \phi(y).$$

From this, $F_y(x,y) = -xy^{-2} + \phi'(y)$. But we must have

$F_y(x,y) = N(x,y) = 2y - xy^{-2}$. Thus $\phi'(y) = 2y$,

$\phi(y) = y^2 + c_0$. Hence $F(x,y) = x^2 + xy^{-1} + y^2 + c_0$.

The family of solutions $F(x,y) = c_1$ is then expressed

as $x^2 + xy^{-1} + y^2 = c$, where $c = c_1 - c_0$.

5. We first multiply the stated equation through by

$x^p y^q$ to obtain $(4x^{p+1}y^{q+2} + 6x^p y^{q+1})dx + (5x^{p+2}y^{q+1} + 8x^{p+1}y^q)dy = 0$. For this equation, we have

$M(x,y) = 4x^{p+1}y^{q+2} + 6x^p y^{q+1}$,

$N(x,y) = 5x^{p+2}y^{q+1} + 8x^{p+1}y^q$,

$M_y(x,y) = 4(q+2)x^{p+1}y^{q+1} + 6(q+1)x^p y^q$, and

$N_x(x,y) = 5(p+2)x^{p+1}y^{q+1} + 8(p+1)x^p y^q$.

In order for the equation to be exact, we must have

$M_y(x,y) = N_x(x,y)$, and hence

$$\begin{cases} 4(q+2) = 5(p+2) \\ 6(q+1) = 8(p+1) \end{cases}, \text{ that is }, \begin{cases} 5p - 4q = -2 \\ 8p - 6q = -2 \end{cases}.$$

Solving these, we find $p = 2$, $q = 3$. Thus the desired

I.F. of the form $x^p y^q$ is $x^2 y^3$.

Now multiplying the original equation through by

this I.F. $x^2 y^3$, we obtain the equivalent exact equation

$$(4x^3 y^5 + 6x^2 y^4)dx + (5x^4 y^4 + 8x^3 y^3)dy = 0.$$

It is easy to check that this is indeed exact. We seek
$F(x,y)$ such that $F_x(x,y) = M(x,y) = 4x^3y^5 + 6x^2y^4$ and
$F_y(x,y) = N(x,y) = 5x^4y^4 + 8x^3y^3$. From the first of
these,

$$F(x,y) = \int M(x,y)\partial x = \int (4x^3y^5 + 6x^2y^4)\partial x$$

$$= x^4y^5 + 2x^3y^4 + \phi(y).$$

From this, $F_y(x,y) = 5x^4y^4 + 8x^3y^3 + \phi'(y)$. But we must
have $F_y(x,y) = N(x,y) = 5x^4y^4 + 8x^3y^3$. Thus $\phi'(y) = 0$,
$\phi(y) = c_0$. Thus $F(x,y) = x^4y^5 + 2x^3y^4 + c_0$, and the
solution $F(x,y) = c_1$ is

$x^4y^5 + 2x^3y^4 = c$ or $x^3y^4(xy + 2) = c$, where

$c = c_1 - c_0$.

    7. For this equation $a_1 = 5$, $b_1 = 2$, $a_2 = 2$,
$b_2 = 1$, so

$$\frac{a_2}{a_1} = \frac{2}{5} \neq \frac{1}{2} = \frac{b_2}{b_1} \qquad .$$

Therefore this is Case 1 of Theorem 2.7. We make the
transformation $\begin{cases} x = X + h \\ y = Y + k \end{cases}$,

where (h,k) is the solution of the system

$$\begin{cases} 5h + 2k + 1 = 0 \\ 2h + k + 1 = 0 \end{cases}$$

The solution of this system is $h = 1$, $k = -3$, and so
the transformation is $\begin{cases} x = X + 1 \\ y = Y - 3 \end{cases}$.

This reduces the given equation to the homogeneous
equation

$$(5X + 2Y) \, dX + (2X + Y) \, dY = 0 \quad .$$

We write this in the form

$$\frac{dY}{dX} = -\frac{5 + 2(Y/X)}{2 + (Y/X)}$$

and let $Y = vX$ to obtain

$$v + X \frac{dv}{dX} = -\frac{5 + 2v}{2 + v} \quad .$$

This reduces to

$$X \frac{dv}{dX} = \frac{5 - v^2}{2 + v} \quad \text{or} \quad \frac{(v + 2)dv}{v^2 + 4v + 5} = -\frac{dX}{X} \quad .$$

Integrating we find

$$\ln|v^2 + 4v + 5| = -2 \ln|X| + \ln |c_1| \quad .$$

or $\quad \ln|v^2 + 4v + 5| = \ln(|c_1| \, X^{-2}) \quad .$

From this, $|v^2 + 4v + 5| = |c_1| \, X^{-2} \quad .$

Now replacing v by Y/X, and simplifying, we obtain

$$|5X^2 + 4XY + Y^2| = |c_1| \quad .$$

Finally replacing X by $x - 1$ and Y by $y + 3$, we obtain
the solutions of the original D.E. in the form

$$|5(x-1)^2 + 4(x-1)(y+3) + (y+3)^2| = |c_1|$$

or
$$|5x^2 + 4xy + y^2 + 2x + 2y + 14| = |c_1| \quad .$$

Assuming $x > 0$, $y > 0$, and writing $c = c_1 - 14$, this

takes the form

$$5x^2 + 4xy + y^2 + 2x + 2y = c.$$

8. Here $a_1 = 3$, $b_1 = -1$, $a_2 = -6$, $b_2 = 2$, and $a_2/a_1 = -2 = b_2/b_1$. Therefore this is Case 2 of Theorem 2.7. We therefore let $z = 3x - y$. Then $dy = 3dx - dz$, and the given D.E. transforms into

$$(z + 1)dx - (2z - 3)(3dx - dz) = 0$$

or $(-5z + 10)dx + (2z - 3)dz = 0$,

which is separable. Separating variables, we obtain

$$dx + \frac{2z - 3}{-5z + 10} \ dz = 0$$

or

$$dx + (-\frac{2}{5} - \frac{1}{5(z - 2)}) \ dz = 0.$$

Integrating, we have

$$x - \frac{2}{5} z - \frac{1}{5} \ln |z - 2| = c_1 \ .$$

or

$$5x - 2z - \ln |z - 2| = 5c_1 \ .$$

Replacing z by $3x - y$ and simplifying, we obtain the solution of the given D.E. in the form

$$x - 2y + \ln|3x - y - 2| = c \ , \quad \text{where } c = -5c_1 \ .$$

12. The D.E. of this problem is of the form (2.52) of Theorem 2.7, with $a_1 = 3$, $b_1 = -1$, $a_2 = 1$, $b_2 = 1$, and hence $a_2/a_1 = 1/3 \neq -1 = b_2/b_1$. Therefore this is Case 1 of the theorem. We let $x = X + h$, $y = Y + k$,

where (h,k) is the solution of the system

$$\begin{cases} 3h - k - 6 = 0 \\ h + k + 2 = 0. \end{cases}$$ The solution os this is h = 1,

k = -3, and so the transformation is $\begin{cases} x = X + 1, \\ y = Y - 3 \end{cases}$.

This reduces the given equation to the homogeneous

equation $(3X - Y)dX + (X + Y)dY = 0$.

We write this in the form

$$\frac{dY}{dX} = \frac{(Y/X) - 3}{(Y/X) + 1}$$

and let Y = vX to obtain

$$v + X \frac{dv}{dX} = \frac{v-3}{v+1}$$.

This reduces to

$$X \frac{dv}{dX} = -\frac{v^2 + 3}{v + 1} \quad \text{or} \quad \frac{(v+1)dv}{v^2 + 3} = -\frac{dX}{X}$$.

Integrating (tables may be needed!), we find

$$\frac{1}{2} \ln(v^2 + 3) + \frac{1}{\sqrt{3}} \text{ arc tan } \frac{v}{\sqrt{3}} = -\ln|X| + c$$

or

$$\ln(v^2 + 3) + \frac{2}{\sqrt{3}} \text{ arc tan } \frac{v}{\sqrt{3}} = -\ln X^2 + c_1$$.

Now replacing v by Y/X, we obtain

$$\ln\left[\frac{Y^2}{X^2} + 3\right] + \frac{2}{\sqrt{3}} \text{ arc tan } \frac{Y}{\sqrt{3}X} = -\ln X^2 + c_1$$.

Combining the ln terms, this takes the form

$$\ln(Y^2 + 3X^2) + \frac{2}{\sqrt{3}} \text{ arc tan } \frac{Y}{\sqrt{3}X} = C_1$$.

Finally replacing X by x - 1 and Y by y + 3, we obtain
the solutions of the original D.E. in the form

$$\ln[3(x-1)^2 + (y+3)^2] + \frac{2}{\sqrt{3}} \text{ arc tan } \frac{y+3}{\sqrt{3}(x-1)} = c_1.$$

We now apply the I.C. $y(2) = -2$ to this, obtaining

$$\ln 4 + \frac{2}{\sqrt{3}} \text{ arc tan } \frac{1}{\sqrt{3}} = c_1$$

or $\ln 4 + (2/\sqrt{3})(\pi/6) = c_1$. Thus we obtain the
solution of the stated I.V. problem as

$$\ln[3(x-1)^2 + (y+3)^2] + \frac{2}{\sqrt{3}} \text{ arc tan } \frac{y+3}{\sqrt{3}(x-1)} = \ln 4 + \frac{\pi}{3\sqrt{3}}.$$

CHAPTER 3

Section 3.1, Page 75

1. **Step 1.** We first find the D.E. of the given family $y = cx^3$. Differentiating, we obtain $\frac{dy}{dx} = 3cx^2$. Eliminating the parameter c between the equation of the given family and its derived equation, we obtain the D.E. of the given family in the form $\frac{dy}{dx} = 3(\frac{y}{x^3})x^2$ or $\frac{dy}{dx} = \frac{3y}{x}$.

Step 2. We now find the D.E. of the orthogonal trajectories by replacing $3y/x$ in the D.E. (of Step 1) by its negative reciprocal $-x/3y$, thus obtaining

$$\frac{dy}{dx} = -\frac{x}{3y} \quad .$$

Step 3. We solve this last D.E. Separating variables, we have $3ydy = -x\,dx$. Integrating, we find

95

$(3y^2/2) = -(x^2/2) + k_1$ or $x^2 + 3y^2 = k^2$, where $k^2 = -2k_1 \geq 0$. This is the family of orthogonal trajectories of the given family of cubics. It is a family of ellipses with centers at the origin and major axes along the x axis.

4. <u>Step 1</u>. We first find the D.E. of the given family $y = e^{cx}$. Differentiating, we obtain $\frac{dy}{dx} = ce^{cx}$. We eliminate the parameter c between the equation of the given family and its derived equation. From the given equation, we have $y > 0$ and $\ln y = cx$. Thus $c = (\ln y)/x$. Substituting in the derived equation we find $\frac{dy}{dx} = (y \ln y)/x$. This is the D.E. of the given family.

      <u>Step 2</u>. We now find the D.E. of the orthogonal trajectories by replacing $(y \ln y)/x$ by its negative reciprocal $-x/y \ln y$, thereby obtaining the D.E.

$$\frac{dy}{dx} = -\frac{x}{y \ln y} \, \circ$$

      <u>Step 3</u>. We solve this last D.E. Separating variables, we have $y \ln y \, dy = -x \, dx$. Integrating (by parts), we find $(y^2 \ln y)/2 - y^2/4$
$$= -x^2/2 + k_1 \text{ or}$$
$$y^2(\ln y - \tfrac{1}{2}) = -x^2 + k.$$
This is the family of orthogonal trajectories of the given family of exponentials.

8. <u>Step 1.</u> We first find the D.E. of the given
family $x^2 = 2y - 1 + ce^{-2y}$. Differentiating, we obtain
$2x = 2\frac{dy}{dx} - 2ce^{-2y} \frac{dy}{dx}$. We eliminate the parameter c
between these two equations as follows. From the given
equation, $c = (x^2 - 2y + 1)e^{2y}$. Substituting this in
the second equation, we get

$$x = \frac{dy}{dx} - (x^2 - 2y + 1) \frac{dy}{dx} \quad \text{or} \quad \frac{dy}{dx} = \frac{x}{2y - x^2}$$

This is the D.E. of the given family.

   <u>Step 2.</u> We now find the D.E. of the orthogonal
trajectories by replacing $x/(2y - x^2)$ by its negative
reciprocal $(x^2 - 2y)/x$ thereby obtaining the D.E.

$$\frac{dy}{dx} = \frac{x^2 - 2y}{x} \quad .$$

   <u>Step 3.</u> We solve this last D.E. Writing it as
$\frac{dy}{dx} + \frac{2}{x} y = x$, we recognize that it is a linear equation
in standard form with $P(x) = \frac{2}{x}$ and $Q(x) = x$. An I.F.
is $e^{\int P(x)dx} = e^{\int (2/x)dx} = e^{2\ln|x|} = x^2$.
Multiplying through by this, we have $x^2 \frac{dy}{dx} + 2xy = x^3$
or $\frac{d}{dx}[x^2y] = x^3$. Integrating, we find $x^2y = x^4/4 + k$.
This is the equation of the one-parameter family of
orthogonal trajectories.

12. We first find the D.E. of the given family of
parabolas $y = c_1 x^2 + K$. Differentiating, we obtain
$dy/dx = 2c_1 x$. Eliminating the parameter c between the
given equation and the derived equation, we obtain the
D.E. of the family of parabolas in the form
$dy/dx = (2y - 2K)/x$. It will be convenient to rewrite
this slightly, as follows: $dy/dx = (4y - 4K)/2x$. Then
the D.E. of the family of orthogonal trajectories of the
given family of parabolas is

$$dy/dx = 2x/(4K - 4y) \ . \qquad\qquad (\ast)$$

We now find the D.E. of the given family of ellipses
$x^2 + 2y^2 - y = c_2$. Differentiating, we obtain
$2x + 4y \, dy/dx - dy/dx = 0$, from which we at once obtain
the D.E. of the family of ellipses, that is,

$$dy/dx = 2x/(1 - 4y) \ . \qquad\qquad (\ast\ast)$$

Comparing ($\ast$) and ($\ast\ast$) we see that the two given
families are orthogonal provided $K = {}^1/4$.

17. Step 1. We first find the D.E. of the given
family $x + y = cx^2$. Differentiating, we obtain
$1 + dy/dx = 2cx$. Eliminating the parameter c between
the given equation and the derived equation, we obtain
the D.E. of the given family in the form $dy/dx =$
$(x + 2y)/x$.

Step 2'. We replace $f(x,y) = (x + 2y)/x$ in this D.E. by

$$\frac{f(x,y) + \tan \alpha}{1 - f(x,y)\tan \alpha} = \frac{f(x,y) + 2}{1 - 2f(x,y)} = \frac{(x+2y)/x + 2}{1 - 2(x+2y)/x} = -\frac{3x + 2y}{x + 4y} .$$

We have the D.E. $\frac{dy}{dx} = -\frac{3x + 2y}{x + 4y}$ of the desired oblique trajectories.

Step 3. We now solve this D.E. Writing it in the form $\frac{dy}{dx} = -\frac{3 + 2(y/x)}{1 + 4(y/x)}$, we see that it is homogeneous. We let $y = vx$ to obtain $v + x \, dv/dx =$ $-(3 + 2v)/(1 + 4v)$ or $\frac{(4v + 1)dv}{4v^2 + 3v + 3} = -\frac{dx}{x}$. Integrating this (we recommend tables), we find $(1/2)\ln|4v^2 + 3v + 3|$ $-(1/2)(2/\sqrt{39})$ arc tan $(8v + 3)/\sqrt{39} = -\ln x + (1/2)c$, where the final 1/2 is for convenience. We multiply through by 2 and replace v by y/x to obtain

$$\ln \left| \frac{4y^2 + 3xy + 3x^2}{x^2} \right| - (2/\sqrt{39}) \text{ arc tan}[(8y+3x)/\sqrt{39}x]$$

$= -\ln x^2 + c$. Thus we find the desired family of oblique trajectories in the form

$\ln|3x^2 + 3xy + 4y^2| - (2/\sqrt{39})$ arc tan$[(3x + 8y)/\sqrt{39}x] = c.$

Section 3.2, Page 85

1. We choose the positive x axis vertically down-ward along the path of the stone and the origin at the

point from which the body fell. The forces acting on
the stone are: (1) F_1, its weight, 4 lbs., which acts
downward and so is positive, and (2) F_2, the air
resistance, numerically equal to v/2, which acts
upward and so is the negative quantity - v/2. Newton's
Second Law gives m dv/dt = F_1 + F_2. Using g = 32 and
m = w/g = 4/32 = 1/8, this becomes

$$\frac{1}{8} \frac{dv}{dt} = 4 - v/2 \ .$$

The initial condition is v(0) = 0.

The D.E. is separable. We write it as

$$\frac{dv}{4 - v/2} = 8 \ dt \ .$$

Integrating we find - 2 ln |4 - v/2| = 8t + c_0 which
reduces to |4 - v/2| = c_1 e^{-4t} . Applying the I.C.
v(0) = 0 to this, we find c_1 = 4. Thus the velocity at
time t is given by v = 8 - $8e^{-4t}$. Writing this as
dx/dt = 8 - $8e^{-4t}$ and integrating, we find x = 8t + $2e^{-4t}$
+ c_2. Applying the I.C. x(0) = 0, we find c_2 = -2.
Thus the distance fallen is x = 8t + $2e^{-4t}$ - 2. Thus
the answers to part (a) are: v = 8(1 - e^{-4t}) and
x = 2(4t + e^{-4t} - 1).

To answer part (b), we simply let t = 5 in these
expressions for v and x. Thus v(5) = 8(1 - e^{-20})$\approx$ 8 ft/sec
and x(5) = 2(19 + e^{-20})$\approx$ 38 feet.

3. We choose the positive x axis vertically upward
along the path of the ball and the origin at the ground
level. The forces acting on the ball are: (1) F_1, its
weight, (3/4) lb, which acts downward and so is negative
and (2) F_2, the air resistance, numerically equal to
v/64, which also acts downward and so is the negative
quantity -v/64. Newton's Second Law gives m dv/dt =
$F_1 + F_2$. Using g = 32 and m = w/g = (3/4)/32 = 3/128,
this becomes

$$\frac{3}{128} \frac{dv}{dt} = - \frac{3}{4} - \frac{v}{64} \quad .$$

The initial condition is v(0) = 20.

The D.E. is separable. We write it as

$$\frac{3dv}{2v + 96} = - dt \quad .$$

Integrating we find (3/2) ln $|2v + 96|$ = -t + c_0 which
reduces to $|v + 48|$ = c_1 $e^{-2t/3}$. Applying the I.C.
v(0) = 20 to this, we find c_1 = 68. Thus the velocity
at time t is given by v = 68 $e^{-2t/3}$ - 48. From this we
obtain x = -102 $e^{-2t/3}$ - 48t+c_2. Applying the I.C.
x(0) = 6, we obtain c_2 = 108. Thus we have the distance
x = -102$e^{-2t/3}$ - 48t + 108.

The question asks how high the ball will rise. It
will stop rising and start falling when the velocity
v = 0. Thus we set 68$e^{-2t/3}$ - 48 = 0, obtaining

$e^{-2t/3}$ = 12/17 and t $\approx$ 0.5225 (seconds). For this value of t, we find x $\approx$ 10.92, which is the height above the ground that the ball will rise.

4. The forces acting on the ship are: (1) F_1, the constant propeller thrust of 100,000 lb., which moves the ship forward and so is positive, and (2) F_2, the resistance, numerically equal to 8000v, which acts against the forward motion of the ship and so is the negative quantity -8000v. Newton's Second Law gives m dv/dt = F_1 + F_2. Using g = 32 and m = w/g = (32000 tons)(2000 lbs/ton)/32 = 2,000,000, this becomes

$$2,000,000 \frac{dv}{dt} = 100,000 - 8000v.$$

The initial condition is v(0) = 0.

The D.E. is separable. We write it as

$$\frac{dv}{dt} = \frac{1}{20} - \frac{v}{250} \text{ or } \frac{dv}{25 - 2v} = \frac{dt}{500} .$$

Integrating we find $\ln|25 - 2v|$ = -t/250 + c_0 which reduces to $|25 - 2v|$ = $c_1 e^{-t/250}$. Applying the I.C. v(0) = 0 to this, we find c_1 = 25. Thus the velocity at time t is given by

$$v = \frac{25}{2} (1 - e^{-t/250}) \qquad (\ast)$$

This is the answer to part (a). The answer to part (b) is then

$\lim\limits_{t\to\infty} v = \lim\limits_{t\to\infty} \frac{25}{2}(1 - e^{-t/250}) = 12.5$ (ft/sec). To

answer part (c), let $v = (0.80)(12.5)$ in (✱) and solve

for t. We obtain $0.80 = 1 - e^{-t/250}$ and hence

$e^{-t/250} = 1/5$. Thus $-t/250 = \ln(1/5)$, from which

$t = 402$ (sec.).

8. We choose the positive x-axis horizontally along

the given direction of motion. The forces acting on

the boat and rider are: (1) F_1, the constant force of

12 lb., which acts in the given direction and so is

positive, and (2) F_2, the resistance force, numerically

equal to 2v, which acts opposite to the given direction

and so is the negative quantity - 2v. Newton's Second

Law gives $m\ dv/dt = F_1 + F_2$. Using $g = 32$ and

$m = w/g = (150 + 170)/32 = 10$, this becomes

$$10\,\frac{dv}{dt} = 12 - 2v.$$

The initial velocity is 20 m.p.h. = 88/3 ft/sec. Thus

the I.C. is $v(0) = 88/3$.

The D.E. is separable. We write it as

$$\frac{5\ dv}{6 - v} = dt.$$

Integrating we find $5\ \ln|v - 6| = -t + c_0$ which reduces

to $|v - 6| = c\ e^{-t/5}$. Applying the I.C. $v(0) = 88/3$ to

this, we find $c = 70/3$. Thus the velocity at time t

is given by $v = 6 + (70/3)\ e^{-t/5}$.   (✱)

To answer part (a), we let $t = 15$ in (✱) to obtain $v(15) = 6 + (70/3) e^{-3} \approx 7.16$ ft/sec. To answer part (b), let $v = (1/2)(88/3) = 44/3$ in (✱) and solve for t. We have, $6 + (70/3)e^{-t/5} = (44)(3)$, from which $e^{-t/5} = 13/35$. From this, $-t/5 \approx -0.99$ and $t \approx 4.95$ (seconds).

10. We choose the positive x-axis vertically upward along the path of the shell with the origin at the earth's surface. The forces acting on the shell are: (1) F_1, its weight, 1 lb., which acts downward and so is the negative quantity -1; and (2) F_2, the air resistance numerically equal to $10^{-4} v^2$, which also acts downward (against the rising shell) and so is the negative quantity $-10^{-4} v^2$. Newton's Second Law gives $m \, dv/dt = F_1 + F_2$. Using $g = 32$ and $m = w/g = 1/32$, this becomes

$$\frac{1}{32} \frac{dv}{dt} = -1 - 10^{-4}v^2 \ .$$

The initial condition is $v(0) = 1000$.

The D.E. is separable. We write it as

$$\frac{1}{32} \frac{dv}{dt} = - \frac{10^4 + v^2}{10^4} \quad \text{or} \quad \frac{dv}{v^2 + 10^4} = - \frac{32}{10^4} dt \ .$$

Integrating we find $(1/100)$ arc tan $(v/100) = -32t/10^4 + c$ or arc tan$(v/100) = - 32t/100 + c$.

Application of the initial condition at once gives

c = arc tan 10. Thus we obtain the solution in the form

arc tan $(v/100)$ = arc tan 10 - 32t/100 .

Taking the tan of each side and multiplying by 100 gives

v = 100 tan (arc tan 10 - 0.32t) .

This is the answer to part (a). To answer part (b),
note that the shell will stop rising when v = 0. **Setting**
v = 0, we at once have arc tan 10 - 0.32t = 0, and
thus t = arc tan 10/0.32 $\approx$ 4.60 (sec.).

13. We choose the positive x-axis horizontally
along the given direction of motion and the origin at
the point at which the man stops pushing. The forces
acting on the loaded sled as it continues are: (1) F_1,
the air resistance, numerically equal to 3v/4, which
acts opposite to the direction of motion and so is given
by -3v/4, and (2) F_2, the frictional force, having
numerical value μN = (0.04)(80), which also acts
opposite to the direction of motion and so is given by
-(0.04)(80) = -16/5. Newton's Second Law gives
m dv/dt = F_1 + F_2. Using g = 32 and m = w/g = 80/32 =
5/2, this becomes

$$\frac{5}{2}\frac{dv}{dt} = -\frac{3v}{4} - \frac{16}{5} \ .$$

The initial velocity is 10 ft/sec., so the I.C. is
$v(0)$ = 10.

The D.E. is separable. We write it as

$$\frac{dv}{15v + 64} = - \frac{dt}{50} .$$

Integrating, we find $\ln|15v + 64| = -3t/10 + c_0$ which reduces to $|15v + 64| = c\, e^{-3t/10}$. Applying the I.C. $v(0) = 10$ to this, we find $c = 214$. Thus the velocity is given by

$$v = (214/15)\, e^{-3t/10} - 64/15 .$$

From this, integration and simplification give $x = -(428/9)\, e^{-3t/10} - 64t/15 + c_1$. Application of the condition $x(0) = 0$ then gives $c_1 = 428/9$. Hence the distance x is given by

$$x = (428/9)(1 - e^{-3t/10}) - 64t/15 .$$

To answer the stated question, note that the sled will continue until the velocity $v = 0$. Thus we set $v = 0$ and solve for t, finding $e^{-3t/10} = 32/107$ from which $t \approx 4.02$. We now evaluate x at $t \approx 4.02$ to determine the distance which the sled will continue. We find

$x(4.02) \approx (428/9)(1 - 32/107) - 64(4.02)/15 \approx 16.18$ feet.

15. We choose the positive x direction down the slide with the origin at the top. The forces acting on the case are: (1) F_1, the component of its weight parallel to the slide; (2) F_2, the frictional force;

and F_3, the air resistance. The case weighs 24 lbs.,
and the component parallel to the slide has numerical
value $24 \sin 45° = 24/\sqrt{2}$. Since this acts in the
positive (downward) direction along the slide,
$F_1 = 24/\sqrt{2}$. The frictional force F_2 has numerical
value μN, where μ is the coefficient of friction
and N is the normal force. Here $\mu = 0.4$ and the
magnitude of N is $24 \cos 45° = 24/\sqrt{2}$. Since the
force F_2 acts in the negative (upward) direction along
the slide, we have $F_2 = -(0.4)(24/\sqrt{2})$. Finally, the air
resistance F_3 has numerical value $v/3$. Since this also
acts in the negative direction, we thus have $F_3 = -v/3$.
Newton's Second Law now gives $m\, dv/dt = F_1 + F_2 + F_3$.
With $g = 32$, $m = w/g = 24/32 = 3/4$ and the above forces,
this becomes

$(3/4)\,(dv/dt) = (24/\sqrt{2}) - (0.4)(24/\sqrt{2}) - v/3$.

The initial condition is $v(0) = 0$.

The D.E. is separable. We write it as

$(3/4)\,(dv/dt) = (-5v + 108\sqrt{2})/15$ or $\dfrac{dv}{5v-108\sqrt{2}} = -\dfrac{4}{45}\,dt$.

Integrating we find $\ln|5v - 108\sqrt{2}| = -4t/9 + c_0$ or

$|5v - 108\sqrt{2}| = c\, e^{-4t/9}$. Applying the initial condition,
we at once find $c = 108\sqrt{2}$. Thus the velocity is given
by

$$v = (108\sqrt{2}/5)\,(1 - e^{-4t/9}). \qquad (\ast)$$

To answer (a), simply let t = 1 in (✱). We obtain
$v(1) = (108\sqrt{2}/5)(1 - e^{-4}/9) \approx 10.96$ (ft/sec). To
answer (b), more work is required. We first integrate
(✱) to find the distance x from the top of the slide.
We have

$$x = (108\sqrt{2}/5)t + (243\sqrt{2}/5)e^{-4t/9} + c.$$

Since x = 0 at t = 0, we find c = $-243\sqrt{2}/5$, and hence
$$x = (108\sqrt{2}/5)t + (243\sqrt{2}/5)e^{-4t/9} - 243\sqrt{2}/5.$$
The slide being 30 ft. long, we let x = 30 in this
to determine the time t at which the case reaches the
bottom. That is, we must find t such that
$$30 = (108\sqrt{2}/5)t + (243\sqrt{2}/5)e^{-4t/9} - 243\sqrt{2}/5.$$
We simplify this so that it takes the form
$$68.72\, e^{-4t/9} = -30.54t + 98.72.$$
A little trial-and-error calculation with a hand
calculator shows that the two sides of this are
approximately equal for t = 2.49. This is the time at
which the case reaches the bottom. Letting t = 2.49 in
(✱), we find the velocity at that time to be
approximately 20.46 (ft/sec).

16. We choose the positive x direction down the
hill with the origin at the starting point. The forces
acting on the boy and sled are: (1) F_1, the component
of their weight parallel to the hill; (2) F_2, the

frictional force; and F_3, the air resistance. The boy
and sled weigh 72 lbs., and the component parallel to
the hill has numerical value 72 sin 30° = 36. Since
this acts in the positive (downward) direction on the
hill, F_1 = 36. The frictional force F_2 has numerical
value μN, where μ > 0 is the coefficient of
friction and N is the normal force. The magnitude of
N is 72 cos 30° = $36\sqrt{3}$; the μ is an unknown which will
be determined in due course from the given data of the
problem. Since the force F_2 acts in the negative
(upward) direction on the hill, we have F_2 = $-\mu(36\sqrt{3})$.
Finally, the air resistance F_3 has numerical value 2v.
Since this also acts in the negative direction, we thus
have F_3 = -2v. Newton's Second Law now gives
m dv/dt = $F_1 + F_2 + F_3$. With g = 32,
m = w/g = 72/32 = 9/4 and the above forces, this becomes
$$(9/4)(dv/dt) = 36 - \mu\, 36\sqrt{3} - 2v.$$
The initial condition is v(0) = 0. Another condition
is also given, that is, v(5) = 10; and this extra
condition will be sufficient   for us to eventually
determine μ .

The D.E. is separable. We write it as
$$dv/dt = (8/9)(18 - \mu\, 18\sqrt{3} - v)$$
or

$$\frac{dv}{v + 18(\mathcal{M}\sqrt{3} - 1)} = -\frac{8}{9} dt .$$

Integrating we find $\ln|v + 18(\mathcal{M}\sqrt{3} - 1)| = -8t/9 + c_0$

or $|v + 18(\mathcal{M}\sqrt{3} - 1)| = c\, e^{-8t/9} .$

Applying the initial condition, we at once find that
$c = |18(\mathcal{M}\sqrt{3} - 1)|$. Thus we obtain the solution in
the form $v + 18(\mathcal{M}\sqrt{3} - 1) = 18(\mathcal{M}\sqrt{3} - 1)\, e^{-8t/9}$ or
$v = 18(1 - \mathcal{M}\sqrt{3})(1 - e^{-8t/9})$. Now we apply the extra
condition $v(5) = 10$ to this to determine $\mathcal{M}$. We have
$10 = 18(1 - \mathcal{M}\sqrt{3})(1 - e^{-40/9})$ and hence
$1 - \mathcal{M}\sqrt{3} = (5/9)(0.988)^{-1}$, from which we obtain
$\mathcal{M} = 0.25$.

17. This problem has two parts: (A), <u>before</u> the
object reaches the surface of the lake; and (B), <u>after</u>
it passes beneath the surface. We consider (A) first.
We take the positive x-axis vertically downward with
the origin at the point of release of the object. The
forces acting on the body are: (1) F_1, its weight,
32 lbs., which acts downward and so is positive; and
(2) F_2, the air resistance, numerically equal to 2v,
which acts upward and so is the negative quantity $-2v$.
Applying Newton's Second Law, with $m = w/g = 32/32 = 1$,
we at once obtain the D.E.

$$dv/dt = 32 - 2v \quad .$$

The initial condition is $v(0) = 0$. The D.E. is
separable. We write it as $\dfrac{dv}{32 - 2v} = dt$. Integrating
we find $\ln|32 - 2v| = -2t + c_0$ or $|32 - 2v| = c_1 e^{-2t}$.
Applying the initial condition we find $c = 32$. With
this, we at once have

$$v = 16(1 - e^{-2t}) \quad . \qquad\qquad (\ast)$$

This gives the velocity at each time t before the
object reaches the surface of the lake. To solve
problem (B), we will need to know the velocity at the
instant when the object reaches the surface. To find
out when this is, we need to know distance fallen as a
function of time. This is found by integrating ($\ast$).
We at once obtain

$$x = 16t + 8e^{-2t} + k \quad .$$

Since $x(0) = 0$, we have $k = -8$, and hence the distance
fallen (<u>before</u> striking the water) is given by

$$x = 16t + 8e^{-2t} - 8 \quad .$$

Since the point of release was 50 feet above the water,
if we let $x = 50$ in this, it will determine the time at
which the object hits the surface. Thus we must solve

$$50 = 16t + 8e^{-2t} - 8$$

for t. We write this in the form $58 - 16t = 8e^{-2t}$. A
little trial-and-error calculation with a hand

calculator leads to the approximate solution t = 3.62.
The velocity at this instant is then found by letting
t = 3.62 in (✻). The result is approximately 15.99
(ft/sec.).

We now turn to problem (B). We again take the
positive x-axis vertically downward, but now we take
the origin at the point where the object hits the
surface of the lake. The forces now acting on the body
are given by F_1 = 32, F_2 = -6v, and F_3 = -8. The last
two (the water resistance and the buoyancy) have
negative signs since they act upward. Newton's Second
Law leads to the D.E.

$$dv/dt = 32 - 6v - 8 .$$

By part (A), the velocity of the object at the surface
of the lake is 15.99 (ft/sec). We take this as the I.C.
here: v(0) = 15.99.

Separating variables, the D.E. becomes $\frac{dv}{24 - 6v}$ = dt.
Integrating, we find $\ln|24 - 6v|$ = -6t + c_1 or
$|24 - 6v|$ = ce^{-6t} . Applying the I.C. gives c = 71.94.
Thus we obtain $-(24 - 6v) = 71.94e^{-6t}$ and hence
$v = 4 + 11.99e^{-6t}$. This is the velocity after the
object passes beneath the surface. We want to know
what this is 2 sec. after. Hence we let t = 2 in this
to obtain $v \approx$ 4.00 (ft/sec.).

Section 3.3, Page 96

    1. Let x be the amount of radioactive nuclei
present after t years. Then x satisfies the D.E.
$dx/dt = -kx$, where $k > 0$. Letting x_0 denote the
amount initially present, we have the I.C. $x(0) = x_0$.
Also, since 10 % of the original number have undergone
disintegration in 200 years, 90 % remain, and so we
have the additional condition $x(200) = 9x_0/10$. The
D.E. is separable; and separating variables and
integrating, we at once obtain $x = c\, e^{-kt}$. Application
of the I.C. immediately gives $x_0 = c$. Hence we have
$x = x_0 e^{-kt}$. Now apply the additional condition to
determine k. We have $9x_0/10 = x_0 e^{-200k}$, which reduces
to $e^{200k} = 10/9$. Thus $e^k = (10/9)^{1/200}$, and
$k = (1/200)\ln(10/9)$. Thus the solution $x = x_0(e^k)^{-t}$
takes the form
$$x = x_0(10/9)^{-t/200}. \qquad (\ast)$$
To answer question (a), we let $t = 1000$ in ($\ast$). We
find $x(1000) = x_0(10/9)^{-5} \approx 0.5905\, x_0$. Thus the
answer is 59.05 % . To answer question (b), we let
$x = x_0/4$ in ($\ast$). We have $x_0/4 = x_0(10/9)^{-t/200}$, and
hence $(10/9)^{t/200} = 4$. Thus $t/200\ln(10/9) = \ln 4$,
from which $t = 200[\ln 4/\ln(10/9)] \approx 2630.55$. Thus
$t \approx 2631$ years.

4.   In the first printings of the Third Edition of
the text, this problem is incorrectly stated. The
second sentence should read, "at the end of one hour,
two-thirds kilogram of the first chemical remains, while
at the end of four hours, only one-third kilogram
remains."

Let x = the amount of the first chemical present.
Then x satisfies the D.E. $dx/dt = -kx$, where k > 0.
Two conditions are given: $x(1) = 2/3$ and $x(4) = 1/3$.
The D.E. is separable; and separating variables and
integrating, we at once obtain $x = ce^{-kt}$. Applying
the two conditions, we obtain respectively $ce^{-k} = 2/3$
and $ce^{-4k} = 1/3$. These two equations will determine c
and k. Dividing the first by the second gives $e^{3k} = 2$,
from which $e^{k} = 2^{1/3}$. Then from the second equation for
c and k, $c = (1/3)e^{4k} = 2^{4/3}/3$. Thus the solution of
the D.E. which satisfies the two given conditions is
$x = (2^{4/3}/3)(e^{k})^{-t} = (2^{4/3}/3)(2^{1/3})^{-t}$ or
$x = 2^{(4 - t)/3}/3$.   (✳)

To answer question (a), let t = 7 in (✳). This
gives $x(7) = 2^{-1}/3 = 1/6$ (kg). Now compare this with
the original amount, which is $x(0) = 2^{4/3}/3$ (kg). We
have $x(7)/x(0) = (1/6)/(2^{4/3}/3) = 1/4\sqrt[3]{2} \approx 0.1984$.
Thus 19.8 % of the first chemical remains at the end of

seven hours.

To answer question (b), we first note that one tenth of the first chemical is $x(0)/10 = 2^{4/3}/30$ (kg). We thus let $x = 2^{4/3}/30$ in (✱) and solve for t. We have $2^{(4-t)/3}/3 = 2^{4/3}/30$. From this $[(4-t)/3][\ln 2] = \ln 2^{4/3}/10$; and solving for t, we find $t \approx 9.97$. Thus the answer is 9 hours, 58 minutes.

5. Let x = the population at time t. Then we at once have the D.E. $dx/dt = kx$. Letting x_0 denote the population at the start of the given 40 year period, we have the I.C. $x(0) = x_0$. Since the population doubles in 40 years, we have the additional condition $x(40) = 2x_0$. The solution of the D.E. is $x = c\, e^{kt}$. Applying the I.C. to this we at once have $c = x_0$, and hence $x = x_0\, e^{kt}$. Now applying the additional condition to this, we have $2x_0 = x_0\, e^{40k}$, from which $e^{40k} = 2$ and $k = \ln 2/40 \approx 0.0173$. Thus the solution of the D.E. which satisfies the two given conditions is $x = x_0 e^{0.0173t}$. To answer the stated question, we let $x = 3x_0$ in this, from which $e^{0.0173t} = 3$ and $0.0173t = \ln 3$. From this we find $t \approx 63.5$. Thus the population triples in approximately 63.5 years.

9. The population x satisfies the D.E.
$$dx/dt = 3x/100 - 3x^2/10^8 \quad .$$

The I.C. is x(1980) = 200,000. The D.E. is separable.
We write it in the form

$$\frac{10^6 \, dx}{x(x - 10^6)} = -\frac{3 \, dt}{100} \, .$$

To integrate the left member, we use partial fractions.
Thus we obtain

$$\left[\frac{1}{x - 10^6} - \frac{1}{x} \right] dx = -\frac{3}{100} \, dt \, .$$

Integrating we find $\ln|x - 10^6| - \ln|x| = -3t/100 + c_0$
or $\ln|(x - 10^6)/x| = -3t/100 + c_0$. From this we obtain
$(x - 10^6)/x = ce^{-3t/100}$. We apply the I.C. x(1980) =
200,000 = (2)(10^5) to this. We have
$[(2)(10^5) - 10^6]/2(10)^5 = ce^{-3(198)/10}$, or $c = -4e^{59.4}$.
Thus we obtain the solution in the form $(x - 10^6)/x =$
$-4e^{59.4 - 3t/100}$. We solve this for x, obtaining
$x + 4x \, e^{59.4 - 3t/100} = 10^6$ and hence

$$x = \frac{10^6}{1 + 4e^{59.4 - 3t/100}} \qquad (*)$$

This is the answer to part (a). To answer part (b),
we let t = 2000 in ($*$). We find x(2000) =
$10^6/[1 + 4e^{0.6}] \approx 312,966$. To answer (c), simply find
$\lim_{t \to \infty} x$, where x is given by ($*$). Since $\lim_{t \to \infty} e^{59.4 - 3t/100}$
= 0, we have $\lim_{t \to \infty} x = 10^6/1 = 1,000,000$.

11. Let x = the population at time t. Since 100 people leave the island every year, the D.E. (3.50) is correctly modified by subtracting 100 from its right member. Thus we have the D.E.

$$dx/dt = (1/400)x - 10^{-8} x^2 - 100.$$

We also have the I.C. x(1980) = 20,000. The D.E. is separable. We write it in the successive forms

$$\frac{dx}{dt} = \frac{(10)^6 x - 4 x^2 - 4(10)^{10}}{4(10)^8} ,$$

$$\frac{dx}{[x^2 - 25(10)^4 x + (10)^{10}]} = - \frac{dt}{(10)^8} ,$$

$$\frac{dx}{[x - 5(10)^4][x - 20(10)^4]} = - \frac{dt}{(10)^8} .$$

We now apply partial fractions to the left member and multiply through by $(15)(10)^4$ to obtain

$$\left[\frac{1}{x - 20(10)^4} - \frac{1}{x - 5(10)^4}\right] dx = - \frac{15\, dt}{(10)^4} .$$

Integrating and simplification give

$$\ln\left[\frac{x - 20(10)^4}{x - 5(10)^4}\right] = - 15t/(10)^4 + c_0$$

or

$$\frac{x - 20(10)^4}{x - 5(10)^4} = c\, e^{-15t/(10)^4} .$$

We apply the I.C. $x(1980) = 20,000 = (20)(10)^3$ to this.

With $x = (20)(10)^3$, the left member reduces thus:

$$\frac{(20)(10)^3(1 - 10)}{(5)(10)^3(4 - 10)} = 6.$$

Thus we find $6 = c\ e^{-(15)(1980)/(10)^4}$, and so

$c = 6\ e^{(15)(1980)/(10)^4}$. Thus the solution takes the

form

$$\frac{x - (20)(10)^4}{x - 5(10)^4} = 6\ e^{15(1980 - t)/(10)^4}.$$

We must solve this for x. After some algebraic manipu-

lations, we obtain the desired result

$$x = \frac{(10)^5[3\ e^{15(1980 - t)/(10)^4} - 2]}{6\ e^{15(1980 - t)/(10)^4} - 1}.$$

12. This falls into two problems: (A) Before the
cats arrived, and (B) after the cats arrived. We let
x = the number of mice on the island at time t and
proceed to solve problem (A). During this time, in the
absence of cats killing mice, the D.E. is simply
$dx/dt = kx$, with solution $x = c\ e^{kt}$. Choosing January
1, 1960 as time $t = 0$, the I.C. is $x(0) = 50,000$.
Then, measuring t in years, Jan. 1, 1970 is time $t = 10$;
and we have the additional condition $x(10) = 100,000$.
Applying the I.C. $x(0) = 50,000$, we at once find

$c = 50,000$ and hence $x = 50,000 \, e^{kt}$. The additional
condition now gives $e^{10k} = 2$, from which $k = \ln 2/10$.
With this, the solution takes the form
$x = 50,000 \, e^{t \ln 2/10}$. But this solution only holds
until 1970 and so will not help us to answer the stated
question. It is the number k, which represents the
natural rate of population increase, which we need here
to solve problem (B) and answer the stated question.

Turning to problem (B), since the cats kill 1000
mice/month = 12,000 mice /year, the D.E. of problem
(A) must be modified by subtracting 12,000 from its
right member. Thus we have $dx/dt = kx - 12,000$, where
$k = \ln 2/10$ was found in problem (A). The D.E. is
separable, and we write it in the form $dx/(kx - 12,000)$
$= dt$. Integrating we obtain $kx = 12,000 + c \, e^{kt}$. Now
we have an I.C. here, for on Jan. 1, 1970, there were
100,000 mice. Taking this date as time $t = 0$ for problem (B), we thus have the I.C. $x(0) = 100,000$. Application of this gives $c = 100,000 \, k - 12,000$. With this
value of c and $k = \ln 2/10$, we obtain the solution of
problem (B):

$x = (10/\ln 2)[12,000 + (10,000 \ln 2 - 12,000)e^{t \ln 2/10}]$

The answer to the stated question is now found by letting
$t = 1$ in this expression. We have

$x(1) = (14.427)[12,000 + (6931 - 12,000)(1.0718)]$
       $94,742.$

15. Let x = the amount of salt at time t. We use
the basic equation (3.55), dx/dt = IN - OUT. We find
IN = (3 lb./gal.)(4 gal./min.) = 12 (lb./min.); and
OUT = (C lb./gal.)(4 gal./min.) where C lb./gal. is the
concentration. Since the rates of inflow and outflow
are the same, there is always 100 gal. in the tank and
so this is simply $x/100$. Thus we have OUT = $4x/100$
(lb./min.). Hence the D.E. of the problem is
dx/dt = 12 - $4x/100$. The I.C. is $x(0)$ = 20.

The D.E. is separable. We write it as
$dx/dt = (300 - x)/25$ or $dx/(300 - x) = dt/25$.
Integrating, we find $\ln|300 - x| = -t/25 + c_0$ or
$|300 - x| = c\, e^{-t/25}$. Applying the I.C., we find
$280 = c$; and since $300 - x > 0$ initially, we take
$|300 - x| = 300 - x$. Thus we obtain the solution in
the form $300 - x = 280\, e^{-t/25}$ or

$$x = 300 - 280\, e^{-t/25} \qquad (\ast)$$

To answer question (a), let t = 10 in ($\ast$). We
find $x \approx 112.31$ (lb.). To answer question (b), let
x = 160 in ($\ast$). We have $280\, e^{-t/25} = 140$ or
$e^{-t/25} = 0.5$. Then $-t/25 = \ln 0.5$, from which we find

$t \approx 17.33$ (min.).

19. Let x = the amount of salt at time t. We use
the basic equation dx/dt = IN - OUT. We find IN =
(30 gm/liter)(4 liters/min) = 120(gm/min); and OUT =
(C gm/liter)(5/2 liters/min), where C gm/liter is the
concentration. The rate of inflow is 4 liters /min and
that of outflow is 5/2 liters/min, so there is a net
gain of 3/2 liters /min of fluid in the tank. Hence
at the end of t minutes the amount of fluid in the tank
is 300 + 3t/2 liters. Thus the concentration at time t
is x/(300 + 3t/2) gm/liter, and so OUT = 5x/2(300 +
3t/2) = 5x/(3t + 600) (gm/min). Hence the D.E. of the
problem is dx/dt = 120 - 5x/(3t + 600). The I.C. is
x(0) = 50.

The D.E. is linear. We write it in the standard
form

$$\frac{dx}{dt} + \frac{5}{3t + 600} x = 120 \quad .$$

An I.F. is

$$e^{\int [(5/(3t + 600)]dt} = e^{(5/3) \ln (3t + 600)} =$$

$(3t + 600)^{5/3}$. Multiplying through by this we have

$(3t + 600)^{5/3}$ dx/dt + $5(3t + 600)^{2/3}$ x = $120(3t + 600)^{5/3}$

or

$$\frac{d}{dt} [(3t + 600)^{5/3} x] = 120(3t + 600)^{5/3}.$$

Integration gives $(3t + 600)^{5/3} x = 15(3t + 600)^{5/3} + c$
or
$$x = 15(3t + 600) + c(3t + 600)^{5/3}.$$

Applying the I.C. $x(0) = 50$ gives $50 = (15)(600) +$
$c(600)^{-5/3}$, from which $c = -8950(600)^{5/3}$. Thus the
solution is given by   (✱)

$$x = 15(3t + 600) - (8950)(600)^{5/3} (3t + 600)^{-5/3}.$$

The stated question asks for x at the instant when
the tank overflows. Since the amount of fluid increases
at the rate of 3/2 liter/min and the 500 liter tank
originally had 300 liters in it, this time t is given
by $3t/2 = 200$ or $t = 400/3$. We thus let $t = 400/3$ in
(✱) and obtain $x(400/3) = 1500(1000) -$
$(8950)(600)^{5/3} (1000)^{5/3} = 15000 - 8950(3/5)^{5/3}$
$\approx 11, 179.96$ (grams).

20. Let x = the amount of salt at time t. We use
the basic equation $dx/dt = IN - OUT$. We find IN =
(50 gm/liter)(5 liters/min) = 250/gm/min ; and OUT =
(C gm/liter)(7 liters /min), where C gm/liter is the
concentration. The rate of inflow is 5 liters/min.
and that of outflow is 7 liters/min., so there is a
net loss of 2 liters/min of fluid in the tank. Hence
at the end of t minutes the amount of fluid in the tank
is $200 - 2t$ liters. Thus the concentration at time t is

$x/(200 - 2t)$ gm/liter, and so OUT = $7x/(200 - 2t)$
(gm./min). Hence the D.E. is $dx/dt = 250 - 7x/(200-2t)$.
The I.C. is $x(0) = 40$.

   The D.E. is linear. We write it in the standard
form

$$\frac{dx}{dt} + \frac{7}{200 - 2t} x = 250 \quad .$$

An I.F. is $e^{\int [7/(200 - 2t)]dt} = e^{-(7/2)\ln(200 - 2t)}$

$= (200 - 2t)^{-7/2}$. Multiplying through by this, we
have $(200 - 2t)^{-7/2} dx/dt + 7(200-2t)^{-9/2} x = 250(200 - 2t)^{-7/2}$ or

$$\frac{d}{dt} [(200 - 2t)^{-7/2} x] = 250(200 - 2t)^{-7/2}.$$

Integration gives $(200 - 2t)^{-7/2} x = 50(200 - 2t)^{-5/2} + c$
or

$$x = 50 (200 - 2t) + c(200 - 2t)^{7/2} \quad .$$

Applying the I.C. $x(0) = 40$, we find
$40 = 10,000 + c(200)^{7/2}$, from which $c = -9960(200)^{-7/2}$.
Thus the solution is given by

$$x = 50(200 - 2t) - 9960(1 - t/100)^{7/2}. \quad (\ast)$$

   The tank will be half full when it contains 100
liters. Since it initially contained 200 liters and it
loses 2 liters/min, it will contain 100 liters after 50
min. Thus we let $t = 50$ in ($\ast$). We find

$x(50) = 5000 - 9960(1/2)^{7/2} \approx 4119.65$ (gm).

23. Let x = the temperature of the body at time t.
We have the D.E. $dx/dt = k(x - 50)$, the I.C.
$x(0) = 80$, and the additional condition $x(5) = 70$.
The D.E. is separable. We write it in the form
$dx/(x-50) = k \, dt$. Integrating, we find $\ln|x - 50| =$
$kt + c_0$ and hence $|x - 50| = c \, e^{kt}$. Since $x \geq 50$,
$|x - 50| = x - 50$, so we have $x = 50 + c \, e^{kt}$.
Application of the I.C. gives $c = 30$, so $x = 50 + 30 \, e^{kt}$.
Application of the additional condition then gives
$70 = 50 + 30e^{5k}$ or $e^{5k} = 2/3$, from which
$k = (1/5) \ln(2/3) = -0.0811$. Thus we obtain the solu-
tion

$$x = 50 + 30 \, e^{-0.0811t} . \qquad (\text{✱})$$

To answer part (a), let $t = 10$ in (✱). We find
$x(10) = 50 + 30e^{-0.811} \approx 63.33°$ F. To answer part (b),
let $x = 60$ in (✱). We have $60 + 30e^{-0.0811t}$ or
$e^{-0.0811t} = 1/3$. From this $-0.0811t = \ln(1/3)$ and
hence $t \approx 13.55$ (min).

CHAPTER 4

Section 4.1B, Page 112

7. (a) One readily verifies by direct substitution into the D.E. that each of these functions is indeed a solution. To show that they are linearly independent, we apply Theorem 4.4. We have

$$W(e^{2x}, e^{3x}) = \begin{vmatrix} e^{2x} & e^{3x} \\ 2e^{2x} & 3e^{3x} \end{vmatrix} = 3e^{5x} - 2e^{5x} = e^{5x} \neq 0$$

for all x, $-\infty < x < \infty$. Thus the two solutions are linearly independent on $-\infty < x < \infty$.

(b) $y = c_1 e^{2x} + c_2 e^{3x}$.

(c) To satisfy the condition $y(0) = 2$, we let $x = 0$, $y = 2$ in the general solution of part (b). We have $c_1 + c_2 = 2$ (✳). Now we differentiate the general solution, obtaining $y' = 2c_1 e^{2x} + 3c_2 e^{3x}$. To satisfy the condition $y'(0) = 3$, we let $x = 0$, $y = 3$ in this derived equation. We have $2c_1 + 3c_2 = 3$.   (✳✳). Solving the two equations (✳) and (✳✳) for c_1 and c_2,

we find $c_1 = 3$, $c_2 = -1$. Substituting these values of
c_1 and c_2 into the general solution of part (b), we
have the desired particular solution $y = 3e^{2x} - e^{3x}$.
This is unique by Theorem 4.1; and it is defined on
$-\infty < x < \infty$.

10. (a) One readily verifies by direct substitution into the D.E. that each of these functions is
indeed a solution. To show that they are linearly
independent, we apply Theorem 4.4. We have

$$W(x^2, 1/x^2) = \begin{vmatrix} x^2 & 1/x^2 \\ 2x & -2/x^3 \end{vmatrix} = -2/x - 2/x = -4/x \neq 0$$

for all x, $0 < x < \infty$. Thus the two solutions are
linearly independent on $0 < x < \infty$.

(b) $y = c_1 x^2 + c_2/x^2$.

(c) To satisfy the condition $y(2) = 3$, we let
x = 2, y = 3 in the general solution of part (b). We
have $4c_1 + c_2/4 = 3$ (✱). Now we differentiate the
general solution, obtaining $y' = 2c_1 x - 2c_2 x^{-3}$. To
satisfy the condition $y'(2) = -1$, we let x = 2,
$y' = -1$ in this derived equation. We have
$4c_1 - c_2/4 = -1$ (✱✱). Solving the two equations
(✱) and (✱✱) for c_1 and c_2, we find $c_1 = 1/4$,
$c_2 = 8$. Substituting these values of c_1 and c_2 into

the general solution of part (b), we have the desired
particular solution $y = (1/4) x^2 + 8/x^2$. This is
unique by Theorem 4.1; and it is defined on $0 < x < \infty$.

12. We use Theorem 4.4. We have

$$W(e^{-x}, e^{3x}, e^{4x}) = \begin{vmatrix} e^{-x} & e^{3x} & e^{4x} \\ -e^{-x} & 3e^{3x} & 4e^{4x} \\ e^{-x} & 9e^{3x} & 16e^{4x} \end{vmatrix}$$

$$= e^{-x}e^{3x}e^{4x} \begin{vmatrix} 1 & 1 & 1 \\ -1 & 3 & 4 \\ 1 & 9 & 16 \end{vmatrix}$$

$$= e^{6x} \left\{ \begin{vmatrix} 3 & 4 \\ 9 & 16 \end{vmatrix} - \begin{vmatrix} -1 & 4 \\ 1 & 16 \end{vmatrix} + \begin{vmatrix} -1 & 3 \\ 1 & 9 \end{vmatrix} \right\}$$

$$= e^{6x}[12 + 20 - 12] = 20e^{6x} \neq 0$$

for all x, $-\infty < x < \infty$. Thus the three solutions are
linearly independent on $-\infty < x < \infty$. The general
solution is $y = c_1 e^{-x} + c_2 e^{3x} + c_3 e^{4x}$.

Section 4.1D, Page 123

1. Let $y = xv$. Then $\frac{dy}{dx} = x\frac{dv}{dx} + v$ and

$\frac{d^2x}{dx^2} = x\frac{d^2v}{dx^2} + 2\frac{dv}{dx}$. Substituting these into the given

D.E., we obtain

$$x^2(x\frac{d^2v}{dx^2} + 2\frac{dv}{dx}) - 4x(x\frac{dv}{dx} + v) + 4xv = 0$$

or $x^3\frac{d^2v}{dx^3} - 2x^2\frac{dv}{dx} = 0$ or $x\frac{d^2v}{dx^2} - 2\frac{dv}{dx} = 0$.

Letting $w = dv/dx$, we obtain $x\, dw/dx - 2w = 0$ or
$dw/w = 2\, dx/x$. Integrating, we find
$\ln|w| = 2\ln|x| + \ln|c|$ or $w = cx^2$. We choose $c = 1$,
recall $dv/dx = w$, and integrate to obtain $v = x^3/3$.
Now forming $y = xv$, we obtain $y = x^4/3$. This or any
nonzero constant multiple thereof serves as the desired
linearly independent solution. Choosing multiple 3, we
have $y = x^4$. The general solution is then

$y = c_1x + c_2x^4$.

5. Let $y = e^{2x}v$. Then $\frac{dy}{dx} = e^{2x}\frac{dv}{dx} + 2e^{2x}v$ and

$\frac{d^2y}{dx^2} = e^{2x}\frac{d^2v}{dx^2} + 4e^{2x}\frac{dv}{dx} + 4e^{2x}v$. Substituting this

into the given D.E., we obtain

$$(2x + 1)(e^{2x}\frac{d^2v}{dx^2} + 4e^{2x}\frac{dv}{dx} + 4e^{2x}v)$$

$$- 4(x + 1)(e^{2x} \frac{dv}{dx} + 2e^{2x} v) + 4e^{2x} v = 0$$

or

$$(2x + 1)e^{2x} \frac{d^2v}{dx^2} + 4xe^{2x} \frac{dv}{dx} = 0$$

or

$$(2x + 1) \frac{d^2v}{dx^2} + 4x \frac{dv}{dx} = 0 \ .$$

Letting $w = dv/dx$, we obtain $(2x + 1)\ dw/dx + 4xw = 0$ or $dw/w = - 4x\ dx/(2x + 1)$. Using long division on the right member, we rewrite this as $\frac{dw}{w} = \left(-2 + \frac{2}{2x+1} \right) dx$. Integrating, we have $\ln|w| = -2x + \ln|2x + 1| + \ln|c|$ or $w = c(2x + 1)e^{-2x}$. We choose $c = 1$ and recall $w = dv/dx$ to obtain $dv/dx = (2x + 1)e^{-2x}$. Integrating this, we obtain $v = -(x+1)e^{-2x}$. Now forming $y = e^{2x}v$, we obtain $y = -(x+1)$. This or any nonzero constant multiple thereof serves as the desired linearly independent solution. Choosing multiple -1, we have $y = x + 1$. The general solution is then $y = c_1 e^{2x} + c_2(x + 1)$,

Section 4.2, Page 134

1. The auxiliary equation is $m^2 - 5m + 6 = 0$. Its roots are 2 and 3 (real and distinct). The G. S. is $y = c_1 e^{2x} + c_2 e^{3x}$.

4.   The auxiliary equation is $3m^2 - 14m - 5 = 0$.
Its roots are 5 and $- 1/3$ (real and distinct).   The
G.S. is $y = c_1 e^{5x} + c_2 e^{-x/3}$.

6.   The auxiliary equation is $m^3 - 3m^2 - m + 3 = 0$,
which can be written $m^2(m - 3) - (m - 3) = 0$ or
$(m + 1)(m - 1)(m - 3) = 0$.   Hence its roots are 1, -1,
and 3 (real and distinct).   The G.S. is $y = c_1 e^x +$
$c_2 e^{-x} + c_3 e^{3x}$.

7.   The auxiliary equation is $m^2 - 8m + 16 = 0$.
Its roots are 4, 4 (real, double root).   The G. S. is
$y = (c_1 + c_2 x)\ e^{4x}$ or $y = c_1 e^{4x} + c_2 x e^{4x}$.

9.   The auxiliary equation is $m^2 - 4m + 13 = 0$.
Solving it, obtain

$$m = \frac{4 \pm \sqrt{16 - 52}}{2} = 2 \pm 3i \ .$$

So the roots are the conjugate complex numbers
$2 \pm 3i$.   The G. S. is $y = e^{2x}(c_1 \sin 3x + c_2 \cos 3x)$.

12.   The auxiliary equation is $4m^2 + 1 = 0$.   Its
roots are $m = \pm (1/2)i$.   The G. S. is
$y = c_1 \sin (1/2)x + c_2 \cos (1/2)x$.

14.   The auxiliary equation is $4m^3 + 4m^2 - 7m + 2 = 0$
Observe by inspection that -2 is a root.   Then by

synthetic division,

$$
\begin{array}{r|rrrr}
-2 & 4 & 4 & -7 & 2 \\
   &   & -8 & 8 & -2 \\
\hline
   & 4 & -4 & 1 & 0
\end{array}
\quad ,
$$

find the factorization $(m + 2)(4m^2 - 4m + 1) = 0$ and
hence $(m + 2)(2m - 1)^2 = 0$. Thus the roots are -2
(real simple root) $1/2$, $1/2$ (real double root). The
G. S. is $y = c_1 e^{-2x} + (c_2 + c_3 x)e^{(1/2)x}$ or

$$y = c_1 e^{-2x} + c_2 e^{(1/2)x} + c_3 x e^{(1/2)x} .$$

16. The auxiliary equation is $m^3 + 4m^2 + 5m + 6 = 0$
Observe by inspection that -3 is a root. Then by
synthetic division,

$$
\begin{array}{r|rrrr}
-3 & 1 & 4 & 5 & 6 \\
   &   & -3 & -3 & -6 \\
\hline
   & 1 & 1 & 2 & 0
\end{array}
$$

find the factorization $(m + 3)(m^2 + m + 2) = 0$. From
$m^2 + m + 2 = 0$, obtain

$$m = \frac{-1 \pm \sqrt{1 - 8}}{2} = -\frac{1}{2} \pm \frac{\sqrt{7}}{2} i .$$

Thus the roots are the real simple root -3 and the
conjugate complex roots $-1/2 \pm (\sqrt{7}/2)i$. The G. S. is
$y = c_1 e^{-3x} + e^{-x/2}[c_2 \sin (\sqrt{7}/2)x + c_3 \cos(\sqrt{7}/2)x] .$

19. The auxiliary equation is $m^5 - 2m^4 + m^3 = 0$.
Factoring gives $m^3(m^2 - 2m + 1) = 0$ or $m^3(m-1)^2 = 0$.
Thus the roots are 0, 0, 0 [triple root, from factor
m^3], 1, 1 [double root, from factor $(m - 1)^2$]. The
G. S. is

$$y = (c_1 + c_2 x + c_3 x^2)e^{0x} + (c_4 + c_5 x)e^x$$

that is,

$$y = c_1 + c_2 x + c_3 x^2 + (c_4 + c_5 x)e^x$$

or

$$y = c_1 + c_2 x + c_3 x^2 + c_4 e^x + c_5 x e^x \ .$$

22. The auxiliary equation is
$$m^4 + 6m^3 + 15m^2 + 20m + 12 = 0.$$
Observe by inspection that -2 is a root. Then by
synthetic division,

-2	1	6	15	20	12
		-2	-8	-14	-12
	1	4	7	6	0

find the factorization $(m + 2)(m^3 + 4m^2 + 7m + 6) = 0$.
Now observe by inspection that -2 is also a root of the
reduced cubic $m^3 + 4m^2 + 7m + 6 = 0$. Then by
synthetic division,

$$\begin{array}{c|cccc} -2 & 1 & 4 & 7 & 6 \\ & & -2 & -4 & -6 \\ \hline & 1 & 2 & 3 & \boxed{0} \end{array}$$

find the factorization $(m + 2)(m^2 + 2m + 3) = 0$
of the reduced cubic. Hence the auxiliary equation
has the factored form $(m + 2)^2(m^2 + 2m + 3) = 0$. The
factor $(m + 2)^2$ yields the real double root -2, -2.
The factor $m^2 + 2m + 3 = 0$ gives the conjugate complex
roots

$$m = \frac{-2 \pm \sqrt{4 - 12}}{2} = \frac{-2 \pm 2\sqrt{2}\, i}{2} = -1 \pm \sqrt{2}\, i \ .$$

Thus the roots are -2, -2, $-1 \pm \sqrt{2}\, i$. The G. S. is

$$y = (c_1 + c_2 x)e^{-2x} + e^{-x}(c_3 \sin \sqrt{2}x + c_4 \cos \sqrt{2}x) \quad \text{or}$$

$$y = c_1 e^{-2x} + c_2 x e^{-2x} + e^{-x}(c_3 \sin \sqrt{2}x + c_4 \cos \sqrt{2}x).$$

25. The auxiliary equation is $m^2 - m - 12 = 0$.
Its roots are 4, -3 (real and distinct). The G. S. is
$y = c_1 e^{4x} + c_2 e^{-3x}$. (A). From this,
$y' = 4c_1 e^{4x} - 3c_2 e^{-3x}$ (B). Apply condition
$y(0) = 3$ to (A) to obtain $c_1 + c_2 = 3$. Apply condition
$y'(0) = 5$ to (B) to obtain $4c_1 - 3c_2 = 5$. From these
two equations in c_1 and c_2, find $c_1 = 2$, $c_2 = 1$.
Thus the solution of stated I. V. P. is $y = 2e^{4x} + e^{-3x}$.

29. The auxiliary equation is $m^2 + 6m + 9 = 0$. Its roots are $-3, -3$ (real double root). The G. S. is $y = (c_1 + c_2 x) e^{-3x}$. (A). From this, $y' = (-3c_1 + c_2 - 3c_2 x)e^{-3x}$ (B). Apply condition $y(0) = 2$ to (A) to obtain $c_1 = 2$. Apply condition $y'(0) = -3$ to (B) to obtain $-3c_1 + c_2 = -3$. From this, find $c_2 = 3$. Thus the solution of the stated I. V. P. is $y = (2 + 3x) e^{-3x}$.

33. The auxiliary equation is $m^2 - 4m + 29 = 0$. Solving, obtain the conjugate complex roots

$$m = \frac{4 \pm \sqrt{16 - 116}}{2} = \frac{4 \pm 10i}{2} = 2 \pm 5i \ .$$

The G. S. is $y = e^{2x} (c_1 \sin 5x + c_2 \cos 5x)$ (A). From this, $y' = e^{2x}[(2c_1 - 5c_2) \sin 5x + (5c_1 + 2c_2) \cos 5x]$. (B) Apply condition $y(0) = 0$ to (A) to obtain

$$e^0(c_1 \sin 0 + c_2 \cos 0) = 0 \quad \text{or} \quad c_2 = 0$$

Apply condition $y'(0) = 5$ to (B) to obtain
$$e^0[(2c_1 - 5c_2) \sin 0 + (5c_1 + 2c_2) \cos 0] = 5$$
or
$$5c_1 + 2c_2 = 5.$$

From this, $c_1 = 1$.   Thus the solution of the stated
I. V. P. is $y = e^{2x} \sin 5x$.

38.   The auxiliary equation is $4m^2 + 4m + 37 = 0$.
Solving, obtain the conjugate complex roots

$$m = \frac{-4 \pm \sqrt{16 - 16(37)}}{8} = \frac{-4 \pm 24i}{8} = -\frac{1}{2} \pm 3i$$

The G.S. is $y = e^{-x/2} (c_1 \sin 3x + c_2 \cos 3x)$.   (A)

From this, $y' = e^{-x/2}[(-c_1/2 - 3c_2) \sin 3x +$

$(3c_1 - c_2/2) \cos 3x]$.   (B)   Apply condition

$y(0) = 2$ to (A) to obtain $c_2 = 2$.   Apply condition
$y'(0) = -4$ to (B) to obtain $3c_1 - c_2/2 = -4$.   From this,
$c_1 = -1$.   Thus the solution of the stated I. V. P. is
$y = e^{-x/2}(- \sin 3x + 2 \cos 3x)$.

42.   The auxiliary equation is $m^3 - 5m^2 + 9m - 5 = 0$.
Observe by inspection that 1 is a root.   Then by
synthetic division,

1	1	-5	9	-5
		1	-4	5
	1	-4	5	0

find the factorization $(m - 1)(m^2 - 4m + 5) = 0$.   From
$m^2 - 4m + 5 = 0$, obtain

$$m = \frac{4 \pm \sqrt{16 - 20}}{2} = 2 \pm i .$$

Thus the roots are the real simple root i and the conjugate complex roots $2 \pm i$. The G. S. is

$$y = c_1 e^x + e^{2x}(c_2 \sin x + c_3 \cos x). \qquad (A)$$

From this,

$$y' = c_1 e^x + e^{2x}[(2c_2 - c_3) \sin x + (c_2 + 2c_3) \cos x],$$
$$\qquad\qquad (B)$$

$$y'' = c_1 e^x + e^{2x}[(3c_2 - 4c_3) \sin x + (4c_2 + 3c_3) \cos x]$$
$$\qquad\qquad (C)$$

Apply condition $y(0) = 0$ to (A) to obtain $c_1 + c_3 = 0$. Apply condition $y'(0) = 1$ to (B) to obtain $c_1 + c_2 + 2c_3 = 1$. Apply condition $y''(0) = 6$ to (C) to obtain $c_1 + 4c_2 + 3c_3 = 6$. The solution of these three equations in c_1, c_2, c_3 is $c_1 = 1$, $c_2 = 2$, $c_3 = -1$. Thus the solution of the stated I. V. P. is $y = e^x + e^{2x} (2 \sin x - \cos x)$.

45. Since sin x is a solution of the D. E., $m = \pm i$ must be roots of the auxiliary equation $m^4 + 2m^3 + 6m^2 + 2m + 5 = 0$, and hence $(m - i)(m + i) = m^2 + 1$ must be a factor of $m^4 + 2m^3 + 6m^2 + 2m + 5$. By long division, find the other factor is $m^2 + 2m + 5$. Hence in factored form

the auxiliary equation is

$$(m^2 + 1)(m^2 + 2m + 5) = 0 \quad .$$

The factor $m^2 + 2m + 5 = 0$ gives

$$m = \frac{-2 \pm \sqrt{4 - 20}}{2} = -1 \pm 2i \quad .$$

Thus the roots of the auxiliary equation are the two
pairs of conjugate complex numbers $\pm i$ and $-1 \pm 2i$.
The G. S. of the D. E. is

$$y = c_1 \sin x + c_2 \cos x + e^{-x}(c_3 \sin 2x + c_4 \cos 2x).$$

46. Since $e^x \sin 2x$ is a solution of the D. E.,
$1 \pm 2i$ must be roots of the auxiliary equation
$m^4 + 3m^3 + m^2 + 13m + 30 = 0$, and hence
$[m - (1 + 2i)][m - (1 - 2i)] = m^2 - 2m + 5$ must be a
factor of $m^4 + 3m^3 + m^2 + 13m + 30 = 0$. By long
division, find the other factor $m^2 + 5m + 6$. Hence in
factored form the auxiliary equation is
$(m^2 - 2m + 5)(m^2 + 5m + 6) = 0$. The factor
$m^2 + 5m + 6 = 0$ gives $m = -2, -3$. Thus the roots of
the auxiliary equation are the real distinct numbers -2
and -3 and the conjugate complex pair $1 \pm 2i$. Thus
the G. S. is

$$y = c_1 e^{-2x} + c_2 e^{-3x} + e^x(c_3 \sin 2x + c_4 \cos 2x) \quad .$$

Section 4.3, Page 151

Exercises 1 through 24 follow the pattern of Examples 4.36, 4.37 and 4.38 on pages 145-150 of the text. The five steps of the method outline on pages 143-144 are indicated in each solution.

1. The corresponding homogeneous D. E. is $y'' - 3y' + 2y = 0$. The auxiliary equation is $m^2 - 3m + 2 = 0$, with roots 1, 2. The complementary function is $y_c = c_1 e^x + c_2 e^{2x}$. The NH term is a constant multiple of the UC function given by x^2.

Step 1: Form the UC set of x^2. It is
$$S_1 = \left\{ x^2, x, 1 \right\} .$$

Step 2: This step does not apply, since there is only one UC set present.

Step 3: An examination of the complementary function shows that none of the functions in S_1 is a solution of the corresponding homogeneous D. E. Hence S_1 does not need revision.

Step 4: Thus the original set S_1 remains. Form a linear combination of its three members.

Step 5: Thus we take
$$y_p = Ax^2 + Bx + C$$
as a particular solution. Then

$$y_p' = 2Ax + B, \quad y_p'' = 2A \quad .$$

We substitute in the D. E., obtaining

$$2A - 3(2Ax + B) + 2(Ax^2 + Bx + C) = 4x^2$$

or

$$2Ax^2 + (-6A + 2B)x + (2A - 3B + 2C) = 4x^2 \quad .$$

We equate coefficients of like terms on both sides of this to obtain

$$2A = 4, \quad -6A + 2B = 0, \quad 2A - 3B + 2C = 0.$$

From these, we find $A = 2$, $B = 6$, $C = 7$. Thus we obtain the particular integral

$$y_p = 2x^2 + 6x + 7 \quad .$$

The G. S. of the D. E. is

$$y = c_1 e^x + c_2 e^{2x} + 2x^2 + 6x + 7 \quad .$$

4. The corresponding homogeneous D. E. is $y'' + 2y' + 2y = 0$. The auxiliary equation is $m^2 + 2m + 2 = 0$ with roots $-1 \pm i$. The complementary function is $y_c = e^{-x} (c_1 \sin x + c_2 \cos x)$. The NH term is a constant multiple of the UC function given by $\sin 4x$.

Step 1: Form the UC set of $\sin 4x$. It is $$S_1 = \left\{ \sin 4x, \cos 4x \right\} \quad .$$

Step 2: This step does not apply.

Step 3:   An examination of the complementary function shows that none of the functions in S_1 is a solution of the corresponding homogeneous D. E.   Hence S_1
does not need revision.

Step 4:   Thus the original set S_1 remains.   Form a
linear combination of its two members.

Step 5:   Thus we take y_p = A sin 4x + B cos 4x as a
particular solution.   Then

$$y_p' = 4A \cos 4x - 4B \sin 4x \quad ,$$

$$y_p'' = -16A \sin 4x - 16B \cos 4x \quad .$$

We substitute into the D. E., obtaining

-16A sin 4x - 16B cos 4x + 8A cos 4x - 8B sin 4x

+ 2A sin 4x + 2B cos 4x = 10 sin 4x

or

(-14A - 8B) sin 4x + (8A - 14B) cos 4x = 10 sin 4x.

We equate coefficients of like terms on both sides of
this to obtain

-14A - 8B = 10,   8A - 14B = 0   .

From these, we find A = -7/13, B = -4/13.   Thus we
obtain the particular integral

$$y_p = -(7/13) \sin 4x - (4/13) \cos 4x \quad .$$

The G. S. of the D. E. is

$$y = e^{-x}(c_1 \sin x + c_2 \cos x) - (7/13) \sin 4x - (4/13) \cos 4x.$$

8. The corresponding homogeneous D. E. is
$y'' + 2y' + 10y = 0$. The auxiliary equation is
$m^2 + 2m + 10 = 0$ with roots $-1 \pm 3i$. The complementary
function is $y_c = e^{-x}(c_1 \sin 3x + c_2 \cos 3x)$. The NH
term is a constant multiple of the UC function given
by xe^{-2x}.

Step 1: Form the UC set of xe^{-2x}. It is
$$S_1 = \left\{ xe^{-2x}, \ e^{-2x} \right\} .$$

Step 2: This step does not apply.

Step 3: An examination of the complementary function shows that none of the functions in S_1 is a solution of the corresponding homogeneous D. E. Hence S_1
does not need revision.

Step 4: Thus the original set S_1 remains. Form a
linear combination of its two members.

Step 5: Thus we take $y_p = Axe^{-2x} + Be^{-2x}$ as a
particular solution. Then

$$y_p' = -2Axe^{-2x} + (A - 2B)\, e^{-2x},$$

$$y_p'' = 4Axe^{-2x} + (-4A + 4B)e^{-2x} .$$

We substitute into the D. E., obtaining

$$4Axe^{-2x} + (-4A + 4B)e^{-2x} - 4Axe^{-2x} + (2A - 4B)e^{-2x}$$

$$+ 10Axe^{-2x} + 10Be^{-2x} = 5xe^{-2x}$$

or
$$10Axe^{-2x} + (-2A + 10B)e^{-2x} = 5xe^{-2x} .$$

We equate coefficients of like terms on both sides of
this to obtain $10A = 5$, $-2A + 10B = 0$. From these,
we find $A = 1/2$, $B = 1/10$. Thus we obtain the
particular integral

$$y_p = xe^{-2x}/2 + e^{-2x}/10 .$$

The G. S. of the D. E. is

$$y = e^{-x}(c_1 \sin 3x + c_2 \cos 3x) + xe^{-2x}/2 + e^{-2x}/10 .$$

9. The corresponding homogeneous D. E. is
$y''' + 4y'' + y' - 6y = 0$. The auxiliary equation is
$m^3 + 4m^2 + m - 6 = 0$. By inspection note that $m = 1$ is
a root. From this (by synthetic division or otherwise)
we obtain the factored form $(m - 1)(m + 2)(m + 3)$. Thus
the roots of the auxiliary equation are $m = 1, -2, -3$.
The complementary function is $y_c = c_1 e^x + c_2 e^{-2x} + c_3 e^{-3x}$
The NH term is a linear combination of the UC functions
x^2 and 1.

Step 1: Form the UC set of each of these two
functions: $S_1 = \left\{ x^2, x, 1 \right\}$, $S_2 = \left\{ 1 \right\}$.

Step 2:   Set S_2 is completely included in S_1, so S_2 is omitted, leaving just S_1.

Step 3:   An examination of the complementary function shows that none of the functions in S_1 is a solution of the corresponding homogeneous D. E.   Hence S_1 does not need revision.

Step 4:   Thus the original set S_1 remains.   We form a linear combination of its three members.

Step 5:   Thus we take $y_p = Ax^2 + Bx + C$.   Then $y_p' = 2Ax + B$, $y_p'' = 2A$, $y_p''' = 0$.   We substitute in the D.E., obtaining

$$0 + 4(2A) + (2Ax + B) - 6(Ax^2 + Bx + C) = -18x^2 + 1$$

or

$$-6Ax^2 + (2A - 6B)x + (8A + B - 6C) = -18x^2 + 1 .$$

We equate coefficients of like terms on both sides of this to obtain $-6A = -18$, $2A - 6B = 0$, $8A + B - 6C = 1$. From these, we find $A = 3$, $B = 1$, $C = 4$.   Thus we obtain the particular solution

$$y_p = 3x^2 + x + 4 .$$

The G. S. of the D. E. is

$$y = c_1 e^x + c_2 e^{-2x} + c_3 e^{-3x} + 3x^2 + x + 4 .$$

11.   The corresponding homogeneous D. E. is
$y''' + y'' + 3y' - 5y = 0$. The auxiliary equation is
$m^3 + m^2 + 3m - 5 = 0$. By inspection note that $m = 1$ is
a root. From this (by synthetic division or otherwise)
we obtain the factored form $(m - 1)(m^2 + 2m + 5) = 0$.
From this, the roots of the auxiliary equation are 1 and
$-1 \pm 2i$. The complementary function is
$y_c = e^x + e^{-x}(c_1 \sin 2x + c_2 \cos 2x)$. The NH term is a
linear combination of the UC functions given by $\sin 2x$,
x^2, x, and 1.

Step 1: Form the UC set of each of these four
functions: $S_1 = \{ \sin 2x, \cos 2x \}$, $S_2 = \{ x^2, x, 1 \}$
$S_3 = \{ x, 1 \}$, $S_4 = \{ 1 \}$.

Step 2: Sets S_3 and S_4 are completely included in
S_2; so S_3 and S_4 are omitted, leaving the two sets S_1
and S_2.

Step 3: An examination of the complementary function shows that none of the functions in S_1 or S_2 is a
solution of the corresponding homogeneous D. E. Hence
neither S_1 nor S_2 needs revision.

Step 4: Thus the original sets S_1 and S_2 remain.
We form a linear combination of their five members.

Step 5:   Thus we take

$$y_p = A \sin 2x + B \cos 2x + Cx^2 + Dx + E$$

as a particular solution. Then:

$$y_p' = 2A \cos 2x - 2B \sin 2x + 2Cx + D,$$

$$y_p'' = -4A \sin 2x - 4B \cos 2x + 2C,$$

$$y_p''' = -8A \cos 2x + 8B \sin 2x.$$

We substitute in the D. E., obtaining

$$(-8A \cos 2x + 8B \sin 2x) + (-4A \sin 2x - 4B \cos 2x + 2C)$$

$$+ 3(2A \cos 2x - 2B \sin 2x + 2Cx + D)$$

$$- 5(A \sin 2x + B \cos 2x + Cx^2 + Dx + E)$$

$$= 5 \sin 2x + 10x^2 - 3x + 7$$

or

$$(-9A + 2B) \sin 2x + (-2A - 9B) \cos 2x$$

$$-5Cx^2 + (6C - 5D)x + (2C + 3D - 5E)$$

$$= 5 \sin 2x + 10x^2 - 3x + 7 \ .$$

We equate coefficients of like terms on both sides of
this to obtain

$$-9A + 2B = 5, \quad -2A - 9B = 0, \quad -5C = 10,$$

$$6C - 5D = -3, \quad 2C + 3D - 5E = 7 \ .$$

From these, we find $A = -9/17$, $B = 2/17$, $C = -2$,
$D = -9/5$, $E = -82/25$. Thus we obtain the particular

integral

$$y_p = -\frac{9 \sin 2x}{17} + \frac{2 \cos 2x}{17} - 2x^2 - \frac{9x}{5} - \frac{82}{25} \ .$$

The G. S. of the D. E. is

$$y = e^x + e^{-x}(c_1 \sin 2x + c_2 \cos 2x)$$

$$-\frac{9 \sin 2x}{17} + \frac{2 \cos 2x}{17} - 2x^2 - \frac{9x}{5} - \frac{82}{25} \ .$$

13. The corresponding homogeneous D. E. is $y'' + y' - 6y = 0$. The auxiliary equation is $m^2 + m - 6 = 0$, with roots 2, -3. The complementary function is $y_c = c_1 e^{2x} + c_2 e^{-3x}$. The NH term is a linear combination of the UC functions e^{2x}, e^{3x}, x and 1.

Step 1: Form the UC set of each of these four functions: $S_1 = \left\{ e^{2x} \right\}$, $S_2 = \left\{ e^{3x} \right\}$, $S_3 = \left\{ x, 1 \right\}$ $S_4 = \left\{ 1 \right\}$.

Step 2: Set S_4 is completely included in S_3; so S_4 is omitted, leaving the three sets S_1, S_2, S_3.

Step 3: Observe that the only member e^{2x} of $S_1 = \left\{ e^{2x} \right\}$ is included in the complementary function and so is a solution of the corresponding homogeneous D. E. Thus we multiply the member e^{2x} of S_1 by x to obtain the revised set

$$S_1' = \left\{ xe^{2x} \right\} ,$$

whose only member is not a solution of the homogeneous
D. E.

Step 4: We now have the three sets

$$S_1' = \left\{ xe^{2x} \right\} , \quad S_2 = \left\{ e^{3x} \right\} , \quad S_3 = \left\{ x, 1 \right\} .$$

We form a linear combination of their four elements.

Step 5: Thus we take

$$y_p = Axe^{2x} + Be^{3x} + Cx + D$$

as a particular solution. Then

$$y_p' = 2Axe^{2x} + Ae^{2x} + 3Be^{3x} + C ,$$

$$y_p'' = 4Axe^{2x} + 4Ae^{2x} + 9Be^{3x} .$$

We substitute in the D. E., obtaining

$$(4Axe^{2x} + 4Ae^{2x} + 9Be^{3x}) + (2Axe^{2x} + Ae^{2x} + 3Be^{3x} + C)$$

$$- 6(Axe^{2x} + Be^{3x} + Cx + D) = 10e^{2x} - 18e^{3x} - 6x - 11$$

or

$$5Ae^{2x} + 6Be^{3x} - 6Cx + (C - 6D) = 10e^{2x} - 18e^{3x} - 6x - 11.$$

We equate coefficients of like terms on both sides of
this to obtain

$$5A = 10, \quad 6B = -18, \quad -6C = -6, \quad C - 6D = -11.$$

From this, we find $A = 2$, $B = -3$, $C = 1$, $D = 2$. We

thus obtain the particular integral

$$y_p = 2xe^{2x} - 3e^{3x} + x + 2.$$

The G. S. of the D. E. is

$$y = c_1e^{2x} + c_2e^{-3x} + 2xe^{2x} - 3e^{3x} + x + 2 \ .$$

16. The corresponding homogeneous D. E. is
$y''' - 2y'' - y' + 2y = 0$. The auxiliary equation is
$m^3 - 2m^2 - m + 2 = 0$, that is $(m^2 - 1)(m - 2) = 0$.
From this, the roots are 1, -1, 2. The complementary
function is $y_c = c_1e^x + c_2e^{-x} + c_3e^{2x}$. The NH term is
a linear combination of the UC functions given by
e^{2x} and e^{3x}.

Step 1: Form the UC set of each of these two
functions: $S_1 = \left\{ e^{2x} \right\}$, $S_2 = \left\{ e^{3x} \right\}$.

Step 2: Neither set is identical with nor included
in the other, so both are retained.

Step 3: Observe that the only member e^{2x} of $S_1 = \left\{ e^{2x} \right\}$
is included in the complementary function, and so is a
solution of the corresponding homogeneous D. E. Thus
we multiply the member e^{2x} of S_1 by x to obtain the
revised set $S_1' = \left\{ xe^{2x} \right\}$, whose only member is not a
solution of the homogeneous D. E.

Step 4: We now have the two sets $S_1' = \left\{ xe^{2x} \right\}$, $S_2 = \left\{ e^{3x} \right\}$. We form a linear combination of their two elements.

Step 5: Thus we take $y_p = Axe^{2x} + Be^{3x}$ as a particular solution. Then $y_p' = 2Axe^{2x} + Ae^{2x} + 3Be^{3x}$,

$$y_p'' = 4Axe^{2x} + 4Ae^{2x} + 9Be^{3x},$$

$$y_p''' = 8Axe^{2x} + 12Ae^{2x} + 27Be^{3x},$$

We substitute in the D. E., obtaining

$$(8Axe^{2x} + 12Ae^{2x} + 27e^{3x}) - 2(4Axe^{2x} + 4Ae^{2x} + 9Be^{3x})$$

$$- (2Axe^{2x} + Ae^{2x} + 3Be^{3x}) + 2(Axe^{2x} + Be^{3x})$$

$$= 9e^{2x} - 8e^{3x}$$

or

$$3Ae^{2x} + 8Be^{3x} = 9e^{2x} - 8e^{3x} \quad .$$

We equate coefficients of like terms on both sides of this to obtain $3A = 9$, $8B = -8$, and hence $A = 3$, $B = -1$. Thus we obtain the particular integral $y_p = 3xe^{2x} - e^{3x}$. The G. S. of the D. E. is

$$y = c_1e^x + c_2e^{-x} + c_3e^{2x} + 3xe^{2x} - e^{3x} \quad .$$

17. The corresponding homogeneous D. E. is $y''' + y' = 0$. The auxiliary equation is $m^3 + m = 0$, with roots $m = 0, \pm i$. The complementary function is

$y_c = c_1 + c_2 \sin x + c_3 \cos x$. The NH term is a linear combination of the UC functions given by x^2 and $\sin x$.

Step 1: Form the UC set of each of these two UC functions: $S_1 = \{x^2, x, 1\}$, $S_2 = \{\sin x, \cos x\}$.

Step 2: Neither set is identical with nor included in the other, so both are retained.

Step 3: Observe that the member 1 of S_1 is included in the complementary function (in the $c_1 = c_1 \cdot 1$ term) and so is a solution of the corresponding homogeneous D. E. Thus we multiply each member of set S_1 by x to obtain the revised set $S_1' = \{x^3, x^2, x\}$, which has no members which are solutions of the homogeneous D. E. Similarly, both members of S_2 are included in the complementary function. Thus we multiply each member of S_2 by x to obtain the revised set $S_2' = \{x \sin x, x \cos x\}$, which has no members which are solutions of the homogeneous D. E.

Step 4: We now have the two sets $S_1' = \{x^3, x^2, x\}$ $S_2' = \{x \sin x, x \cos x\}$. We form a linear combination of their five elements.

Step 5: Thus we take $y_p = Ax^3 + Bx^2 + Cx + Dx \sin x + Ex \cos x$.

Then $y_p' = 3Ax^2 + 2Bx + C + Dx \cos x - Ex \sin x + D \sin x$
$$+ E \cos x,$$

$y_p'' = 6Ax + 2B - Dx \sin x - Ex \cos x + 2D \cos x$
$$- 2E \sin x,$$

$y_p''' = 6A - Dx \cos x + Ex \sin x - 3D \sin x - 3E \cos x.$

We substitute in the D. E., obtaining

$(6A - Dx \cos x + Ex \sin x - 3D \sin x - 3E \cos x)$
$$+ (3Ax^2 + 2Bx + C + Dx \cos x - Ex \sin x + D \sin x$$
$$+ E \cos x) = 2x^2 + 4 \sin x$$

or

$3Ax^2 + 2Bx + (6A + C) - 2D \sin x - 2E \cos x =$
$$2x^2 + 4 \sin x.$$

We equate coefficients of like terms on both sides of this to obtain $3A = 2$, $2B = 0$, $6A + C = 0$, $-2D = 4$, $-2E = 0$, and hence $A = 2/3$, $B = 0$, $C = -4$, $D = -2$, $E = 0$. Thus we obtain the particular integral $y_p = (2/3)x^3 - 4x - 2x \sin x$. The G. S. of the D. E. is $y = c_1 + c_2 \sin x + c_3 \cos x + (2/3)x^3 - 4x - 2x \sin x.$

18. The corresponding homogeneous D. E. is $y^{iv} - 3y''' + 2y'' = 0$. The auxiliary equation is $m^4 - 3m^3 + 2m^2 = 0$ or $m^2(m-1)(m-2) = 0$, with roots 0, 0 (real double root), 1, 2. The complementary function

is $y_c = c_1 + c_2 x + c_3 e^x + c_4 e^{2x}$. The NH term is a linear combination of the UC functions given by e^{-x}, e^{2x}, and x.

Step 1: Form the UC set of each of these three UC functions: $S_1 = \{ e^{-x} \}$, $S_2 = \{ e^{2x} \}$, $S_3 = \{ x, 1 \}$.

Step 2: No set is identical with nor included in any other, so each is retained.

Step 3: Observe that the member e^{2x} of S_2 is included in the complementary function and so is a solution of the corresponding homogeneous D. E. Thus we multiply the member of S_2 by x to obtain the revised set $S_2' = \{ xe^{2x} \}$, whose member is not a solution of the homogeneous D. E. Next observe that both members x and 1 of S_3 are included in the complementary function. Thus we multiply each member of S_3 by x^2 to obtain the revised set $S_3' = \{ x^3, x^2 \}$, whose members are not solutions of the homogeneous D. E. (note that multiplication by x, instead of x^2, is not sufficient here).

Step 4: We now have the three sets

$$S_1 = \{ e^{-x} \}, \quad S_2' = \{ xe^{2x} \}, \quad S_3' = \{ x^3, x^2 \}.$$

We form a linear combination of their four elements.

Step 5: Thus we take $y_p = Ae^{-x} + Bxe^{2x} + Cx^3 + Dx^2$
as a particular solution. Then

$$y_p' = -Ae^{-x} + 2Bxe^{2x} + Be^{2x} + 3Cx^2 + 2Dx,$$

$$y_p'' = Ae^{-x} + 4Bxe^{2x} + 4Be^{2x} + 6Cx + 2D,$$

$$y_p''' = -Ae^{-x} + 8Bxe^{2x} + 12Be^{2x} + 6 C,$$

$$y_p^{iv} = Ae^{-x} + 16Bxe^{2x} + 32Be^{2x} .$$

We substitute into the D. E., obtaining

$$(Ae^{-x} + 16Bxe^{2x} + 32Be^{2x}) - 3(-Ae^{-x} + 8Bxe^{2x} + 12Be^{2x}$$
$$+ 6C) + 2(Ae^{-x} + 4Bxe^{2x} + 4Be^{2x} + 6Cx + 2D)$$
$$= 3e^{-x} + 6e^{2x} - 6x \qquad \text{or}$$

$$6Ae^{-x} + 4Be^{2x} + 12Cx + (-18C + 4D) = 3e^{-x} + 6e^{2x} - 6x .$$

We equate coefficients of like terms on both sides of
this to obtain $6A = 3$, $4B = 6$, $12C = -6$, $-18C + 4D = 0$,
and hence $A = 1/2$, $B = 3/2$, $C = -1/2$, $D = -9/4$. Thus
we obtain the particular integral $y_p = (1/2)e^{-x} +$
$(3/2)xe^{2x} - (1/2)x^3 - (9/4)x^2$. The G. S. of the D. E.
is

$$y = c_1 + c_2x + c_3e^x + c_4e^{2x} + (1/2)e^{-x} + (3/2)xe^{2x}$$
$$- (1/2)x^3 - (9/4)x^2 .$$

21. The corresponding homogeneous D. E. is
$y'' + y = 0$. The auxiliary equation is $m^2 + 1 = 0$,
with roots $\pm i$. The complementary function is

$y_c = c_1 \sin x + c_2 \cos x$. The NH term is the UC function $x \sin x$.

Step 1:   Form the UC set of this UC function.   It is $S_1 = \left\{ x \sin x, \; x \cos x, \; \sin x, \; \cos x \right\}$.

Step 2:   This step does not apply.

Step 3:   Observe that the members $\sin x$ and $\cos x$ of S_1 are included in the complementary function and so are solutions of the corresponding homogeneous D. E. Thus we multiply each member of S_1 by x to obtain the revised set

$$S_1' = \left\{ x^2 \sin x, \; x^2 \cos x, \; x \sin x, \; x \cos x \right\}$$

whose members are not solutions of the homogeneous D. E.

Step 4:   We now have the set S_1'.   We form a linear combination of its four elements $x^2 \sin x$, $x^2 \cos x$, $x \sin x$, and $x \cos x$.

Step 5:   Thus we take

$$y_p = Ax^2 \sin x + Bx^2 \cos x + Cx \sin x + Dx \cos x$$

as a particular solution.   Then

$$y_p' = Ax^2 \cos x - Bx^2 \sin x + (2A - D)x \sin x$$
$$+ (2B + C)x \cos x + C \sin x + D \cos x \qquad \text{and}$$
$$y_p'' = -Ax^2 \sin x - Bx^2 \cos x + (-4B - C)x \sin x$$
$$+ (4A - D)x \cos x + (2A - 2D)\sin x + (2B + 2C)\cos x.$$

We substitute into the D. E., obtaining

$-4B$ x sin x + 4Ax cos x + $(2A - 2D)$ sin x

  + $(2B + 2C)$ cos x = x sin x .

We equate coefficients of like terms on both sides of this to obtain $-4B = 1$, $4A = 0$, $2A - 2D = 0$, $2B + 2C = 0$. From this, $A = 0$, $B = -1/4$, $C = 1/4$, $D = 0$. Thus we obtain the particular integral $y_p = -(1/4) x^2$ cos x + $(1/4)$ x sin x. The G. S. of the D. E. is

$$y = c_1 \sin x + c_2 \cos x - (1/4)x^2 \cos x + (1/4)x \sin x.$$

24. The corresponding homogeneous D. E. is $y^{iv} - 5y''' + 7y'' - 5y' + 6y = 0$. The auxiliary equation is $m^4 - 5m^3 + 7m^2 - 5m + 6 = 0$. By inspection we find that $m = 2$ is a root. Then by synthetic division, we have $(m - 2)(m^3 - 3m^2 + m - 3) = 0$ or $(m - 2)(m - 3)(m^2 + 1) = 0$. Thus the roots are $m = 2, 3, \pm i$. The complementary function is $y_c = c_1 e^{2x} + c_2 e^{3x} + c_3 \sin x + c_4 \cos x$. The NH term is a linear combination of the UC functions given by sin x and sin 2x.

Step 1: Form the UC set of each of these two UC functions: $S_1 = \{\sin x, \cos x\}$, $S_2 = \{\sin 2x, \cos 2x\}$.

Step 2: Neither set is identical with nor included

in the other, so both are retained.

Step 3:   Observe that both members of S_1 are included in the complementary function and so are solutions of the corresponding homogeneous D. E.   Thus we multiply each member of S_1 by x to obtain the revised set $S_1' = \left\{ x \sin x, \ x \cos x \right\}$, whose members are not solutions of the homogeneous D. E.

Step 4:   We now have the two sets

$$S_1' = \left\{ x \sin x, \ x \cos x \right\}, \ S_2 = \left\{ \sin 2x, \ \cos 2x \right\} .$$

We form a linear combination of their four elements.

Step 5:   Thus we take

$$y_p = Ax \sin x + Bx \cos x + C \sin 2x + D \cos 2x$$

as a particular solution.   Then

$y_p' = Ax \cos x - Bx \sin x + A \sin x + B \cos x +$
$2C \cos 2x - 2D \sin 2x,$ $y_p'' = -Ax \sin x - Bx \cos x +$
$2A \cos x - 2B \sin x - 4C \sin 2x - 4D \cos 2x,$

$y_p''' = -Ax \cos x + Bx \sin x - 3A \sin x - 3B \cos x -$
$8C \cos 2x + 8D \sin 2x,$ $y_p^{iv} = Ax \sin x + Bx \cos x -$
$4A \cos x + 4B \sin x + 16C \sin 2x + 16D \cos 2x.$

We substitute into the D. E., obtaining

$(Ax \sin x + Bx \cos x - 4A \cos x + 4B \sin x + 16C \sin 2x +$
$16D \cos 2x) - 5(-Ax \cos x + Bx \sin x - 3A \sin x -$
$3B \cos x - 8C \cos 2x + 8D \sin 2x) + 7(-Ax \sin x -$
$Bx \cos x + 2A \cos x - 2B \sin x - 4C \sin 2x - 4D \cos 2x)$
$- 5(Ax \cos x - Bx \sin x + A \sin x + B \cos x + 2C \cos 2x -$
$2D \sin 2x) + 6(Ax \sin x + Bx \cos x + C \sin 2x + D \cos 2x)$
$= 5 \sin x - 12 \sin 2x$

or $(10A - 10B) \sin x + (10A + 10B) \cos x$

$+ (-6C - 30D) \sin 2x + (30C - 6D) \cos 2x$

$= 5 \sin x - 12 \sin 2x$.

We equate coefficients of like terms on both sides of
this to obtain $10A - 10B = 5$, $10A + 10B = 0$,
$-6C - 30D = -12$, $30C - 6D = 0$. The equations in A and B
are equivalent to $2A - 2B = 1$, $A + B = 0$, from which
$A = 1/4$, $B = -1/4$. The equations in C and D are
equivalent to $C + 5D = 2$, $5C - D = 0$, from which
$C = 1/13$, $D = 5/13$. Thus we obtain the particular
integral

$y_p = (1/4)x \sin x - (1/4)x \cos x$

$+ (1/13) \sin 2x + (5/13) \cos 2x.$

The G. S. of the D. E. is

$y = c_1 e^{2x} + c_2 e^{3x} + c_3 \sin x + c_4 \cos x + (1/4)x \sin x -$
$(1/4)x \cos x + (1/13) \sin 2x + (5/13) \cos 2x.$

27. The auxiliary equation of the corresponding homogeneous D. E. is $m^2 - 8m + 15 = 0$, with roots 3, 5. The complementary function is $y_c = c_1 e^{3x} + c_2 e^{5x}$. The UC set of UC function xe^{2x} is $S_1 = \left\{ xe^{2x}, e^{2x} \right\}$. This does not need revision, so we take $y_p = Axe^{2x} + Be^{2x}$ as a particular integral. Then $y_p' = 2Axe^{2x} + (A + 2B)e^{2x}$ and $y_p'' = 4Axe^{2x} + (4A + 4B)e^{2x}$. We substitute in the D. E., obtaining

$$4Axe^{2x} + (4A + 4B)e^{2x} - 8[2Axe^{2x} + (A + 2B)e^{2x}]$$
$$+ 15(Axe^{2x} + Be^{2x}) = 9xe^{2x} \qquad \text{or}$$
$$3Axe^{2x} + (-4A + 3B)e^{2x} = 9xe^{2x} \ .$$

We equate coefficients of like terms on both sides of this to obtain $3A = 9$, $-4A + 3B = 0$. From these, we find $A = 3$, $B = 4$. Thus we obtain the particular integral $y_p = 3xe^{2x} + 4e^{2x}$. The G. S. of the D. E. is

$$y = c_1 e^{3x} + c_2 e^{5x} + 3xe^{2x} + 4e^{2x} \ .$$

We apply the I. C. $y(0) = 5$ to this. We find $c_1 + c_2 + 4 = 5$ or $c_1 + c_2 = 1$. ($*$) We next differentiate the G. S. to obtain $y' = 3c_1 e^{3x} + 5c_2 e^{5x} + 6xe^{2x} + 11e^{2x}$. We apply the I. C. $y'(0) = 10$ to this. We find $3c_1 + 5c_2 + 11 = 10$ or $3c_1 + 5c_2 = -1$. ($**$) From ($*$) and ($**$) we find $c_1 = 3$, $c_2 = -2$. Thus we obtain the particular solution

$$y = 3e^{3x} - 2e^{5x} + 3xe^{2x} + 4e^{2x} \quad .$$

32. The auxiliary equation of the corresponding homogeneous D. E. is $m^2 - 10m + 29 = 0$, with roots $5 \pm 2i$. The complementary function is $y_c = e^{5x}(c_1 \sin 2x + c_2 \cos 2x)$. The UC set of the UC function e^{5x} is $S_1 = \left\{ e^{5x} \right\}$. This does not need revision, so we take $y_p = Ae^{5x}$ as a particular integral. Then $y_p' = 5Ae^{5x}$ and $y_p'' = 25Ae^{5x}$. We substitute in the D. E., obtaining

$$25Ae^{5x} - 10(5Ae^{5x}) + 29Ae^{5x} = 8e^{5x}$$

or $\quad 4Ae^{5x} = 8e^{5x}.$ We at once see that $A = 2$ and hence obtain the particular integral $y_p = 2e^{5x}$. The G. S. of the D. E. is

$$y = e^{5x}(c_1 \sin 2x + c_2 \cos 2x) + 2e^{5x} \quad .$$

We apply the I. C. $y(0) = 0$ to this. We find $c_2 + 2 = 0$, so $c_2 = -2$. We next differentiate the G. S. to obtain

$$y' = e^{5x}[(5c_1 - 2c_2)\sin 2x + (2c_1 + 5c_2) \cos 2x] + 10e^{5x}.$$

We apply the I. C. $y'(0) = 8$ to this. We find $2c_1 + 5c_2 + 10 = 8$ or $2c_1 + 5c_2 = -2$. Since $c_2 = -2$, this gives $c_1 = 4$. Thus we obtain the particular solution $\quad y = 2e^{5x}(2 \sin 2x - \cos 2x + 1) \quad .$

39. The auxiliary equation of the corresponding homogeneous D. E. is $m^3 - 4m^2 + m + 6 = 0$ or $(m + 1)(m - 3)(m - 2) = 0$, with roots -1, 2, 3. The complementary function is $y_c = c_1 e^{-x} + c_2 e^{2x} + c_3 e^{3x}$. The UC sets of the UC functions in the right member of the D. E. are $S_1 = \left\{ xe^x, e^x \right\}$, $S_2 = \left\{ e^x \right\}$ and $S_3 = \left\{ \sin x, \cos x \right\}$. Since $S_1 \supset S_2$, we omit S_2, retaining S_1 and S_3. Neither of these need revision, so we take a linear combination of their four members as a particular integral. That is, we take $y_p = Axe^x + Be^x + C \sin x + D \cos x$. Then $y_p' = Axe^x + (A + B)e^x + C \cos x - D \sin x$, $y_p'' = Axe^x + (2A + B)e^x - C \sin x - D \cos x$, $y_p''' = Axe^x + (3A + B)e^x - C \cos x + D \sin x$.

We substitute into the D. E., obtaining
$Axe^x + (3A + B)e^x - C \cos x + D \sin x - 4[Axe^x + (2A + B)e^x - C \sin x - D \cos x] + [Axe^x + (A + B)e^x + C \cos x - D \sin x] + 6[Axe^x + Be^x + C \sin x + D \cos x] = 3xe^x + 2e^x - \sin x$ or $4Axe^x + (-4A + 4B)e^x + 10C \sin x + 10D \cos x = 3xe^x + 2e^x - \sin x$. We equate coefficients of like terms on both sides of this to obtain $4A = 3$, $-4A + 4B = 2$, $10C = -1$, $10D = 0$. From these, we find $A = 3/4$, $B = 5/4$, $C = -1/10$, $D = 0$. Thus we obtain the particular integral,

$$y_p = (3/4)xe^x + (5/4)e^x - (1/10) \sin x \quad .$$

The G. S. of the D. E. is

$$y = c_1 e^{-x} + c_2 e^{2x} + c_3 e^{3x} + (3/4)x e^x$$
$$+ (5/4)e^x - (1/10) \sin x \quad .$$

We apply the I. C. $y(0) = 33/40$ to this. We find

$$c_1 + c_2 + c_3 + 5/4 = 33/40 \text{ or } c_1 + c_2 + c_3 = -17/40 \; (\ast \,)$$

We next differentiate the G. S. to obtain

$$y' = -c_1 e^{-x} + 2c_2 e^{2x} + 3c_3 e^{3x} + (3/4)x e^x$$
$$+ 2e^x - (1/10) \cos x \quad .$$

We apply the I. C. $y'(0) = 0$ to this. We find

$$-c_1 + 2c_2 + 3c_3 + 2 - (1/10) = 0 \quad \text{or}$$

$$-c_1 + 2c_2 + 3c_3 = -19/10 \quad . \qquad\qquad (\ast\ast\,)$$

We differentiate once more, obtaining

$$y'' = c_1 e^{-x} + 4c_2 e^{2x} + 9c_3 e^{3x} + (3/4)x e^x$$
$$+ (11/4)e^x + (1/10) \sin x \quad .$$

We apply the I. C. $y''(0) = 0$ to this. We find

$$c_1 + 4c_2 + 9c_3 + (11/4) = 0 \qquad \text{or}$$

$$c_1 + 4c_2 + 9c_3 = -11/4 \qquad\qquad (\ast\ast\ast).$$

The three equations $(\ast)$, $(\ast\ast\,)$, $(\ast\ast\ast)$ determine
c_1, c_2, c_3. Adding $(\ast)$ and $(\ast\ast)$, we have
$3c_2 + 4c_3 = -93/40$. Adding $(\ast\ast\,)$ and $(\ast\ast\ast)$, we
have $6c_2 + 12c_3 = -93/20$. Solving these two resulting
equations in c_2 and c_3, we find $c_2 = -31/40$, $c_3 = 0$.

Then (✳) gives $c_1 = -c_2 - c_3 - 17/40 = 7/20$. Thus we obtain the desired particular solution

$$y = (7/20)e^{-x} - (31/40)e^{2x} + (3/4)xe^{x}$$
$$+ (5/4)e^{x} - (1/10) \sin x \ .$$

44. The auxiliary equation of the corresponding homogeneous D. E. is $m^2 - 6m + 9 = 0$, with roots 3, 3 (double root). The complementary function is $y_c = (c_1 + c_2 x)e^{3x}$. The NH member is a linear combination of the three UC functions $x^4 e^{x}$, $x^3 e^{2x}$, $x^2 e^{3x}$. The UC sets of these functions are

$$S_1 = \left\{ x^4 e^{x},\ x^3 e^{x},\ x^2 e^{x},\ xe^{x},\ e^{x} \right\} \ ,$$
$$S_2 = \left\{ x^3 e^{2x},\ x^2 e^{2x},\ xe^{2x},\ e^{2x} \right\} ,\ \text{and}$$
$$S_3 = \left\{ x^2 e^{3x},\ xe^{3x},\ e^{3x} \right\} \ ,\ \text{respectively.}$$

None is completely contained in any other, so each is retained. The members xe^{3x} and e^{3x} of S_3 are included in the complementary function and so are solutions of the corresponding homogeneous D. E. Thus we multiply each member of S_3 by x^2 to obtain the revised set $S_3' = \left\{ x^4 e^{3x},\ x^3 e^{3x},\ x^2 e^{3x} \right\}$, whose members are not solutions of the homogeneous D. E. (Note that multiplication by x, instead of x^2, is not sufficient here.) We now have the three sets S_1, S_2, and S_3'.

We form a linear combination of their twelve members.
Thus:

$$y_p = Ax^4e^x + Bx^3e^x + Cx^2e^x + Dxe^x + Ee^x$$

$$+ Fx^3e^{2x} + Gx^2e^{2x} + Hxe^{2x} + Ie^{2x}$$

$$+ Jx^4e^{3x} + Kx^3e^{3x} + Lx^2e^{3x} \quad .$$

45. The auxiliary equation of the corresponding
homogeneous D. E. is $m^2 + 6m + 13 = 0$, with roots
$-3 \pm 2i$. The complementary function is
$y_c = e^{-3x}(c_1 \sin 2x + c_2 \cos 2x)$. The NH member is a
linear combination of the two UC functions $xe^{-3x} \sin 2x$
and $x^2e^{-2x} \sin 3x$. The UC sets of these functions are
$$S_1 = \left\{ xe^{-3x} \sin 2x, \ xe^{-3x} \cos 2x, \ e^{-3x} \sin 2x, \right.$$
$$\left. e^{-3x} \cos 2x \right\}$$

and
$$S_2 = \left\{ x^2e^{-2x} \sin 3x, \ x^2e^{-2x} \cos 3x, \ xe^{-2x} \sin 3x, \right.$$
$$\left. xe^{-2x} \cos 3x, \ e^{-2x} \sin 3x, \ e^{-2x} \cos 3x \right\} \quad ,$$

respectively. Neither is completely contained in the
other, so each is retained.

The members $e^{-3x} \sin 2x$ and $e^{-3x} \cos 2x$ of S_1 are
included in the complementary function and so are solutions of the corresponding homogeneous D. E. Thus we
multiply each member of S_1 by x to obtain the revised
set

$$S_1' = \left\{ x^2 e^{-3x} \sin 2x, \ x^2 e^{-3x} \cos 2x, \ xe^{-3x} \sin 2x, \ xe^{-3x} \cos 2x \right\},$$

whose members are not solutions of the homogeneous D.E.

We now have the two sets S_1' and S_2. We form a linear combination of their ten members. Thus:

$$\begin{aligned}
y_p = \ & Ax^2 e^{-3x} \sin 2x + Bx^2 e^{-3x} \cos 2x \\
& + Cxe^{-3x} \sin 2x + Dxe^{-3x} \cos 2x \\
& + Ex^2 e^{-2x} \sin 3x + Fx^2 e^{-2x} \cos 3x \\
& + Gxe^{-2x} \sin 3x + Hxe^{-2x} \cos 3x \\
& + Ie^{-2x} \sin 3x + Je^{-2x} \cos 3x.
\end{aligned}$$

50. The auxiliary equation of the corresponding homogeneous D. E. is $m^6 + 2m^5 + 5m^4 = 0$, with roots 0, 0, 0, 0 (four-fold root) and $-1 \pm 2i$. The complementary function is $y_c = c_1 + c_2 x + c_3 x^2 + c_4 x^3 + e^{-x}(c_5 \sin 2x + c_6 \cos 2x)$. The NH member is a linear combination of the three UC functions x^3, $x^2 e^{-x}$, and $e^{-x} \sin 2x$. The UC sets of these functions are $S_1 = \left\{ x^3, x^2, x, 1 \right\}$, $S_2 = \left\{ x^2 e^{-x}, xe^{-x}, e^{-x} \right\}$, and $S_3 = \left\{ e^{-x} \sin 2x, e^{-x} \cos 2x \right\}$. None is completely contained in any other, so all three are retained.

All four members of S_1 are contained in the complementary function and so are solutions of the correspon-

ding homogeneous D. E. Thus we multiply each member
of S_1 by the lowest positive integral power of x so
that the resulting revised set will contain no members
that are solutions of the corresponding homogeneous D.
E. By actual trial, we see that multiplying by x, x^2,
or x^3 will <u>not</u> accomplish this. Rather, we must multiply each member of S_1 by x^4, obtaining

$$S_1' = \left\{ x^7, x^6, x^5, x^4 \right\}$$

whose members are not solutions of the homogeneous D. E.

Also, both members of S_3 are included in the complementary function and so are solutions of the corresponding homogeneous D. E. Thus we multiply each member of
S_3 by x to obtain the revised set

$$S_3' = \left\{ xe^{-x} \sin 2x, xe^{-x} \cos 2x \right\},$$

whose members are not solutions of the homogeneous D. E.

We now have the three sets S_1', S_2, and S_3'. We
form a linear combination of their nine members. Thus:

$$y_p = Ax^7 + Bx^6 + Cx^5 + Dx^4 + Ex^2e^{-x} + Fxe^{-x} + Ge^{-x}$$
$$+ Hxe^{-x} \sin 2x + Ixe^{-x} \cos 2x.$$

53. The auxiliary equation of the corresponding
homogeneous D. E. is $m^4 + 3m^2 - 4 = 0$ or

$(m^2 + 4)(m^2 - 1) = 0$, with roots 1, -1, $\pm$ 2i. The complementary function is $y_c = c_1 e^x + c_2 e^{-x} + c_3 \sin 2x + c_4 \cos 2x$. The NH member is a linear combination of $\cos^2 x$ and cosh x. Since $\cos^2 x = (1 + \cos 2x)/2$ and cosh x = $(e^x + e^{-x})/2$, we see that the NH member is in fact the linear combination

$$\tfrac{1}{2} + \tfrac{1}{2} \cos 2x - \tfrac{1}{2} e^x - \tfrac{1}{2} e^{-x}$$

of the four UC functions 1, $\cos 2x$, e^x, and e^{-x}. The UC sets of these are $S_1 = \{1\}$, $S_2 = \{\cos 2x, \sin 2x\}$, $S_3 = \{e^x\}$, and $S_4 = \{e^{-x}\}$, respectively.

Both members $\cos 2x$ and $\sin 2x$ of S_2 are contained in the complementary function and so are solutions of the corresponding homogeneous D. E. Thus we multiply each member of S_2 by x to obtain the revised set $S_2' = \{x \cos 2x, x \sin 2x\}$, whose members are not solutions of the homogeneous D. E. Thus we replace S_2 by the revised set S_2'. The situation is exactly the same for each of S_3 and S_4, so we replace S_3 by the revised set $S_3' = \{xe^x\}$ and S_4 by the revised set $S_4' = \{xe^{-x}\}$. Thus we have the sets $S_1 = \{1\}$, $S_2' = \{x \sin 2x, x \cos 2x\}$, $S_3' = \{xe^x\}$, and $S_4' = \{xe^{-x}\}$. We form a linear combination of their five members. Thus

$$y_p = A + Bx \sin 2x + Cx \cos 2x + Dxe^x + Exe^{-x} \quad .$$

Alternatively, we regard the NH member of the D. E. as a linear combination of the UC function $\cos^2 x = (\cos x)(\cos x)$ and the UC combination $\cosh x = (e^x + e^{-x})/2$. The UC sets of these are respectively $S_1 = \left\{ \sin^2 x, \cos^2 x, \sin x \cos x \right\}$ and $S_2 = \left\{ \sinh x, \cosh x \right\}$. One finds by direct substitution that the member $\sin x \cos x$ of S_1 is a solution of the corresponding homogeneous D. E. Thus we multiply each member of S_1 by x to obtain the revised set $S_1' = \left\{ x \sin^2 x, x \cos^2 x, x \sin x \cos x \right\}$, whose members are not solutions of the homogeneous D. E. So we replace S_1 by the revised set S_1'. In like manner, one finds $\cosh x$ is a solution of the homogeneous D. E.; and so we replace S_2 by the revised set $S_2' = \left\{ x \sinh x, x \cosh x \right\}$. So we now have the two revised sets S_1' and S_2'. We form a linear combination of their five members. Thus:

$$y_p = Ax \sin^2 x + Bx \cos^2 x + Cx \sin x \cos x + Dx \sinh x$$
$$+ Ex \cosh x.$$

54. The auxiliary equation of the corresponding homogeneous D. E. is $m^4 + 10m^2 + 9 = 0$ or

$(m^2 + 9)(m^2 + 1) = 0$ with roots $\pm i$, $\pm 3i$. The complementary function is $y_c = c_1 \sin x + c_2 \cos x + c_3 \sin 3x$
$+ c_4 \cos 3x$. The NH member is the UC function
$\sin x \sin 2x$. The UC set of this function is
$S_1 = \{\sin x \sin 2x,\ \sin x \cos 2x,\ \cos x \sin 2x,$
  $\cos x \cos 2x\}$. One finds by direct substitution
that the member $\sin x \sin 2x$ of S_1 is a solution of the
corresponding homogeneous D. E.   Thus we multiply each
member of S_1 by x to obtain the revised set $S_1' =$
$\{x \sin x \sin 2x,\ x \sin x \cos 2x,\ x \cos x \sin 2x,$
$x \cos x \cos 2x\}$, whose members are not solutions of
the homogeneous D. E. So we replace S_1 by the revised
set S_1'. We form a linear combination of its four
members. Thus:

$y_p = Ax \sin x \sin 2x + Bx \sin x \cos 2x + Cx \cos x \sin 2x$
$\qquad + Dx \cos x \cos 2x.$ $\qquad\qquad$ (✱)

Alternatively, using the trigonometric identity
$\sin u \sin v = (1/2)[\cos(u - v) - \cos(u + v)]$ with
$u = 2x$, $v = x$, we see that the NH member is equal to
$(1/2)[\cos x - \cos 3x]$ and thus is a linear combination
of the simple UC functions $\cos x$ and $\cos 3x$. The UC
sets of these two functions are $S_2 = \{\cos x,\ \sin x\}$
and $S_3 = \{\cos 3x,\ \sin 3x\}$, respectively.

Both members of S_2 are contained in the complementary function and so are solutions of the corresponding
homogeneous D. E. Thus we multiply each member of S_2
by x to obtain the revised set $S_2' = \{$ x cos x, x sin x$\}$
whose members are not solutions of the homogeneous
D. E. Thus we replace S_2 by the revised set S_2'. The
situation is exactly the same for the set S_3, so we
replace S_3 by the revised set $S_3' = \{$x cos 3x, x sin 3x$\}$.
We now have the two revised sets S_2' and S_3'. We form
a linear combination of their four members. Thus:

y_p = Ex cos x + Fx sin x + Gx cos 3x + Hx sin 3x (✳✳)

The student should convince himself that the y_p given by
(✳) can be expressed in the form given by (✳✳), and
vice versa.

Section 4.4, Page 162

6. The complementary function is defined by
$y_c(x) = c_1 \sin x + c_2 \cos x$. We assume a particular
integral of the form

$$y_p(x) = v_1(x) \sin x + v_2(x) \cos x. \qquad (1)$$

Then $y_p'(x) = v_1(x) \cos x - v_2(x) \sin x + v_1'(x) \sin x$
$+ v_2'(x) \cos x$. We impose the condition

$$v_1'(x) \sin x + v_2'(x) \cos x = 0, \qquad (2)$$

leaving $y_p'(x) = v_1(x) \cos x - v_2(x) \sin x$. Then from this

$$y_p''(x) = -v_1(x) \sin x - v_2(x) \cos x + v_1'(x) \cos x$$
$$-v_2'(x) \sin x. \qquad (3)$$

Substituting (1) and (3) into the given D. E., we obtain

$$v_1'(x) \cos x - v_2'(x) \sin x = \tan x \sec x. \qquad (4)$$

We now have conditions (2) and (4) from which to determine $v_1'(x)$ and $v_2'(x)$:

$$\begin{cases} \sin x \ \ v_1'(x) + \cos x \ \ v_2'(x) = 0, \\ \cos x \ \ v_1'(x) - \sin x \ \ v_2'(x) = \tan x \sec x. \end{cases}$$

Solving, we find

$$v_1'(x) = \frac{\begin{vmatrix} 0 & \cos x \\ \tan x \sec x & -\sin x \end{vmatrix}}{\begin{vmatrix} \sin x & \cos x \\ \cos x & -\sin x \end{vmatrix}} = \frac{-\cos x \tan x \sec x}{-1}$$

$$= \tan x \ .$$

$$v_2'(x) = \frac{\begin{vmatrix} \sin x & 0 \\ \cos x & \tan x \sec x \end{vmatrix}}{\begin{vmatrix} \sin x & \cos x \\ \cos x & -\sin x \end{vmatrix}} = \frac{\sin x \tan x \sec x}{-1}$$

$$= -\tan^2 x = 1 - \sec^2 x \ .$$

Integrating, we find $v_1(x) = -\ln|\cos x|$,

$v_2(x) = x - \tan x$. Substituting into $y_p(x) = v_1(x) \sin x$ + $v_2(x) \cos x$, we find $y_p(x) = -\sin x \ln|\cos x|$ + $(x - \tan x) \cos x = -\sin x \ln|\cos x| + x \cos x - \sin x$.

The G. S. of the D. E. is

$$y = c_1 \sin x + c_2 \cos x - \sin x \ln|\cos x| + x \cos x$$
$$- \sin x ,$$

which may be more simply written as

$$y = c_0 \sin x + c_2 \cos x - \sin x \ln|\cos x| + x \cos x,$$

where $c_0 = c_1 - 1$.

7. The auxiliary equation $m^2 + 4m + 5 = 0$ has the conjugate complex roots

$$m = \frac{-4 \pm \sqrt{16 - 20}}{2} = -2 \pm i, \quad \text{so the complementary func-}$$

tion is $y_c(x) = c_1 e^{-2x} \sin x + c_2 e^{-2x} \cos x$. We assume a particular integral of the form

$$y_p(x) = v_1(x)e^{-2x} \sin x + v_2(x) e^{-2x} \cos x. \qquad (1)$$

Then $y_p'(x) = v_1(x)e^{-2x} \cos x - 2v_1(x)e^{-2x} \sin x$

$+ v_1'(x)e^{-2x} \sin x - v_2(x)e^{-2x} \sin x - 2v_2(x)e^{-2x} \cos x$

$+ v_2'(x)e^{-2x} \cos x$. We impose the condition

$$v_1'(x)e^{-2x} \sin x + v_2'(x)e^{-2x} \cos x = 0, \qquad (2)$$

leaving $y_p'(x) = v_1(x)e^{-2x} \cos x - 2v_1(x)e^{-2x} \sin x$

$$- v_2(x)e^{-2x} \sin x - 2v_2(x)e^{-2x} \cos x. \quad (3)$$

Then, differentiating this and collecting like terms, we find

$$y_p''(x) = 3v_1(x)e^{-2x} \sin x - 4v_1(x)e^{-2x} \cos x$$

$$+ v_1'(x)[e^{-2x} \cos x - 2e^{-2x} \sin x] + 3v_2(x)e^{-2x} \cos x$$

$$+ 4v_2(x)e^{-2x} \sin x - v_2'(x)[e^{-2x} \sin x + 2e^{-2x} \cos x].$$

$$(4)$$

Then substituting (1), (3), and (4) into the given D.E. and collecting like terms, we obtain

$$v_1'(x)[e^{-2x} \cos x - 2e^{-2x} \sin x]$$

$$- v_2'(x)[e^{-2x} \sin x + 2e^{-2x} \cos x]$$

$$= e^{-2x} \sec x.$$

$$(5)$$

We now have conditions (2) and (5) from which to determine $v_1'(x)$ and $v_2'(x)$:

$$e^{-2x} \sin x \; v_1'(x) + e^{-2x} \cos x \; v_2'(x) = 0 \quad,$$

$$e^{-2x}(\cos x - 2 \sin x) \; v_1'(x)$$

$$+ e^{-2x}(-\sin x - 2 \cos x) \; v_2'(x) = e^{-2x} \sec x.$$

Solving, we find:

$$v_1'(x) = \frac{\begin{vmatrix} 0 & e^{-2x} \cos x \\ e^{-2x} \sec x & e^{-2x}(-\sin x - 2 \cos x) \end{vmatrix}}{\begin{vmatrix} e^{-2x} \sin x & e^{-2x} \cos x \\ e^{-2x}(\cos x - 2 \sin x) & e^{-2x}(-\sin x - 2 \cos x) \end{vmatrix}}$$

$$= \frac{-e^{-4x} \cos x \sec x}{e^{-4x}(-1)} = 1 \quad.$$

$$v_2{}'(x) = \frac{\begin{vmatrix} e^{-2x} \sin x & 0 \\ e^{-2x}(\cos x - 2 \sin x) & e^{-2x} \sec x \end{vmatrix}}{\begin{vmatrix} e^{-2x} \sin x & e^{-2x} \cos x \\ e^{-2x}(\cos x - 2 \sin x) & e^{-2x}(-\sin x - 2 \cos x) \end{vmatrix}}$$

$$= \frac{e^{-4x} \sec x \sin x}{e^{-4x}(-1)} = -\tan x \quad .$$

Integrating, we find $v_1(x) = x$, $v_2(x) = \ln|\cos x|$.
Substituting into $y_p(x) = v_1(x)e^{-2x} \sin x$
$+ v_2(x)e^{-2x} \cos x$, we find $y_p(x) = xe^{-2x} \sin x$
$+ [\ln|\cos x|]e^{-2x} \cos x$. The G. S. of the D. E. is

$$y = e^{-2x}(c_1 \sin x + c_2 \cos x) + xe^{-2x} \sin x$$

$$+ [\ln|\cos x|] \, e^{-2x} \cos x.$$

12. The complementary function is defined by
$y_c(x) = c_1 \sin x + c_2 \cos x$. We assume a particular
integral of the form

$$y_p (x) = v_1(x) \sin x + v_2(x) \cos x. \qquad (1)$$

Then $y_p{}'(x) = v_1(x) \cos x - v_2(x) \sin x + v_1{}'(x) \sin x$
$+ v_2{}'(x) \cos x$. We impose the condition

$$v_1{}'(x) \sin x + v_2{}'(x) \cos x = 0, \qquad (2)$$

leaving $y_p{}'(x) = v_1(x) \cos x - v_2(x) \sin x$. Then from
this,

$$y_p''(x) = -v_1(x) \sin x - v_2(x) \cos x + v_1'(x) \cos x$$
$$- v_2'(x) \sin x \ . \tag{3}$$

Substituting (1) and (3) into the given D. E., we obtain

$$v_1'(x) \cos x - v_2'(x) \sin x = \tan^3 x. \tag{4}$$

We now have conditions (2) and (4) from which to determine $v_1'(x)$ and $v_2'(x)$:

$$\begin{cases} \sin x \ v_1'(x) + \cos x \ v_2'(x) = 0, \\ \cos x \ v_1'(x) - \sin x \ v_2'(x) = \tan^3 x \ . \end{cases}$$

Solving, we find

$$v_1'(x) = \frac{\begin{vmatrix} 0 & \cos x \\ \tan^3 x & -\sin x \end{vmatrix}}{\begin{vmatrix} \sin x & \cos x \\ \cos x & -\sin x \end{vmatrix}} = \frac{-\tan^3 x \cos x}{-1}$$

$$= \frac{(1 - \cos^2 x) \sin x}{\cos^2 x} = [(\cos x)^{-2} - 1]\sin x,$$

$$v_2'(x) = \frac{\begin{vmatrix} \sin x & 0 \\ \cos x & \tan^3 x \end{vmatrix}}{\begin{vmatrix} \sin x & \cos x \\ \cos x & -\sin x \end{vmatrix}} = \frac{\tan^3 x \sin x}{-1} = -\frac{\sin^4 x}{\cos^3 x}$$

$$= -\frac{(1 - \cos^2 x)^2}{\cos^3 x} = -\frac{1}{\cos^3 x} + \frac{2}{\cos x} - \cos x$$

$$= -\sec^3 x + 2 \sec x - \cos x \ .$$

Integrating (we recommend using tables), we find

$v_1(x) = (\cos x)^{-1} + \cos x,$

$v_2(x) = -\frac{1}{2} \tan x \sec x + \frac{3}{2} \ln|\sec x + \tan x| - \sin x$.

Substituting into $y_p(x) = v_1(x) \sin x + v_2(x) \cos x$,

we find $y_p(x) = [(\cos x)^{-1} + \cos x] \sin x$

$+ \left[\frac{1}{2} \tan x \sec x + \frac{3}{2} \ln|\sec x + \tan x| - \sin x \right] \cos x$

$= \frac{\tan x}{2} + \frac{3}{2} \cos x \ln|\sec x + \tan x|$.

The G. S. of the D. E. is

$y = c_1 \sin x + c_2 \cos x + \frac{\tan x}{2} + \frac{3}{2} \cos x \ln|\sec x + \tan x|.$

15. The complementary function is defined by
$y_c(x) = c_1 \sin x + c_2 \cos x.$ We assume a particular
integral of the form

$$y_p(x) = v_1(x) \sin x + v_2(x) \cos x. \qquad (1)$$

Then $y_p'(x) = v_1(x) \cos x - v_2(x) \sin x + v_1'(x) \sin x$
$+ v_2'(x) \cos x.$ We impose the condition

$$v_1'(x) \sin x + v_2'(x) \cos x = 0 , \qquad (2)$$

leaving $y_p'(x) = v_1(x) \cos x - v_2(x) \sin x.$ Then from
this

$$y_p''(x) = -v_1(x) \sin x - v_2(x) \cos x + v_1'(x) \cos x$$
$$- v_2'(x) \sin x. \qquad (3)$$

Substituting (1) and (3) into the given D.E., we obtain

$v_1'(x) \cos x - v_2'(x) \sin x = 1/(1 + \sin x).$ (4)

We now have conditions (2) and (4) from which to determine $v_1'(x)$ and $v_2'(x)$:

$$\begin{cases} \sin x \ v_1'(x) + \cos x \ v_2'(x) = 0 \ , \\ \cos x \ v_1'(x) - \sin x \ v_2'(x) = 1/(1 + \sin x) \ . \end{cases}$$

Solving, we find

$$v_1'(x) = \frac{\begin{vmatrix} 0 & \cos x \\ 1/(1 + \sin x) & -\sin x \end{vmatrix}}{\begin{vmatrix} \sin x & \cos x \\ \cos x & -\sin x \end{vmatrix}} = \frac{\cos x}{1 + \sin x} \quad ,$$

$$v_2'(x) = \frac{\begin{vmatrix} \sin x & 0 \\ \cos x & 1/(1 + \sin x) \end{vmatrix}}{\begin{vmatrix} \sin x & \cos x \\ \cos x & -\sin x \end{vmatrix}} = -\frac{\sin x}{1 + \sin x} \quad .$$

Integrating, we find

$v_1(x) = \ln(1 + \sin x),$

$v_2(x) = - \int \frac{\sin x \ dx}{1 + \sin x} = \int (-1 + \frac{1}{1 + \sin x}) \ dx$

$$= -x - \frac{\cos x}{1 + \sin x} \quad .$$

Substituting into $y_p(x) = v_1(x) \sin x + v_2(x) \cos x$, we find $y_p(x) = \sin x[\ln(1 + \sin x)] - x \cos x - \frac{\cos^2 x}{1 + \sin x}.$

The G.S. of the D.E. is

$$y = c_1 \sin x + c_2 \cos x + \sin x[\ln(1 + \sin x)]$$
$$- x \cos x - \frac{\cos^2 x}{1 + \sin x} .$$

16. The auxiliary equation $m^2 - 2m + 1 = 0$ has roots 1, 1 (double root), so the complementary function is $y_c(x) = (c_1 + c_2 x)e^x$, which we rewrite slightly as $y_c(x) = c_1 e^x + c_2 x e^x$. We assume a particular integral of the form $y_p(x) = v_1(x)e^x + v_2(x) x e^x$.   (1)

Then $y_p'(x) = v_1(x)e^x + v_2(x)[(x + 1)e^x]$
$$+ v_1'(x)e^x + v_2'(x) x e^x .$$

We impose the condition $v_1'(x)e^x + v_2'(x)x e^x = 0$ (2)

leaving $y_p'(x) = v_1(x)e^x + v_2(x)[(x + 1)e^x]$.   (3)

Then $y_p''(x) = v_1(x)e^x + v_2(x)[(x + 2)e^x]$
$$+ v_1'(x)e^x + v_2'(x)[(x + 1)e^x].$$ (4)

Then substituting (1), (3), and (4) into the given D.E., we obtain

$$v_1(x)e^x + v_2(x)[(x + 2)e^x] + v_1'(x)e^x + v_2'(x)[(x+1)e^x]$$
$$- 2 \left\{ v_1(x)e^x + v_2(x)[(x + 1)e^x] \right\}$$
$$+ v_1(x)e^x + v_2(x)x e^x = e^x \sin^{-1} x$$

or

$$v_1'(x)e^x + v_2'(x)[(x + 1)e^x] = e^x \sin^{-1} x.$$ (5)

We now have conditions (2) and (5) from which to
determine $v_1'(x)$ and $v_2'(x)$:

$$\begin{cases} e^x \, v_1'(x) + xe^x \, v_2'(x) = 0 \quad, \\ e^x \, v_1'(x) + (x + 1)e^x \, v_2'(x) = e^x \sin^{-1} x \; . \end{cases}$$

Dividing out $e^x \neq 0$, these simplify to

$$\begin{cases} v_1'(x) + x \, v_2'(x) = 0, \\ v_1'(x) + (x + 1) \, v_2'(x) = \sin^{-1} x \quad. \end{cases}$$

Subtracting the first from the second, we find

$$v_2'(x) = \sin^{-1} x \; ;$$

and then the first equation gives

$$v_1'(x) = -xv_2'(x) = -x \sin^{-1} x \quad.$$

Integrating (using tables), we find

$$v_1(x) = -(x^2/2) \sin^{-1} x + (1/4) \sin^{-1} x -(x/4)\sqrt{1 - x^2} \quad,$$

$$v_2(x) = x \sin^{-1} x + \sqrt{1 - x^2}.$$

Substituting into $y_p(x) = v_1(x)e^x + v_2(x)xe^x$, we find

$$y_p(x) = -(x^2 \, e^x/2) \sin^{-1} x + (1/4) \, xe^x \sin^{-1} x$$

$$-(1/4)xe^x\sqrt{1 - x^2} + x^2e^x \sin^{-1} x + xe^x \sqrt{1 - x^2}$$

$$= \frac{e^x \sin^{-1} x}{4} + \frac{x^2 \, e^x \sin^{-1} x}{2} + \frac{3xe^x\sqrt{1 - x^2}}{4} \quad.$$

The G. S. of the D. E. is

$$y = (c_1 + c_2 x)e^x + \frac{e^x \sin^{-1} x}{4} + \frac{x^2 e^x \sin^{-1} x}{2}$$

$$+ \frac{3xe^x \sqrt{1 - x^2}}{4} \; .$$

17. The auxiliary equation $m^2 + 3m + 2 = 0$ has the roots $-1, -2$, so the complementary function is $y_c(x) = c_1 e^{-x} + c_2 e^{-2x}$. We assume a particular integral of the form $y_p(x) = v_1(x)e^{-x} + v_2(x)e^{-2x}$.   (1)

Then $y_p'(x) = -v_1(x)e^{-x} - 2v_2(x)e^{-2x} + v_1'(x)e^{-x}$
$$+ v_2'(x)e^{-2x} \; .$$

We impose the condition $v_1'(x)e^{-x} + v_2'(x)e^{-2x} = 0$ (2)

leaving $y_p'(x) = -v_1(x)e^{-x} - 2v_2(x) e^{-2x}.$ (3)

Then from this, $y_p''(x) = v_1(x)e^{-x} + 4v_2(x)e^{-2x}$
$$- v_1'(x)e^{-x} - 2v_2'(x)e^{-2x}. \qquad (4)$$

Substituting (1), (3), and (4) into the given D.E., we obtain $v_1(x)e^{-x} + 4v_2(x)e^{-2x} - v_1'(x)e^{-x} - 2v_2'(x)e^{-2x}$
$+ 3[-v_1(x)e^{-x} - 2v_2(x)e^{-2x}] + 2[v_1(x)e^{-x} + v_2(x)e^{-2x}]$
$$= e^{-x}/x \qquad \text{or}$$

$$-v_1'(x) \; e^{-x} - 2v_2'(x)e^{-2x} = e^{-x}/x. \qquad (5)$$

We now have conditions (2) and (5) from which to determine $v_1'(x)$ and $v_2'(x)$:

$$\begin{cases} e^{-x} v_1'(x) + e^{-2x}v_2'(x) = 0 \quad, \\ -e^{-x} v_1'(x) - 2e^{-2x} v_2'(x) = e^{-x}/x \quad. \end{cases}$$

Solving, we find

$$v_1'(x) = \frac{\begin{vmatrix} 0 & e^{-2x} \\ e^{-x}/x & -2e^{-2x} \end{vmatrix}}{\begin{vmatrix} e^{-x} & e^{-2x} \\ -e^{-x} & -2e^{-2x} \end{vmatrix}} = \frac{1}{x} \quad,$$

$$v_2'(x) = \frac{\begin{vmatrix} e^{-x} & 0 \\ -e^{-x} & e^{-x}/x \end{vmatrix}}{\begin{vmatrix} e^{-x} & e^{-2x} \\ -e^{-x} & -2e^{-2x} \end{vmatrix}} = -\frac{e^{x}}{x} \quad.$$

Integrating, we find $v_1(x) = \ln|x|$. We also have
$v_2(x) = -\int (e^{x}/x)dx$. Since $\int (e^{x}/x)dx$ cannot be
expressed in closed form in terms of a finite number of
elementary functions, we simply leave it as indicated.
Substituting into $y_p(x) = v_1(x)e^{-x} + v_2(x)e^{-2x}$, we find
$y_p(x) = e^{-x} \ln|x| - e^{-2x} \int (e^{x}/x) \, dx$. The G. S. of
the D. E. is
$y = c_1 e^{-x} + c_2 e^{-2x} + e^{-x} \ln|x| - e^{-2x} \int (e^{x}/x) \, dx$.

18. The auxiliary equation $m^2 - 2m + 1 = 0$ has the
roots 1, 1 (double root), so the complementary function

is $y_c(x) = (c_1 + c_2 x)e^x$, which we rewrite slightly as
$y_c(x) = c_1 e^x + c_2 x e^x$. We assume a particular integral
of the form $y_p(x) = v_1(x)e^x + v_2(x) xe^x$.   (1)

Then $y_p'(x) = v_1(x)e^x + v_2(x)[(x + 1)e^x]$

$$+ v_1'(x)e^x + v_2'(x)xe^x .$$

We impose the condition $v_1'(x)e^x + v_2'(x)xe^x = 0$,   (2)

leaving $y_p'(x) = v_1(x)e^x + v_2(x)[(x + 1)e^x]$.   (3)

Then $y_p''(x) = v_1(x)e^x + v_2(x)[(x + 2)e^x]$

$$+ v_1'(x)e^x + v_2'(x)[(x + 1)e^x].$$ (4)

Then substituting (1), (3), and (4) into the given D.E.,
we obtain

$$v_1(x)e^x + v_2(x)[(x + 2)e^x] + v_1'(x)e^x + v_2'(x)[(x+1) e^x]$$

$$- 2 \left\{ v_1(x)e^x + v_2(x)[(x + 1)e^x]\right\}$$

$$+ v_1(x)e^x + v_2(x)xe^x = x \ln x$$

or $v_1'(x)e^x + v_2'(x)[(x + 1)e^x] = x \ln x.$ (5)

We now have conditions (2) and (5) from which to determine $v_1'(x)$ and $v_2'(x)$:

$$\begin{cases} e^x v_1'(x) + xe^x v_2'(x) = 0 \ , \\ e^x v_1'(x) + (x + 1)e^x v_2'(x) = x \ln x \ . \end{cases}$$

Dividing out $e^x \neq 0$, these simplify to

$$v_1'(x) + x\, v_2'(x) = 0 \quad,$$

$$v_1'(x) + (x + 1)\, v_2'(x) = e^{-x}\, x \ln x \quad.$$

Subtracting the first from the second, we find

$$v_2'(x) = e^{-x}\, x \ln x \quad;$$

and then the first equation gives

$$v_1'(x) = -e^{-x}\, x^2 \ln x \quad.$$

Thus we have $v_1(x) = -\displaystyle\int \frac{x^2 \ln x}{e^x}\, dx$ $\quad$ and

$v_2(x) = \displaystyle\int \frac{x \ln x}{e^x}\, dx$; and since neither of these

integrals can be expressed in closed form in terms of a
finite number of elementary functions, we simply leave
them as indicated. Substituting into
$y_p(x) = v_1(x)e^x + v_2(x)xe^x$, we find

$$y_p(x) = -e^x \int \frac{x^2 \ln x}{e^x}\, dx + xe^x \int \frac{x \ln x}{e^x}\, dx. \quad \text{The G. S.}$$

of the D. E. is

$$y = (c_1 + c_2 x)e^x - e^x \int \frac{x^2 \ln x}{e^x}\, dx + xe^x \int \frac{x \ln x}{e^x}\, dx \quad.$$

21. Since $x + 1$ and x^2 are linearly independent
solutions of the corresponding homogeneous equation,

the complementary function is defined by

$y_c(x) = c_1(x + 1) + c_2x^2$. We assume a particular

integral of the form $y_p(x) = v_1(x)(x + 1) + v_2(x)x^2$. (1)

Then $y_p'(x) = v_1(x) + 2v_2(x)x + v_1'(x)(x+1) + v_2'(x)x^2$.

We impose the condition $v_1'(x)(x+1) + v_2'(x)x^2 = 0$, (2)

leaving $y_p'(x) = v_1(x) + 2v_2(x) x.$ (3)

Then from this, $y_p''(x) = v_1'(x) + 2v_2(x) + 2v_2'(x)x.$ (4)

Substituting (1), (3), and (4) into the given D.E., we

obtain

$(x^2 + 2x)[v_1'(x) + 2v_2(x) + 2v_2'(x)x]$

$\quad - 2(x + 1)[v_1(x) + 2v_2(x)x] + 2[v_1(x)(x+1) + v_2(x)x^2]$

$\qquad = (x + 2)^2 ,$

or $\quad (x^2 + 2x)[v_1'(x) + 2v_2'(x)x] = (x + 2)^2,$

or finally,

$$v_1'(x) + 2v_2'(x)x = (x + 2)/x. \qquad (5)$$

We now have conditions (2) and (5) from which to

determine $v_1'(x)$ and $v_2'(x)$:

$$\begin{cases} (x+1) \, v_1'(x) + x^2 \, v_2'(x) = 0 \ , \\ \quad v_1'(x) + 2x \, v_2'(x) = (x+2)/x \ . \end{cases}$$

Solving, we find

$$v_1'(x) = \frac{\begin{vmatrix} 0 & x^2 \\ (x+2)/x & 2x \end{vmatrix}}{\begin{vmatrix} x+1 & x^2 \\ 1 & 2x \end{vmatrix}} = \frac{-x(x+2)}{x(x+2)} = -1 \quad,$$

$$v_2'(x) = \frac{\begin{vmatrix} x+1 & 0 \\ 1 & (x+2)/x \end{vmatrix}}{\begin{vmatrix} x+1 & x^2 \\ 1 & 2x \end{vmatrix}} = \frac{(x+1)(x+2)/x}{x(x+2)} = \frac{1}{x} + \frac{1}{x^2} \quad.$$

Integrating we find $v_1(x) = -x$, $v_2(x) = \ln|x| - x^{-1}$.

Substituting into $y_p(x) = v_1(x)(x+1) + v_2(x)x^2$, we have

$$y_p(x) = -x(x+1) + \left(\ln|x| - x^{-1}\right)x^2 = -x^2 - 2x + x^2 \ln|x|.$$

The G. S. of the D. E. is

$$y = c_1(x+1) + c_2x^2 - x^2 - 2x + x^2 \ln|x| \quad.$$

24. Since x and $(x + 1)^{-1}$ are linearly independent
solutions of the corresponding homogeneous equation, the
complementary function is defined by
$y_c(x) = c_1x + c_2(x + 1)^{-1}$. We assume a particular
integral of the form $y_p(x) = v_1(x)x + v_2(x)(x+1)^{-1}$. (1)
Then $y_p'(x) = v_1(x) - v_2(x)(x+1)^{-2} + v_1'(x)x$
$\qquad\qquad + v_2'(x)(x+1)^{-1}$.

We impose the condition

$$v_1'(x)x + v_2'(x)(x+1)^{-1} = 0, \qquad (2)$$

leaving $y_p'(x) = v_1(x) - v_2(x)(x+1)^{-2}.$ \qquad (3)

Then from this,

$$y_p''(x) = v_1'(x) + 2v_2(x)(x+1)^{-3} - v_2'(x)(x+1)^{-2}. \quad (4)$$

Substituting (1), (3), and (4) into the given D.E., we obtain

$$(2x + 1)(x+1)[v_1'(x) + 2v_2(x)(x+1)^{-3} - v_2'(x)(x+1)^{-2}]$$

$$+2x[v_1(x) - v_2(x)(x+1)^{-2}] - 2[v_1(x)x + v_2(x)(x+1)^{-1}]$$

$$= (2x + 1)^2 \ ,$$

or

$$(2x + 1)(x+1)[v_1'(x) - v_2'(x)(x+1)^{-2}] = (2x + 1)^2 \ ,$$

or finally,

$$v_1'(x) - v_2'(x)(x+1)^{-2} = (2x + 1)(x+1)^{-1} \qquad (5)$$

We now have conditions (2) and (5) from which to determine $v_1'(x)$ and $v_2'(x)$:

$$\begin{cases} x\, v_1'(x) + (x+1)^{-1}\, v_2'(x) = 0, \\ v_1'(x) - (x+1)^{-2}\, v_2'(x) = (2x + 1)(x+1)^{-1} \end{cases}$$

Solving, we find

$$v_1'(x) = \frac{\begin{vmatrix} 0 & (x+1)^{-1} \\ (2x+1)(x+1)^{-1} & -(x+1)^{-2} \end{vmatrix}}{\begin{vmatrix} x & (x+1)^{-1} \\ 1 & -(x+1)^{-2} \end{vmatrix}}$$

$$= \frac{-(2x+1)(x+1)^{-2}}{-x(x+1)^{-2} - (x+1)^{-1}} = 1 ,$$

$$v_2'(x) = \frac{\begin{vmatrix} x & 0 \\ 1 & (2x+1)(x+1)^{-1} \end{vmatrix}}{\begin{vmatrix} x & (x+1)^{-1} \\ 1 & -(x+1)^{-2} \end{vmatrix}} = \frac{x(2x+1)(x+1)^{-1}}{-x(x+1)^{-2} - (x+1)^{-1}}$$

$$= -x(x+1) .$$

Integrating, we find $v_1(x) = x$, $v_2(x) = -x^3/3 - x^2/2$.
Substituting into $y_p(x) = v_1(x)x + v_2(x)(x+1)^{-1}$, we
have

$$y_p(x) = x^2 - \frac{(2x^3 + 3x^2)(x+1)^{-1}}{6} .$$

The G. S. of the D. E. is

$$y = c_1 x + c_2(x+1)^{-1} + x^2 - \frac{(2x^3 + 3x^2)(x+1)^{-1}}{6} .$$

Section 4.5

1. Let $x = e^t$; then $t = \ln x$; and (as on pages 165 and 166 of the text)

$$x \frac{dy}{dx} = \frac{dy}{dt} \ , \ x^2 \frac{d^2y}{dx^2} = \frac{d^2y}{dt^2} - \frac{dy}{dt} \ .$$

The D.E. transforms into

$$\frac{d^2y}{dt^2} - \frac{dy}{dt} - 3\frac{dy}{dt} + 3y = 0$$

or $\frac{d^2y}{dt^2} - 4\frac{dy}{dt} + 3y = 0$. The auxiliary equation of this is $m^2 - 4m + 3 = 0$, and it has the roots $m = 1, 3$. Thus the general solution of the D. E. in y and t is $y = c_1 e^t + c_2 e^{3t}$. We return to the original independent variable x and replace e^t by x and e^{3t} by x^3. Doing this, we find that the general solution of the given D. E. is $y = c_1 x + c_2 x^3$.

3. Let $x = e^t$; then $t = \ln x$, $x \frac{dy}{dx} = \frac{dy}{dt}$, $x^2 \frac{d^2y}{dx^2} = \frac{d^2y}{dt^2} - \frac{dy}{dt}$. The D. E. transforms into

$$4(\frac{d^2y}{dt^2} - \frac{dy}{dt}) - 4\frac{dy}{dt} + 3y = 0 \text{ or } 4\frac{d^2y}{dt^2} - 8\frac{dy}{dt} + 3y = 0 \ .$$

The auxiliary equation of this is $4m^2 - 8m + 3 = 0$, and it has the roots $m = \frac{1}{2}, \frac{3}{2}$. Thus the general solution of the D.E. in y and t is $y = c_1 e^{t/2} + c_2 e^{3t/2}$. We return to the original independent variable x and re-place $e^{t/2}$ by $x^{1/2}$ and $e^{3t/2}$ by $x^{3/2}$. Doing this, we find that the general solution of the given D.E. is $y = c_1 x^{1/2} + c_2 x^{3/2}$.

6. Let $x = e^t$; then $t = \ln x$, $x \frac{dy}{dx} = \frac{dy}{dt}$, $x^2 \frac{d^2y}{dx^2} = \frac{d^2y}{dt^2} - \frac{dy}{dt}$. The D. E. transforms into

$\frac{d^2y}{dt^2} - \frac{dy}{dt} - 3 \frac{dy}{dt} + 13y = 0$ or $\frac{d^2y}{dt^2} - 4 \frac{dy}{dt} + 13y = 0$. The auxiliary equation of this is $m^2 - 4m + 13 = 0$, and it has the conjugate complex roots $2 \pm 3i$. The general solution of the D.E. in y and t is $y = e^{2t}(c_1 \sin 3t + c_2 \cos 3t)$. We return to the original independent variable x and replace e^{2t} by x^2 and t by $\ln x$. Doing this, we find that the general solution of the given D.E. is $y = x^2[c_1 \sin(3 \ln x) + c_2 \cos(3 \ln x)]$, or $y = x^2[c_1 \sin (\ln x^3) + c_2 \cos(\ln x^3)]$.

9. Let $x = e^t$; then $t = \ln x$, $x \frac{dy}{dx} = \frac{dy}{dt}$, $x^2 \frac{d^2y}{dx^2} = \frac{d^2y}{dt^2} - \frac{dy}{dt}$. The D.E. transforms into

$9(\frac{d^2y}{dt^2} - \frac{dy}{dt}) + 3 \frac{dy}{dt} + y = 0$ or $9 \frac{d^2y}{dt^2} - 6 \frac{dy}{dt} + y = 0$.

The auxiliary equation of this is $9m^2 - 6m + 1 = 0$, and it has the real double root $1/3$. The general solution

of the D.E. in y and t is $y = (c_1 + c_2 t) e^{t/3}$. We
return to the original independent variable x and
replace $e^{t/3}$ by $x^{1/3}$ and t by ln x. Doing this, we
find that the general solution of the given D.E. is
$y = (c_1 + c_2 \ln x) x^{1/3}$.

11. Let $x = e^t$; then $t = \ln x$, $x \frac{dy}{dx} = \frac{dy}{dt}$,

$x^2 \frac{d^2y}{dx^2} = \frac{d^2y}{dt^2} - \frac{dy}{dt}$, and (as on page 167 of the text)

$x^3 \frac{d^3y}{dx^3} = \frac{d^3y}{dt^3} - 3 \frac{d^2y}{dt^2} + 2 \frac{dy}{dt}$. The D. E. transforms into

$\frac{d^3y}{dt^3} - 3 \frac{d^2y}{dt^2} + 2 \frac{dy}{dt} - 3(\frac{d^2y}{dt^2} - \frac{dy}{dt}) + 6 \frac{dy}{dt} - 6y = 0$ or

$\frac{d^3y}{dt^3} - 6 \frac{d^2y}{dt^2} + 11 \frac{dy}{dt} - 6y = 0$. The auxiliary equation of

this is $m^3 - 6m^2 + 11m - 6 = 0$, and its roots are 1, 2,
3. The general solution of the D. E. in y and t is
$y = c_1 e^t + c_2 e^{2t} + c_3 e^{3t}$. We return to the original
independent variable x and replace e^t by x, e^{2t} by x^2,
and e^{3t} by x^3. Doing this, we find that the general
solution of the given D.E. is $y = c_1 x + c_2 x^2 + c_3 x^3$.

14. Let $x = e^t$; then $t = \ln x$, $x \frac{dy}{dx} = \frac{dy}{dt}$,

$x^2 \frac{d^2y}{dx^2} = \frac{d^2y}{dt^2} - \frac{dy}{dt}$. The D.E. transforms into

$\frac{d^2y}{dt^2} - 5\frac{dy}{dt} + 6y = 4e^t - 6$ (1) (note that x in the right member has transformed into e^t; see Remark 1 on page 166 of text). The auxiliary equation of the corresponding homogeneous D.E. is $m^2 - 5m + 6 = 0$, with roots m = 2, 3. Thus the complementary function of (1) is $y_c = c_1 e^{2t} + c_2 e^{3t}$. We find a particular integral by the method of undetermined coefficients. We assume $y_p = Ae^t + B$. Then $y_p' = Ae^t$, $y_p'' = Ae^t$, and substituting into D.E. (1), we quickly have $2Ae^t + 6B = 4e^t - 6$. From this, 2A = 4, 6B = -6, and hence A = 2, B = -1. Thus the particular integral of (1) is $y_p = 2e^t - 1$, and its general solution is $y = c_1 e^{2t} + c_2 e^{3t} + 2e^t - 1$. Returning to the original independent variable x by replacing e^t by x, etc., we find the general solution of the given D.E.: $y = c_1 x^2 + c_2 x^3 + 2x - 1$.

17. Let $x = e^t$; then $t = \ln x$, $x\frac{dy}{dx} = \frac{dy}{dt}$, and $x^2 \frac{d^2y}{dx^2} = \frac{d^2y}{dt^2} - \frac{dy}{dt}$. The D.E. transforms into

$\frac{d^2y}{dt^2} + 4y = 2t\, e^t$. (1) The auxiliary equation of the corresponding homogeneous D.E. is $m^2 + 4 = 0$ with roots m = ± 2i. Thus the complementary function of (1) is $y_c = c_1 \sin 2t + c_2 \cos 2t$. We find a particular integral by the method of undetermined coefficients. We assume $y_p = Ate^t + Be^t$. Then $y_p' = Ate^t + (A+B)e^t$, $y_p'' = Ate^t + (2A + B) e^t$, and substituting into the D.E. (1), we have $5Ate^t + (2A + 5B) e^t = 2te^t$. From this $5A = 2$, $2A + 5B = 0$, and hence $A = 2/5$, $B = -4/25$. Thus the particular integral of (1) is $y_p = (2/5)te^t - (4/25)e^t$, and its general solution is $y = c_1 \sin 2t + c_2 \cos 2t + (2/5)te^t - (4/25)e^t$. Returning to the original independent variable by replacing e^t by x, t by ln x, we find the general solution of the given D.E.:

$$y = c_1 \sin(2 \ln x) + c_2 \cos (2 \ln x)$$
$$+ (2/5)x \ln x - (4/25)x \qquad \text{or}$$

$$y = c_1 \sin(\ln x^2) + c_2 \cos(\ln x^2) + \frac{x \ln x^2}{5} - \frac{4x}{25} \quad .$$

20. Let $x = e^t$; then $t = \ln x$, $x \frac{dy}{dx} = \frac{dy}{dt}$, $x^2 \frac{d^2y}{dx^2} = \frac{d^2y}{dt^2} - \frac{dy}{dt}$. The D.E. transforms into

$\frac{d^2y}{dt^2} - 3 \frac{dy}{dt} - 10y = 0$. The auxiliary equation of this is $m^2 - 3m - 10 = 0$, with roots $m = 5, -2$. The general solution of the D.E. in y and t is $y = c_1 e^{5t} + c_2 e^{-2t}$. Returning to the original independent variable x, we replace e^t by x, etc., and obtain the general solution of the given D.E. in the form $y = c_1 x^5 + c_2 x^{-2}$. (1). Differentiating (1), we find $y' = 5c_1 x^4 - 2c_2 x^{-3}$. (2). Now apply the I.C.'s. Applying $y(1) = 5$ to (1), we have

$c_1 + c_2 = 5$; and applying $y'(1) = 4$ to (2), we have

$5c_1 - 2c_2 = 4$. Solving these two equations in c_1 and

c_2, we find $c_1 = 2$, $c_2 = 3$. Substituting these values

back into (1), we find the particular solution of the

stated I.V.P.: $y = 2x^5 + 3x^{-2}$.

23. Let $x = e^t$; then $t = \ln x$, $x \frac{dy}{dx} = \frac{dy}{dt}$, and

$x^2 \frac{d^2y}{dx^2} = \frac{d^2y}{dt^2} - \frac{dy}{dt}$. The D. E. transforms into

$$\frac{d^2y}{dt^2} - \frac{dy}{dt} - 2y = 4e^t - 8. \quad (1)$$

The auxiliary equation of the corresponding homogeneous

D. E. is $m^2 - m - 2 = 0$ with roots $m = -1, 2$. Thus the

complementary function of (1) is $y_c = c_1 e^{-t} + c_2 e^{2t}$. We

find a particular integral by the method of undetermined

coefficients. We assume $y_p = Ae^t + B$. Then $y_p' = Ae^t$,

$y_p'' = Ae^t$, and substituting into the D.E. (1), we have

$-2Ae^t - 2B = 4e^t - 8$. From this, $A = -2$, $B = 4$. Thus

the particular integral of (1) is $y_p = -2e^t + 4$, and

its general solution is $y = c_1 e^{-t} + c_2 e^{2t} - 2e^t + 4$.

Returning to the original variable by replacing e^t by

x, etc., we find the general solution of the given D.E.

$y = c_1 x^{-1} + c_2 x^2 - 2x + 4$. \quad (2)

Differentiating (2), we find:

$$y' = -c_1 x^{-2} + 2c_2 x - 2. \qquad (3)$$

Now apply the I.C.'s. Applying $y(1) = 4$ to (2), we have
$c_1 + c_2 = 2$; and applying $y'(1) = -1$ to (3), we have
$-c_1 + 2c_2 = 1$. Solving these two equations in c_1 and
c_2, we find $c_1 = 1$, $c_2 = 1$. Substituting these values
back into (2), we find the particular solution of the
stated I.V.P.: $y = x^{-1} + x^2 - 2x + 4$.

CHAPTER 5

Section 5.2, Page 186

1. This is an example of free, undamped motion; and
equation (5.8) applies. Since the 12 lb. weight
stretches the spring 1.5 in = 1/8 ft., Hooke's Law
F = ks gives 12 = k(1/8), so k = 96 lb./ft. Also,
m = w/g = 12/32 = 3/8 (slug). Thus by (5.8) we have
the D.E.

$$\frac{3}{8}\frac{d^2x}{dt^2} + 96x = 0 \quad \text{or} \quad \frac{d^2x}{dt^2} + 256x = 0 \quad . \quad (1)$$

Since the weight was released from rest from a position
2 in. = 1/6 ft. below its equilibrium position, we also
have the I.C. $x(0) = 1/6$, $x'(0) = 0$. The auxiliary
equation of the D.E. (1) is $r^2 + 256 = 0$ with roots
$r = \pm 16i$. The G.S. of the D.E. is

$$x = c_1 \sin 16t + c_2 \cos 16t \quad . \quad (2)$$

Differentiating this, we obtain

$$\frac{dx}{dt} = 16c_1 \cos 16t - 16c_2 \sin 16t. \quad (3)$$

Applying the first I.C. to (2), we find $c_2 = 1/6$; and
applying the second to (3) gives $c_1 = 0$. Thus the solution of the D.E. satisfying the stated I.C.'s is
$x = (1/6) \cos 16t$. The amplitude is 1/6 (ft.); the
period $2\pi/16 = \pi/8$ (sec); and the frequency is
$1/(\pi/8) = 8/\pi$ oscillations/sec.

2.   This is an example of free, undamped motion; and
equation (5.8) applies. Since the 16 lb. weight
stretches the spring 6 in = 1/2 foot, Hooke's Law
$F = ks$ gives $16 = k(1/2)$, so $k = 32$ lb./ft. Also,
$m = w/g = 16/32 = 1/2$ (slug). Thus by (5.8), we have
the D.E.

$$\frac{1}{2} \frac{d^2x}{dt^2} + 32x = 0 \qquad \text{or} \qquad \frac{d^2x}{dt^2} + 64x = 0. \qquad (1)$$

The auxiliary equation corresponding to (1) is
$r^2 + 64 = 0$ with roots $r = \pm 8i$. The G.S. of the D.E.
(1) is

$$x = c_1 \sin 8t + c_2 \cos 8t \ . \qquad (2)$$

Differentiating this, we obtain

$$\frac{dx}{dt} = 8c_1 \cos 8t - 8c_2 \sin 8t \ . \qquad (3)$$

The three cases (a), (b), (c) lead to different I.C.'s.
   In (a), the weight is released from a position
4 in. = 1/3 ft. below its equilibrium position, so we
have the first I.C. $x(0) = 1/3$. Since it is released

with initial velocity of 2 ft./sec., directed <u>downward</u>,
we have the second I.C. $x'(0) = 2$. Applying the first
I.C. to (2), we find $c_2 = 1/3$; and applying the second
to (3), we have $8c_1 = 2$, so $c_1 = 1/4$. Thus we obtain
the particular solution $x = (1/4) \sin 8t + (1/3) \cos 8t$.

In (b), we again have the first I.C. $x(0) = 1/3$.
But here the weight is released with initial velocity of
2 ft./sec., directed <u>upward</u>, so the second I.C. is
$x'(0) = -2$. We again find $c_2 = 1/3$; but applying the
second I.C. to (3) gives $c_1 = -1/4$. Thus we obtain the
particular solution $x = -(1/4) \sin 8t + (1/3) \cos 8t$.

In (c), the weight is released from a position
4 in. $= 1/3$ ft. <u>above</u> its equilibrium position, so we
have the first I.C. $x(0) = -1/3$. Since it is released
with initial velocity of 2 ft./sec., directed <u>downward</u>,
we have the second I.C. $x'(0) = 2$, just as in part (a).
Applying the first I.C. to (2), we find $c_2 = -1/3$, and
just as in part (a), the second I.C. gives $c_1 = 1/4$.
Thus we obtain the particular solution
$x = (1/4) \sin 8t - (1/3) \cos 8t$.

3. Equation (5.8) applies. Since the 4 lb. weight
stretches the spring 6 in. $= 1/2$ ft., Hooke's law $F = ks$
gives $4 = k(1/2)$, so $k = 8$ lb./ft. Also,
$m = w/g = 4/32 = 1/8$ (slug). By (5.8) we have the D.E.

$$\frac{1}{8}\frac{d^2x}{dt^2} + 8x = 0 \qquad \text{or} \qquad \frac{d^2x}{dt^2} + 64x = 0 \quad . \qquad (1)$$

Since the weight was released from its equilibrium

position with an initial velocity of 2 ft./sec.,

directed downward, we have the I.C.'s $x(0) = 0$,

$x'(0) = 2$. The auxiliary equation of the D.E. (1) is

$r^2 + 64 = 0$ with roots $r = \pm 8i$. The G.S. of the D.E.

is $\qquad\qquad x = c_1 \sin 8t + c_2 \cos 8t \quad . \qquad (2)$

Differentiating this, we obtain

$$\frac{dx}{dt} = 8c_1 \cos 8t - 8c_2 \sin 8t \quad . \qquad (3)$$

Applying the first I.C. $x(0) = 0$ to (2), we find $c_2 = 0$;

and applying the second $x'(0) = 2$ to (3), we get

$8c_1 = 2$, so $c_1 = 1/4$.

Thus the solution of the D.E. satisfying the given

I.C.'s is $x = (1/4) \sin 8t$, $\qquad\qquad\qquad (4)$

and its derivative is $dx/dt = 2 \cos 8t$. $\qquad\quad (5)$

These are the displacement and velocity, respectively,

and hence provice the answer to part (a). From the

solution (4), we see that the amplitude is $1/4$ (ft.),

the period is $2\pi/8 = \pi/4$ (sec.), and the frequency is

$1/(\pi/4) = 4/\pi$ oscillations/sec. These are the answers

to (b).

To answer (c), we seek times t at which

$x = 1.5$ in. $= 1/8$ ft. and $dx/dt > 0$. Thus we first

let x = 1/8 in solution (4) and find sin 8t = 1/2. From

this, 8t = $\pi/6$ + 2nπ or   8t = $5\pi/6$ + 2nπ, and hence

t = $\pi/48$ + n$\pi/4$ or   t = $5\pi/48$ + n$\pi/4$,   where n = 0, 1,

2, $\cdots$. We must choose these t for which dx/dt > 0.

From (5), we see that x'($\pi/48$ + n$\pi/4$)

= 2 cos($\pi/6$ + 2nπ) > 0, but x'($5\pi/48$ + n$\pi/4$)

= 2 cos($5\pi/6$ + 2nπ)< 0. Thus the answer to part (c) is

t = $\pi/48$ + n$\pi/4$, (n = 0, 1, 2, $\cdots$). To answer (d), we

seek times t at which x = 1.5 in. = 1/8 ft. and dx/dt < 0

From our work in (c), we see these are given by

t = $5\pi/48$ + n$\pi/4$, (n = 0, 1, 2, $\cdots$).

    6. Equation (5.8) applies, with m = w/g = 8/32

= 1/4 (slug) and the spring constant k > 0 to be

determined. Thus (5.8) becomes

$$\frac{1}{4}\frac{d^2x}{dt^2} + kx = 0 \qquad \text{or} \qquad \frac{d^2x}{dt^2} + 4kx = 0 \ . \qquad (1)$$

Since the weight is released from a position A ft.

<u>below</u> its equilibrium position with an initial velocity

of 3 ft./sec., directed <u>downward</u>, we have the I.C.'s

x(0) = A > 0, x'(0) = 3. The auxiliary equation of D.E.

(1) is r^2 + 4k = 0 with roots r = $\pm$ 2$\sqrt{k}$i. The G.S. of

the D.E. is x = c_1 sin 2$\sqrt{k}$t + c_2 cos 2$\sqrt{k}$t .   (2)

Differentiating we obtain

$$dx/dt = 2\sqrt{k}c_1 \cos 2\sqrt{k}t - 2\sqrt{k}c_2 \sin 2\sqrt{k}t \ . \qquad (3)$$

Applying the first I.C. $x(0) = A$ to (2), we find
$c_2 = A$; and applying the second $x'(0) = 3$ to (3), we
have $2\sqrt{k}\, c_1 = 3$, so $c_1 = 3/2\sqrt{k}$. Thus the solution of
the D.E. satisfying the given I.C.'s is
$x = (3/2\sqrt{k}) \sin 2\sqrt{k}\ t + A \cos 2\sqrt{k}\ t$. We express this
in the form (5.18) of the text. Multiplying and
dividing by

$$c = \sqrt{(3/2\sqrt{k})^2 + A^2} = \sqrt{9 + 4kA^2}\ /\ 2\sqrt{k}\ ,$$

we have $x = c\left[(3/2\sqrt{k}\ c)\sin 2\sqrt{k}\ t + (A/c)\cos 2\sqrt{k}\ t\right]$.
Then letting $A/c = \cos \phi$, $3/2\sqrt{k}\ c = -\sin \phi$, we have
$x = c \cos(2\sqrt{k}\ t + \phi)$. From this, the period is
$2\pi/2\sqrt{k} = \pi/\sqrt{k}$; but the period is given to be $\pi/2$. Thus
$\sqrt{k} = 2$, and hence $k = 4$. The amplitude is
$c = \sqrt{(3/2\sqrt{k})^2 + A^2}\ = \sqrt{(3/4)^2 + A^2}$; but the amplitude
is given to be $\sqrt{10}/2 = \sqrt{5}$. Hence $(3/4)^2 + A^2 = 5$, from
which $A = \sqrt{71}/4$.

7. There are two different D.E.'s of form (5.8)
here, one involving the 8 lb. weight and the other
involving the other weight. Concerning the 8 lb. weight,
$m = w/g = 8/32 = 1/4$ (slug), and the corresponding D.E.
of form (5.8) is

$$\frac{1}{4}\frac{d^2x}{d^2t} + kx = 0 \qquad \text{or} \qquad \frac{d^2x}{dt^2} + 4kx = 0\ ,$$

where k > 0 is the spring constant. The auxiliary
equation is $r^2 + 4k = 0$ with roots $r = \pm 2\sqrt{k}\, i$. The
G.S. of the D.E. is $x = c_1 \sin 2\sqrt{k}\, t + c_2 \cos 2\sqrt{k}\, t$.
From this, the period of the motion is $2\pi/2\sqrt{k}$. But the
period is given as 4. Thus $2\pi/2\sqrt{k} = 4$, from which
$k = \pi^2/16$.

Now let w be the other weight. For this, m = w/g
= w/32 (slugs), and the corresponding D.E. of form
(5.8) is

$$\frac{w}{32}\frac{d^2x}{dt^2} + \frac{\pi^2}{16}\, x = 0 \qquad \text{or} \qquad \frac{d^2x}{dt^2} + \frac{2\pi^2}{w}\, x = 0 \ .$$

The auxiliary equation is $r^2 + 2\pi^2/w = 0$ with roots
$r = \pm \sqrt{2/w}\, \pi\, i$. The G.S. of the D.E. is

$$x = c_1 \sin \sqrt{2/w}\, \pi\, t + c_2 \cos \sqrt{2/w}\, \pi\, t \ .$$

From this the period of the motion is $2\pi/\sqrt{2/w}\, \pi = \sqrt{2w}$.
But the period of this motion is given to be 6. Thus
$\sqrt{2w} = 6$, from which we find w = 18 (lb.) .

Section 5.3, Page 196

1. (a) This is a free damped motion, and Equation
(5.27) applies. Since the 8 lb. weight stretches the
spring 0.4 ft., Hooke's Law F = ks gives 8 = k(0.4), so
k = 20 lb./ft. Also, m = w/g = 8/32 = 1/4 (slug), and
a = 2. Thus equation (5.27) becomes

$$\frac{1}{4} \frac{d^2x}{dt^2} + 2 \frac{dx}{dt} + 20x = 0 \quad . \tag{1}$$

Since the weight is then pulled down 6 in. = 1/2 ft.
below its equilibrium position and released from rest at
t = 0, we have the I.C.'s $x(0) = 1/2$, $x'(0) = 0$. (2)

    (b) The D.E. (1) may be written as $\frac{d^2x}{dt^2} + 8 \frac{dx}{dt}$
+ 80x = 0. The auxiliary equation is
$r^2 + 8r + 80 = 0$, with roots $r = -4 \pm 8i$. The G.S. of
the D.E. is $x = e^{-4t}(c_1 \sin 8t + c_2 \cos 8t)$. (3)
Differentiating this, we find

$$\frac{dx}{dt} = e^{-4t}[(-4c_1 - 8c_2)\sin 8t + (8c_1 - 4c_2)\cos 8t]. \tag{4}$$

Applying the first I.C. (2) to (3), we find $c_2 = 1/2$,
and applying the second to (4), we have $8c_1 - 4c_2 = 0$,
from which $c_1 = c_2/2 = 1/4$. Thus the solution is

$$x = e^{-4t}[(1/4) \sin 8t + (1/2)\cos 8t]. \tag{5}$$

    (c) We first multiply and divide (5) by
$c = \sqrt{(1/4)^2 + (1/2)^2} = \sqrt{5}/4$, obtaining
$x = (\sqrt{5}/4)e^{-4t}[(1/\sqrt{5}) \sin 8t + (2/\sqrt{5}) \cos 8t]$. We can
now write this as $x = (\sqrt{5}/4)e^{-4t} \cos (8t - \phi)$, where ϕ
is such that $\cos \phi = 2/\sqrt{5}$, $\sin \phi = 1/\sqrt{5}$, and hence
$\phi \approx 0.46$ (rad.).

    (d) The period is $2\pi/8 = \pi/4$ (sec.).

3. This is a free damped motion, and Equation
(5.27) applies. Since the 8 lb. weight stretches the spring
6 in. = 1/2 ft., Hooke's Law F = ks gives 8 = k(1/2), so
k = 16 lb./ft. Also, m = w/g = 8/32 = 1/4 (slug), and
a = 4. Thus equation (5.27) becomes

$$\frac{1}{4} \frac{d^2x}{dt^2} + 4 \frac{dx}{dt} + 16x = 0 \quad .$$

The I.C.'s are x(0) = 3/4, x'(0) = 0. The D.E. may be
written in the form

$$\frac{d^2x}{dt^2} + 16 \frac{dx}{dt} + 64x = 0 \quad .$$

The auxiliary equation is r^2 + 16r + 64 = 0 with roots
-8, -8 (double root). The G.S. of the D.E. is
x = $(c_1 + c_2 t) e^{-8t}$. Differentiating this, we find
$\frac{dx}{dt}$ = $(-8c_1 - 8c_2 t + c_2)e^{-8t}$. Applying the first I.C. to
the G.S., we find c_1 = 3/4; and applying the second to
its derivative, we have $-8c_1 + c_2$ = 0, from which
c_2 = 6. Thus we find the solution x = ($\frac{3}{4}$ + 6t) e^{-8t} .

5. This is a free damped motion, and Equation
(5.27) applies. Since a force of 20 lb. would stretch
the spring 6 in. = 1/2 ft., Hooke's Law F = ks gives
20 = k(1/2), so k = 40 lb./ft. The weight is 4 lb., so
m = w/g = 4/32 = 1/8 (slug); and a = 2. Thus equation
(5.27) becomes

$$\frac{1}{8} \frac{d^2x}{dt^2} + 2 \frac{dx}{dt} + 40x = 0 \ .$$

Since the weight is released from rest from a position 8 inches = 2/3 ft. below its equilibrium position, we have the I.C.'s $x(0) = 2/3$, $x'(0) = 0$.

The D.E. may be written $\frac{d^2x}{dt^2} + 16 \frac{dx}{dt} + 320x = 0$. The auxiliary equation is $r^2 + 16r + 320 = 0$, with roots $-8 \pm 16i$. The G.S. of the D.E. is

$$x = e^{-8t}(c_1 \sin 16t + c_2 \cos 16t) \ .$$

Differentiating we obtain

$$\frac{dx}{dt} = e^{-8t}[(-8c_1 - 16c_2) \sin 16t + (16c_1 - 8c_2)\cos 16t] \ .$$

Applying the first I.C. to the G.S., we find $c_2 = 2/3$; and applying the second to the derived equation, we have $16c_1 - 8c_2 = 0$, from which $c_1 = c_2/2 = 1/3$. Thus we have the displacement

$$x = (e^{-8t}/3)(\sin 16t + 2 \cos 16t) \ .$$

To put this in the form (5.32), we multiply and divide by $c = \sqrt{(1)^2 + (2)^2}$ = $\sqrt{5}$, obtaining

$x = (\sqrt{5}e^{-8t}/3)[(1/\sqrt{5})\sin 16t + (2/\sqrt{5}) \cos 16t]$. We can now write this as

$$x = (\sqrt{5}e^{-8t}/3) \cos (16t - \phi), \tag{1}$$

where $\cos \phi = 2/\sqrt{5}$, $\sin \phi = 1/\sqrt{5}$, and hence $\phi \approx 0.46$ (rad.). We have thus answered part (a).

To answer part (b), we see from (1) that the period
is $2\pi/16 = \pi/8$ (sec.). In the notation of page 190 of
the text, the logarithmic decrement is $2\pi b/\sqrt{\lambda^2 - b^2}$.
Here $b = a/2m = 2/(2)(1/8) = 8$ and $\lambda^2 = k/m = 40/(1/8)$
$= 320$. Thus we find the logarithmic decrement is
$(2\pi)8/\sqrt{256} = \pi$.

To answer part (c), we let $x = 0$ in (1) and solve
for t. We have $\cos(16t - \phi) = 0$, so $16t - \phi = \pi/2$, from
which $t = (\phi + \pi/2)/16$. With $\phi = 0.46$, this gives
$t = 0.127$ (sec.).

8. Equation (5.27) applies, with $m = w/g = 10/32$
$= 5/16$ (slug), $a > 0$, and $k = 20$ lb./ft. Thus we have
the D.E.

$$\frac{5}{16} \frac{d^2x}{dt^2} + a \frac{dx}{dt} + 20x = 0 \quad .$$

Writing this as $5 \frac{d^2x}{dt^2} + 16a\frac{dx}{dt} + 320x = 0$, the auxiliary
equation is $5r^2 + 16ar + 320 = 0$. The roots are given
by
$$r = \frac{-16a \pm \sqrt{256a^2 - 6400}}{10} \quad . \tag{1}$$

In part (a), we seek the smallest value of a for
which damping is nonoscillatory. This is the value of
a for which damping is critical. It occurs when the two
roots given by (1) are real and equal. Thus we set
$256a^2 - 6400 = 0$ and find $a^2 = 6400/256$ and hence
$a = 80/16 = 5$.

In part (b), we let a = 5 in (1) and obtain the roots r = -8, -8 (double root). Then the displacement is given by $x = (c_1 + c_2 t)e^{-8t}$. Differentiating this, we have $\frac{dx}{dt} = (-8c_1 - 8c_2 t + c_2)e^{-8t}$. The I.C.'s are $x(0) = 1/2$, $x'(0) = 1$. Applying them to the preceeding expressions for x and dx/dt, we find $c_1 = 1/2$ and $-8c_1 + c_2 = 1$, from which $c_2 = 5$. Thus we obtain the displacement $x = (\frac{1}{2} + 5t)e^{-8t}$.

For part (c), we note that the extrema of the displacement are found by setting $\frac{dx}{dt} = (1 - 40t)e^{-8t}$ equal to zero and solving for t. We at once have $t = 1/40$. Then $x(1/40) = (5/8)e^{-1/5} \approx 0.51$ (sec.). For $t > 1/40$, $1 - 40t < 0$ and $dx/dt < 0$. Also using L'Hospital's Rule, $\lim_{t \to \infty} dx/dt = 0$. Thus the weight approaches its position monotonically.

9. Here $m = w/g = 32/32 = 1$(slug). If there were no resistance, the D.E. would be

$$m \frac{d^2 x}{dt^2} + kx = 0, \quad \text{that is,} \quad \frac{d^2 x}{dt^2} + kx = 0, \text{ where } k > 0.$$

The auxiliary equation of this is $r^2 + k = 0$, with roots $\pm\sqrt{k}i$. The G.S. of this D.E. is then $x = c_1 \sin \sqrt{k}t + c_2 \cos \sqrt{k}t$. The period of this undamped motion is $2\pi/\sqrt{k}$, and hence the natural frequency is $\sqrt{k}/2\pi$. But this is given to be $4/\pi$. Hence we have $\sqrt{k}/2\pi = 4/\pi$, from which $\sqrt{k} = 8$ and $k = 64$. This answers (a).

To answer (b), we take the resistance into account and have the D.E. $m\frac{d^2x}{dt^2} + a\frac{dx}{dt} + kx = 0$, that is,

$\frac{d^2x}{dt^2} + a\frac{dx}{dt} + 64x = 0$. The auxiliary equation of this is $r^2 + ar + 64 = 0$ with roots $r = (-a \pm \sqrt{a^2 - 256})/2$, where $a^2 < 256$. The G.S. of this D.E. is

$x = e^{-at/2}(c_1 \sin \sqrt{256 - a^2}\, t/2 + c_2 \cos \sqrt{256 - a^2}\, t/2)$

The period of the trigonometric factor of this is $2\pi/(\sqrt{256 - a^2}/2) = 4\pi/\sqrt{256 - a^2}$. Hence the frequency of this motion is $\sqrt{256 - a^2}/4\pi$. But the frequency of this damped motion is given as half the natural frequency $4/\pi$ and so is $2/\pi$. Thus $\sqrt{256 - a^2}/4\pi = 2/\pi$. Thus $256 - a^2 = 64$, from which $a = 8\sqrt{3}$.

Section 5.4, Page 205

3. The D.E. is of the form (5.58) of the text, with $m = w/g = 10/32 = 5/16$ (slug), $a = 5$, $k = 20$, and $F(t) = 10 \cos 8t$. Thus we have

$$\frac{5}{16}\frac{d^2x}{dt^2} + 5\frac{dx}{dt} + 20x = 10 \cos 8t$$

or

$$\frac{d^2x}{dt^2} + 16\frac{dx}{dt} + 64x = 32 \cos 8t \ . \tag{1}$$

The I.C.'s are $x(0) = 0$, $x'(0) = 0$. The auxiliary equation of the homogeneous D.E. corresponding to (1) is

$r^2 + 16r + 64 = 0$ or $(r + 8)^2 = 0$, with roots -8, -8
(double root). Thus the complementary function of (1)
is $x_c = (c_1 + c_2t)e^{-8t}$. We use undetermined coefficients to find a particular integral. We let
$x_p = A \cos 8t + B \sin 8t$. Then $x_p' = -8A \sin 8t$
$+ 8B \cos 8t$ and $x_p'' = -64A \cos 8t - 64B \sin 8t$. Substituting into (1) and simplifying, we find
$128B \cos 8t - 128A \sin 8t = 32 \cos 8t$. Hence $-128A = 0$,
$128B = 32$, so $A = 0$, $B = 1/4$. Thus we have the particular integral $x_p = (1/4) \sin 8t$, and the G.S. of D.E.
(1) is $x = (c_1 + c_2t)e^{-8t} + (1/4) \sin 8t$. Differentiating this, we obtain $\frac{dx}{dt} = (-8c_1 -8c_2t + c_2)e^{-8t}$
$+ 2 \cos 8t$. Applying the stated I.C. to these, we have
$c_1 = 0$, $-8c_1 + c_2 + 2 = 0$, and hence $c_2 = -2$. Thus we
obtain the solution $x = -2te^{-8t} + (1/4) \sin 8t$.

    4. The D.E. is of the form (5.58) of the text, with
$m = w/g = 4/32 = 1/8$ (slug), $a = 2$, and $F(t) = 13 \sin 4t$.
Also, by Hooke's Law $F = ks$, we have $4 = k(1/4)$, so
$k = 16$ lb./ft. Thus we have

$$\frac{1}{8} \frac{d^2x}{dt^2} + 2 \frac{dx}{dt} + 16x = 13 \sin 4t,$$

or

$$\frac{d^2x}{dt^2} + 16 \frac{dx}{dt} + 128x = 104 \sin 4t \ . \qquad (1)$$

The I.C.'s are $x(0) = 1/2$ and $x'(0) = 0$. The auxiliary

equation of the homogeneous D.E. corresponding to (1) is
$r^2 + 16r + 128 = 0$ with roots $r = -8 \pm 8i$. Thus the
complementary function of (1) is

$$x = e^{-8t}(c_1 \sin 8t + c_2 \cos 8t).$$

We use undetermined coefficients to find a particular
integral. We let x_p = A sin 4t + B cos 4t. Then
$x_p{}'$ = 4A cos 4t - 4B sin 4t, $x_p{}''$ = -16A sin 4t
- 16B cos 4t. Substituting into (1) and simplifying, we
find (112A - 64B) sin 4t + (64A + 112B) cos 4t
= 104 sin 4t. Thus we have the equations 112A - 64B
= 104 and 64A + 112B = 0. These reduce to
14A - 8B = 13 and 8A + 14B = 0, from which we find
A = 7/10, B = -2/5. Thus we have the particular integral x_p = (7/10) sin 4t - (2/5) cos 4t, and the G.S. of
the D.E. (1) is
$x = e^{-8t}(c_1 \sin 8t + c_2 \cos 8t)$ + (7/10) sin 4t
- (2/5) cos 4t. Differentiating this, we obtain
$\frac{dx}{dt} = e^{-8t}[(-8c_1 - 8c_2) \sin 8t + (8c_1 - 8c_2) \cos 8t]$

+ (14/5) cos 4t + (8/5) sin 4t. Applying the stated
I.C. to these, we have c_2 - 2/5 = 1/2 and
$8c_1 - 8c_2$ + 14/5 = 0. Hence c_2 = 9/10 and c_1 = 11/20.
Thus we obtain the solution

$x = e^{-8t}[(11/20) \sin 8t + (9/10) \cos 8t]$

$\quad + (7/10) \sin 4t - (2/5) \cos 4t$.

This is the answer to part (a). Concerning part (b), we note that the steady-state term is $(7/10) \sin 4t$ $- (2/5) \cos 4t$. The amplitude of this is given by

$c = \sqrt{(7/10)^2 + (2/5)^2} = \sqrt{65}/10.$

7. There are two problems here, one for $0 \leq t \leq \pi$, and the other for $t > \pi$. We first consider that for which $0 \leq t \leq \pi$. Here the D.E. is of the form (5.58), with $m = w/g = 32/32 = 1$ (slug), $a = 4$, $k = 20$, and $F(t) = 40 \cos 2t$. Thus we have

$$\frac{d^2x}{dt^2} + 4 \frac{dx}{dt} + 20x = 40 \cos 2t \ . \qquad (1)$$

The I.C.'s are $x(0) = 0$, $x'(0) = 0$. The auxiliary equation of the homogeneous D.E. corresponding to (1) is $r^2 + 4r + 20 = 0$, with roots $-2 \pm 4i$. Thus the complementary function of (1) is $x = e^{-2t}(c_1 \sin 4t + c_2 \cos 4t)$. We use undetermined coefficients to find a particular integral. We let $x_p = A \sin 2t + B \cos 2t$. Then $x_p' = 2A \cos 2t - 2B \sin 2t$, $x_p'' = -4A \sin 2t - 4B \cos 2t$. Substituting into (1) and simplifying, we find $(16A - 8B)\sin 2t + (8A + 16B)\cos 2t = 40 \cos 2t$. Thus we have $16A - 8B = 0$ and $8A + 16B = 40$, from which we find $A = 1$, $B = 2$. Thus we have the particular integral of (1), x_p = sin 2t + 2 cos 2t; and the G.S. of D.E. (1) is

$$x = e^{-2t}(c_1 \sin 4t + c_2 \cos 4t) + \sin 2t + 2 \cos 2t.$$

Differentiating this, we obtain

$$\frac{dx}{dt} = e^{-2t}[(-2c_1 - 4c_2) \sin 4t + (4c_1 - 2c_2) \cos 4t]$$
$$+ 2 \cos 2t - 4 \sin 2t \quad .$$

Applying the stated I.C. to these, we have

$c_2 + 2 = 0$ and $4c_1 - 2c_2 + 2 = 0$, from which we find $c_1 = -3/2$, $c_2 = -2$. Thus we obtain the solution

$$\tag{2}$$

$$x = e^{-2t}[(-3/2) \sin 4t - 2 \cos 4t] + \sin 2t + 2 \cos 2t,$$

valid for $0 \leq t \leq \pi$.

Now we consider the problem for which $t > \pi$. The D.E. is again of the form (5.58), where m = 1, a = 4, k = 20, but here F(t) = 0. Thus we have

$$\frac{d^2x}{dt^2} + 4 \frac{dx}{dt} + 20x = 0 \quad . \tag{3}$$

Assuming the displacement is continuous at $t = \pi$, the solution of (3) must take the value given by (2) at $t = \pi$. That is, we must impose the I.C.

$$x(\pi) = -2e^{-2\pi} + 2 \tag{4}$$

on the solution of (3). Similarly, assuming the velocity is continuous at $t = \pi$, the derivative of the

solution of (3) **must** take the value given by the
derivative of (2) at $t = \pi$. The derivative of (2) is

$$\frac{dx}{dt} = e^{-2t}[11 \sin 4t - 2 \cos 4t] + 2 \cos 2t - 4 \sin 2t .$$

From this, we see that we must therefore impose the I.C.

$$x'(\pi) = -2e^{-2\pi} + 2 \tag{5}$$

on the solution of (3). The auxiliary equation of (3)
is $r^2 + 4r + 20 = 0$ with roots $-2 \pm 4i$. Thus the G.S.
of (3) is $x = e^{-2t}(k_1 \sin 4t + k_2 \cos 4t)$, and its
derivative is

$$\frac{dx}{dt} = e^{-2t}[(-2k_1 - 4k_2) \sin 4t + (4k_1 - 2k_2) \cos 4t] .$$

Applying the I.C. (4) to this G.S., we have
$k_2 e^{-2\pi} = -2e^{-2\pi} + 2$, from which $k_2 = 2(e^{2\pi} - 1)$.
Applying the I.C. (5) to the derivative of this G.S.,
we have $(4k_1 - 2k_2)e^{-2\pi} = -2e^{-2\pi} + 2$, from which
$k_1 = (3/2)(e^{2\pi} - 1)$. Thus we obtain the solution

$$x = (e^{2\pi} - 1)e^{-2t}[(3/2) \sin 4t + 2 \cos 4t] , \text{ valid for }$$

$t \geq \pi$.

Section 5.5, Page 211

1. (a) Since the 12 lb. weight stretches the spring
6 in. $= 1/2$ ft., Hooke's Law, $F = ks$, gives $12 = k(1/2)$,
so $k = 24$ lb./ft. Then since $m = w/g = 12/32 = 3/8$

(slugs), a = 3, and F(t) = 2 cos ωt, the D.E. is

$$\frac{3}{8}\frac{d^2x}{dt^2} + 3\frac{dx}{dt} + 24x = 2 \cos \omega t . \qquad (1)$$

The I.C.'s are x(0) = 0, x'(0) = 0. The resonance
frequency is given by formula (5.69) of the text. It
is

$$\frac{1}{2\pi} \sqrt{\frac{24}{(3/8)} - \frac{9}{2(3/8)^2}} = \frac{2\sqrt{2}}{\pi} .$$

This is $\omega_1/2\pi$, where ω_1 is the value of ω

for which the forcing function is in resonance with the
system. Thus $\omega_1/2\pi = 2\sqrt{2}/\pi$, and so $\omega_1 = 4\sqrt{2}$.

We now let $\omega = \omega_1 = 4\sqrt{2}$ in (1), obtaining

$$\frac{3}{8}\frac{d^2x}{dt^2} + 3\frac{dx}{dt} + 24x = 2 \cos 4\sqrt{2}\, t$$

or

$$\frac{d^2x}{dt^2} + 8\frac{dx}{dt} + 64x = \frac{16}{3} \cos 4\sqrt{2}\, t. \qquad (2)$$

The auxiliary equation of the corresponding homogeneous
D.E. is $r^2 + 8r + 64 = 0$, with roots $r = -4 \pm 4\sqrt{3}\, i$.
Thus the complementary function of (2) is

$$x_c = e^{-4t}(c_1 \sin 4\sqrt{3}\, t + c_2 \cos 4\sqrt{3}t) .$$

We use undetermined coefficients to find a particular
integral. We let

$$x_p = A \sin 4\sqrt{2}\, t + B \cos 4\sqrt{2}\, t .$$

Differentiating twice and substituting into (1), we

find
$$\begin{cases} 32A - 32\sqrt{2}\,B = 0, \\ 32\sqrt{2}\,A + 32B = 16/3 \;; \end{cases}$$

and from these $A = \sqrt{2}/18$, $B = 1/18$. Thus we obtain the

general solution of (2) in the form

$$x = e^{-4t}(c_1 \sin 4\sqrt{3}\,t + c_2 \cos 4\sqrt{3}\,t)$$

$$+ \frac{\sqrt{2}}{18} \sin 4\sqrt{2}\,t + \frac{1}{18} \cos 4\sqrt{2}\,t \quad .$$

Differentiating, we find

$$\frac{dx}{dt} = e^{-4t}[(-4c_1 - 4\sqrt{3}\,c_2) \sin 4\sqrt{3}\,t + (4\sqrt{3}\,c_1 - 4c_2)$$

$$\cos 4\sqrt{3}\,t] + \frac{4}{9} \cos 4\sqrt{2}\,t - \frac{2\sqrt{2}}{9} \sin 4\sqrt{2}\,t \quad .$$

Applying the I.C.'s to these expressions for x and

dx/dt, we obtain $c_2 + 1/18 = 0$, $4\sqrt{3}\,c_1 - 4c_2 + 4/9 = 0$.

From these, we find $c_1 = -\sqrt{3}/18$, $c_2 = -1/18$. Thus we

obtain the solution of (2) in the form

$$x = e^{-4t}(-\sqrt{3} \sin 4\sqrt{3}\,t - \cos 4\sqrt{3}\,t)/18$$

$$+(\sqrt{2} \sin 4\sqrt{2}\,t + \cos 4\sqrt{2}\,t)/18.$$

(b) In this part m, k, and F(t) are as in (1),

but a = 0, so we have the D.E.

$$\frac{3}{8} \frac{d^2x}{dt^2} + 24x = 2 \cos \omega t \quad . \tag{3}$$

The auxiliary equation of the corresponding homogeneous

equation is $(3/8)r^2 + 24 = 0$ with roots $r = \pm 8i$. Thus the complementary function is

$$x_c = c_1 \sin 8t + c_2 \cos 8t \quad . \tag{4}$$

Undamped resonance occurs when the frequency $\omega/2\pi$ of the impressed force equals the natural frequence $4/\pi$ and hence when $\omega = 8$. Letting $\omega = 8$ in (3), we have the D.E.

$$\frac{3}{8} \frac{d^2x}{dt^2} + 24x = 2 \cos 8t \quad . \tag{5}$$

The complementary function is given by (4). We find a particular integral using undetermined coefficients. We modify the UC set $\left\{\sin 8t, \cos 8t\right\}$ of $\cos 8t$ by multiplying each member by t, obtaining $\left\{t \sin 8t, t \cos 8t\right\}$. Thus we assume $x_p = At \sin 8t + Bt \cos 8t$. Differentiating twice and substituting into (5), we find $A = 1/3$, $B = 0$. Thus we obtain the general solution of (5) in the form

$$x = c_1 \sin 8t + c_2 \cos 8t + t \sin 8t/3 \quad .$$

Differentiating this, we find

$$\frac{dx}{dt} = 8c_1 \cos 8t - 8c_2 \sin 8t + 8t \cos 8t/3 + \sin 8t/3 \quad .$$

The I.C.'s are the same as in part (a). Applying them to these expressions for x and dx/dt, we readily find $c_1 = 0$, $c_2 = 0$. Thus we obtain the solution of (5) in the form $x = t \sin 8t/3$.

Section 5.6, Page 219

1. Let i denote the current in amperes at time t.
The total electromotive force is 40 V. Using the
voltage drop laws 1 and 2 of the text (page 212), we
find the following voltage drops:

1. across the resistor: E_R = Ri = 10i.

2. across the inductor: $E_L = \frac{di}{dt} = 0.2 \frac{di}{dt}$.

Applying Kirchhoff's Law, we have the D.E.

$$0.2 \frac{di}{dt} + 10i = 40 \quad . \tag{1}$$

Since the initial current is 0, the I.C. is

$$i(0) = 0 \quad .$$

The D.E. (1) is a first order linear D.E. In standard

form it is $\frac{di}{dt} + 50i = 200$;   (2)

an I.F. is $e^{\int 50dt} = e^{50t}$. Multiplying (2) through
by this, we obtain

$$e^{50t} \frac{di}{dt} + 50e^{50t} i = 200e^{50t}$$

or

$$\frac{d}{dt}[e^{50t} i] = 200e^{50t} \quad .$$

Integrating and simplifying, we find

$$i = 4 + ce^{-50t} \quad .$$

Applying the I.C. i = 0 at t = 0 to this, we find
c = -4. Thus we obtain the solution

$$x = 4(1 - e^{-50t}) \quad .$$

3.   Let i denote the current and let q denote the charge on the capacitor at time t.   The electromotive force is 100 V.   Using the voltage drop laws 1 and 3, we find the following voltage drops:

1.   across the resistor: $E_R = Ri = 10i$.

2.   across the capacitor: $E_C = \frac{1}{c} q = (10)^4 q/2$.

Applying Kirchhoff's Law, we have the D.E.

$$10i + (10)^4 q/2 = 100 .$$

Since $i = dq/dt$, this reduces to

$$10 \frac{dq}{dt} + (10)^4 q/2 = 100 . \tag{1}$$

Since the charge is initially zero, we have the I.C.

$$q(0) = 0 .$$

The D.E. (10) is a first order linear D.E.   In standard form it is

$$\frac{dq}{dt} + 500 q = 10 , \tag{2}$$

and an I.F. is $e^{\int 500 \, dt} = e^{500t}$.   Multiplying (2) through by this, we obtain

$$e^{500t} \frac{dq}{dt} + 500 e^{500t} q = 10e^{500t}$$

or

$$\frac{d}{dt}[e^{500t} q] = 10e^{500t} .$$

Integrating and simplifying, we find

$$q = 1/50 + ce^{-500t} .$$

Applying the I.C. q = 0 at t = 0 to this, we find

c = -1/50. Thus we obtain the solution

$$q = (1 - e^{-500t})/50 \ .$$

This is the charge. To find the current, we return to

$10i + (10)^4 \ q/2 = 100$, substitute the expression for q

just found, and solve for i. We find i = -500q + 10

and hence $i = 10e^{-500t}$.

    6. Let i denote the current in amperes at time t.

The total electromotive force is $200e^{-100t}$. Using the

voltage drop laws 1, 2, and 3, we find the following

voltage drops:

1. across the resistor: $E_R = Ri = 80i$.

2. across the inductor: $E_L = L \frac{di}{dt} = 0.2 \frac{di}{dt}$.

3. across the capacitor: $E_c = \frac{1}{c} q = 10^6 q/5$.

Applying Kirchhoff's Law, we have the D.E.

$$0.2 \frac{di}{dt} + 80i + 10^6 \ q/5 = 200e^{-100t} \ .$$

Since i = dq/dt, this reduces to

$$0.2 \frac{d^2q}{dt^2} + 80 \frac{dq}{dt} + 10^6 \ q/5 = 200e^{-100t}. \qquad (1)$$

Since the charge q is initially zero, we have the first

I.C.   q(0) = 0 .

Since the current i is initially zero and i = dq/dt, we

have the second I.C.   q'(0) = 0 .

The homogeneous D.E. corresponding to D.E. (1) has the auxiliary equation

$$\frac{1}{5} r^2 + 80r + 200,000 = 0 \ ,$$

and the roots of this are $-200 \pm 979.8i$. Thus the complementary function of D.E. (1) is

$$q_c = e^{-200t}(c_1 \sin 979.8t + c_2 \cos 979.8t) \ .$$

We use undetermined coefficients to find a particular integral. We write $\qquad q_p = Ae^{-100t}$.

Differentiating twice and substituting into (1), we find A = 1/970 $\approx$ 0.0010. Thus the general solution of D.E. (1) is

$$q = e^{-200t}(c_1 \sin 979.8t + c_2 \cos 979.8t) + e^{-100t}/970$$

Differentiating this, we obtain

$$\frac{dq}{dt} = e^{-200t}[(-200c_1 - 979.8c_2) \sin 979.8t$$

$$+ (979.8c_1 - 200c_2) \cos 979.8t] - 10e^{-100t}/97. \quad (2)$$

Applying the I.C.'s to these expressions for q and dq/dt, we have

$$c_2 + 1/970 = 0, \quad 979.8c_1 - 200c_2 - 10/97 = 0 \ ,$$

from these we find that $c_1 = -10/(97)(979.8) \approx -0.0001$ and $c_2 = -1/970 \approx -0.0010$. Thus we obtain

$$q = e^{-200t}(-0.0001 \sin 979.8t - 0.0010 \cos 979.8t)$$

$$+ 0.0010 \ e^{-100t}.$$

Since i = dq/dt, using formula (2) for dq/dt and the

values of c_1 and c_2 determined above, we find

$$i = e^{-200t}(1.0311 \sin 979.8t + 0.1031 \cos 979.8t)$$
$$- 0.1031e^{-100t} \ .$$

CHAPTER 6

Section 6.1, Page 233

3. We assume $y = \sum_{n=0}^{\infty} c_n x^n$. Then

$$\frac{dy}{dx} = \sum_{n=1}^{\infty} n\, c_n x^{n-1}, \quad \frac{d^2 y}{dx^2} = \sum_{n=2}^{\infty} n(n-1)c_n x^{n-2}.$$ Substituting into the D.E., we obtain

$$\sum_{n=2}^{\infty} n(n-1)c_n x^{n-2} + x \sum_{n=1}^{\infty} nc_n x^{n-1} + 2x^2 \sum_{n=0}^{\infty} c_n x^n$$

$$+ \sum_{n=0}^{\infty} c_n x^n = 0$$

or

$$\sum_{n=2}^{\infty} n(n-1)c_n x^{n-2} + \sum_{n=1}^{\infty} nc_n x^n + \sum_{n=0}^{\infty} 2c_n x^{n+2} + \sum_{n=0}^{\infty} c_n x^n = 0.$$

We rewrite the first and third summations so that x has the exponent n in each. Thus we have:

$$\sum_{n=0}^{\infty} (n+2)(n+1)c_{n+2} x^n + \sum_{n=1}^{\infty} nc_n x^n + \sum_{n=2}^{\infty} 2c_{n-2} x^n$$

$$+ \sum_{n=0}^{\infty} c_n x^n = 0 \quad .$$

The common range of these four summations is from 2 to
∞. We write out the individual terms in each that do
not belong to this range. Thus we have:

$(c_0 + 2c_2) + (2c_1 + 6c_3)x$

$$+ \sum_{n=2}^{\infty} [(n+2)(n+1)c_{n+2} + (n+1)c_n + 2c_{n-2}]x^n = 0 .$$

Equating to zero the coefficient of each power of x, we
obtain

$$c_0 + 2c_2 = 0, \qquad 2c_1 + 6c_3 = 0 , \qquad\qquad (1)$$

$$(n+2)(n+1)c_{n+2} + (n+1)c_n + 2c_{n-2} = 0, \quad n \geq 2 . \qquad (2)$$

From (1), we find $c_2 = -c_0/2$, $c_3 = -c_1/3$. From (2), we
find

$$c_{n+2} = - \frac{(n+1)c_n + 2c_{n-2}}{(n+1)(n+2)} , \quad n \geq 2 .$$

Using this, we find $c_4 = -(3c_2 + 2c_0)/12 = -c_0/24$;

$c_5 = -(4c_3 + 2c_1)/20 = -c_1/30$. Substituting these values
into the assumed solution, we have

$$y = c_0 + c_1 x - (c_0/2)x^2 - (c_1/3)x^3 - (c_0/24)x^4$$

$$- (c_1/30)x^5 + \cdots$$

or

$$y = c_0(1 - \frac{x^2}{2} - \frac{x^4}{24} + \cdots) + c_1(x - \frac{x^3}{3} - \frac{x^5}{30} + \cdots) .$$

7. We assume $y = \sum_{n=0}^{\infty} c_n x^n$. Then $\frac{dy}{dx} = \sum_{n=1}^{\infty} n c_n x^{n-1}$,

$\dfrac{d^2 y}{dx^2} = \displaystyle\sum_{n=2}^{\infty} n(n-1)c_n x^{n-2}$. Substituting into the D.E., we

obtain $x^2 \displaystyle\sum_{n=2}^{\infty} n(n-1)c_n x^{n-2} + \sum_{n=2}^{\infty} n(n-1)c_n x^{n-2}$

$$+ x \sum_{n=1}^{\infty} n c_n x^{n-1} + x \sum_{n=0}^{\infty} c_n x^n = 0$$

or

$$\sum_{n=2}^{\infty} n(n-1)c_n x^n + \sum_{n=2}^{\infty} n(n-1)c_n x^{n-2} + \sum_{n=1}^{\infty} n c_n x^n$$

$$+ \sum_{n=0}^{\infty} c_n x^{n+1} = 0 \quad .$$

We rewrite the second and fourth summations so that x has the exponent n in each. Thus we have:

$$\sum_{n=2}^{\infty} n(n-1)c_n x^n + \sum_{n=0}^{\infty} (n+2)(n+1)c_{n+2} x^n + \sum_{n=1}^{\infty} n c_n x^n$$

$$+ \sum_{n=1}^{\infty} c_{n-1} x^n = 0 \quad .$$

The common range of these four summations is from 2 to ∞. We write out the individual terms in each that do not belong to this range. Thus we have:

$2c_2 + (c_0 + c_1 + 6c_3)x$

$$+ \sum_{n=2}^{\infty} \left\{ (n+2)(n+1)c_{n+2} + [n(n-1)+n]c_n + c_{n-1} \right\} x^n = 0.$$

Equating to zero the coefficient of each power of x, we obtain

$$2c_2 = 0 \quad , \qquad c_0 + c_1 + 6c_3 = 0 \quad , \qquad (1)$$

$$(n+2)(n+1)c_{n+2} + [n(n-1)+n]c_n + c_{n-1} = 0, \ n \geq 2. \qquad (2)$$

From (1), we find $c_2 = 0$, $c_3 = -(c_0 + c_1)/6$. From (2), we find

$$c_{n+2} = - \frac{n^2 c_n + c_{n-1}}{(n+1)(n+2)} \quad , \quad n \geq 2 \ .$$

Using this, we find $c_4 = -(4c_2 + c_1)/12 = -c_1/12$;

$c_5 = -(9c_3 + c_2)/20 = (3/40)(c_0 + c_1)$. Substituting

these values into the assumed solution, we have

$$y = c_0 + c_1 x - (c_0 + c_1)x^3/6 - c_1 x^4/12$$

$$+ \ 3(c_0 + c_1)x^5/40 + \cdots$$

or

$$y = c_0(1 - \frac{x^3}{6} + \frac{3x^5}{40} + \cdots) + c_1(x - \frac{x^3}{6} - \frac{x^4}{12} + \frac{3x^5}{40} + \cdots).$$

9. We assume $y = \sum_{n=0}^{\infty} c_n x^n$. Then $\frac{dy}{dx} = \sum_{n=1}^{\infty} nc_n x^{n-1}$,

$$\frac{d^2 y}{dx^2} = \sum_{n=2}^{\infty} n(n-1)c_n x^{n-2}.$$ Substituting into the D.E., we

obtain

$$x^3 \sum_{n=2}^{\infty} n(n-1)c_n x^{n-2} - \sum_{n=2}^{\infty} n(n-1)c_n x^{n-2} + x^2 \sum_{n=1}^{\infty} nc_n x^{n-1}$$

$$+ \ x \sum_{n=0}^{\infty} c_n x^n = 0$$

or

$$\sum_{n=2}^{\infty} n(n-1)c_n x^{n+1} - \sum_{n=2}^{\infty} n(n-1)c_n x^{n-2} + \sum_{n=1}^{\infty} nc_n x^{n+1}$$

$$+ \sum_{n=0}^{\infty} c_n x^{n+1} = 0 \ .$$

We **rewrite** the second summation so that x has the exponent $n + 1$ in it (since this is the exponent of x in all the other summations). Thus we have:

$$\sum_{n=2}^{\infty} n(n-1)c_n x^{n+1} - \sum_{n=-1}^{\infty} (n+3)(n+2)c_{n+3} x^{n+1}$$

$$+ \sum_{n=1}^{\infty} nc_n x^{n+1} + \sum_{n=0}^{\infty} c_n x^{n+1} = 0 \ .$$

The common range of these four summations is from 2 to ∞. We write out the individual terms in each that do not belong to this range. Thus we have:

$$-2c_2 + (c_0 - 6c_3)x + (2c_1 - 12c_4)x^2$$

$$+ \sum_{n=2}^{\infty} \left\{ -(n+3)(n+2)c_{n+3} + [n(n-1)+n+1]c_n \right\} x^n = 0 \ .$$

Equating to zero the coefficients of each power of x, we obtain

$$-2c_2 = 0 \ , \quad c_0 - 6c_3 = 0, \quad 2c_1 - 12c_4 = 0 \ , \qquad (1)$$

$$-(n+3)(n+2)c_{n+3} + [n(n-1)+n+1]c_n = 0, \ n \geq 2. \qquad (2)$$

From (1), we find $c_2 = 0$, $c_3 = c_0/6$, $c_4 = c_1/6$.

From (2), we find

$$c_{n+3} = \frac{(n^2 + 1)c_n}{(n+2)(n+3)} \ , \quad n \geq 2 \ .$$

Using this, we find $c_5 = 5c_2/20 = 0$; $c_6 = 10c_3/30 =$
$c_3/3 = c_0/18$; and $c_7 = 17c_4/42 = 17c_1/252$. Substituting
these values into the assumed solution, we have

$$y = c_0 + c_1 x + c_0 x^3/6 + c_1 x^4/6 + c_0 x^6/18 + 17c_1 x^7/252$$
$$+ \cdots$$

or

$$y = c_0(1 + \frac{x^3}{6} + \frac{x^6}{18} + \cdots) + c_1(x + \frac{x^4}{6} + \frac{17x^7}{252} + \cdots) \ .$$

13. We assume $y = \sum_{n=0}^{\infty} c_n x^n$. Then $\frac{dy}{dx} = \sum_{n=1}^{\infty} nc_n x^{n-1}$,

$\frac{d^2 y}{dx^2} = \sum_{n=2}^{\infty} n(n-1)c_n x^{n-2}$. Substituting into the D.E., we

obtain

$$x^2 \sum_{n=2}^{\infty} n(n-1)c_n x^{n-2} + \sum_{n=2}^{\infty} n(n-1)c_n x^{n-2} + x \sum_{n=1}^{\infty} nc_n x^{n-1}$$

$$+ 2x \sum_{n=0}^{\infty} c_n x^n = 0$$

or

$$\sum_{n=2}^{\infty} n(n-1)c_n x^n + \sum_{n=2}^{\infty} n(n-1)c_n x^{n-2} + \sum_{n=1}^{\infty} nc_n x^n$$

$$+ \sum_{n=0}^{\infty} 2c_n x^{n+1} = 0 \ .$$

We rewrite the second and fourth summations so that x

has the exponent n in each. Thus we have:

$$\sum_{n=2}^{\infty} n(n-1)c_n x^n + \sum_{n=0}^{\infty} (n+2)(n+1)c_{n+2} x^n + \sum_{n=1}^{\infty} nc_n x^n$$

$$+ \sum_{n=1}^{\infty} 2c_{n-1} x^n = 0 .$$

The common range of these four summations is from 2 to
∞. We write out the individual terms in each that do
not belong to this range. Thus we have:

$$2c_2 + (6c_3 + c_1 + 2c_0)x$$

$$+ \sum_{n=2}^{\infty} \left\{ [n(n-1)+n]c_n + (n+2)(n+1)c_{n+2} + 2c_{n-1} \right\} x^n = 0.$$

Equating to zero the coefficient of each power of x, we
obtain

$$2c_2 = 0 , \qquad 2c_0 + c_1 + 6c_3 = 0 , \qquad (1)$$

$$[n(n-1)+n]c_n + (n+2)(n+1)c_{n+2} + 2c_{n-1} = 0 , \quad n \geq 2. \quad (2)$$

From (1), we find $c_2 = 0$, $c_3 = -(2c_0 + c_1)/6$. From (2),
we find

$$c_{n+2} = - \frac{n^2 c_n + 2c_{n-1}}{(n+1)(n+2)} , \quad n \geq 2 .$$

Using this, we find $c_4 = -(4c_2 + 2c_1)/12 = -c_1/6$,

$c_5 = -(9c_3 + 2c_2)/20 = 3(2c_0 + c_1)/40$. Substituting
these values into the assumed solution, we have

$$y = c_0 + c_1 x - (2c_0 + c_1)x^3/6 - c_1 x^4/6$$

$$+ 3(2c_0 + c_1)x^5/40 + \cdots$$

or

$$y = c_0(1 - x^3/3 + 3x^5/20 + \cdots)$$
$$+ c_1(x - x^3/6 - x^4/6 + 3x^5/40 + \cdots).$$

We now apply the I.C. $y(0) = 2$ to this, obtaining

$c_0 = 2$. Differentiating, we find

$$\frac{dy}{dx} = c_0(-x^2 + 3x^4/4 + \cdots)$$
$$+ c_1(1 - x^2/2 - 2x^3/3 + 3x^4/8 + \cdots) .$$

Applying the I.C. $y'(0) = 3$ to this, we have $c_1 = 3$.
Thus we have the solution

$$y = 2(1 - x^3/3 + 3x^5/20 + \cdots)$$
$$+ 3(x - x^3/6 - x^4/6 + 3x^5/40 + \cdots)$$

or

$$y = 2 + 3x - 7x^3/6 - x^4/2 + 21x^5/40 + \cdots .$$

16. We assume $y = \sum_{n=0}^{\infty} c_n(x-1)^n$. Then

$$\frac{dy}{dx} = \sum_{n=1}^{\infty} nc_n(x-1)^{n-1} , \frac{d^2y}{dx^2} = \sum_{n=2}^{\infty} n(n-1)c_n(x-1)^{n-2} .$$

Substituting into the D.E., we obtain

$$x^2 \sum_{n=2}^{\infty} n(n-1)c_n(x-1)^{n-2} + 3x\sum_{n=1}^{\infty} nc_n(x-1)^{n-1}$$

$$- \sum_{n=0}^{\infty} c_n(x-1)^n = 0 .$$

Since the summations involve powers of x-1, we must

express the respective "coefficients" x^2 and $3x$ of the
first two summations in powers of x-1. Thus we have

$$[(x-1)^2+2(x-1)+1]\sum_{n=2}^{\infty} n(n-1)c_n(x-1)^{n-2}$$

$$+ [3(x-1)+3]\sum_{n=1}^{\infty} nc_n(x-1)^{n-1} - \sum_{n=0}^{\infty} c_n(x-1)^n = 0 \quad .$$

or

$$\sum_{n=2}^{\infty} n(n-1)c_n(x-1)^n + \sum_{n=2}^{\infty} 2n(n-1)c_n(x-1)^{n-1}$$

$$+ \sum_{n=2}^{\infty} n(n-1)c_n(x-1)^{n-2} + \sum_{n=1}^{\infty} 3nc_n(x-1)^n$$

$$+ \sum_{n=1}^{\infty} 3nc_n(x-1)^{n-1} - \sum_{n=0}^{\infty} c_n(x-1)^n = 0 \quad .$$

We rewrite the second, third, and fifth summations so
that x-1 has the exponent n in each. Thus we have:

$$\sum_{n=2}^{\infty} n(n-1)c_n(x-1)^n + \sum_{n=1}^{\infty} 2(n+1)nc_{n+1}(x-1)^n$$

$$+ \sum_{n=0}^{\infty} (n+2)(n+1)c_{n+2}(x-1)^n + \sum_{n=1}^{\infty} 3nc_n(x-1)^n$$

$$+ \sum_{n=0}^{\infty} 3(n+1)c_{n+1}(x-1)^n - \sum_{n=0}^{\infty} c_n(x-1)^n = 0 \quad .$$

The common range of these six summations is from 2 to ∞.
We write out the individual terms in each that do not
belong to this range. Thus we have:

$(-c_0 + 3c_1 + 2c_2) + (2c_1 + 10c_2 + 6c_3)(x-1)$

$+ \sum_{n=2}^{\infty} \left\{ [n(n-1)+3n-1]c_n + [2(n+1)n+3(n+1)]c_{n+1} \right.$

$\left. + (n+2)(n+1)c_{n+2} \right\} (x-1)^n = 0 .$

Equating to zero the coefficient of each power of x, we obtain

$$-c_0 + 3c_1 + 2c_2 = 0, \qquad 2c_1 + 10c_2 + 6c_3 = 0, \qquad (1)$$

$$(n^2 + 2n - 1)c_n + (2n+3)(n+1)c_{n+1} + (n+2)(n+1)c_{n+2} = 0,$$

$$n \geq 2 . \qquad (2)$$

From (1), we find $c_2 = (c_0 - 3c_1)/2$, $c_3 = -(c_1 + 5c_2)/3$

$= (-5c_0 + 13c_1)/6$. From (2), we find

$$c_{n+2} = - \frac{(n^2 + 2n-1)c_n + (2n+3)(n+1)c_{n+1}}{(n+1)(n+2)} , n \geq 2 .$$

Using this, we find $c_4 = -(7c_2 + 21c_3)/12$

$= 7(2c_0 - 5c_1)/12$. Substituting these values into the assumed solution, we have

$y = c_0 + c_1(x-1) + (c_0 - 3c_1)(x-1)^2/2$

$+ (-5c_0 + 13c_1)(x-1)^3/6 + 7(2c_0 - 5c_1)(x-1)^4/12 + \cdots$

or

$y = c_0[1+(x-1)^2/2 - 5(x-1)^3/6 + 7(x-1)^4/6 + \cdots]$

$+ c_1[(x-1) - 3(x-1)^2/2 + 13(x-1)^3/6 - 35(x-1)^4/12 + \cdots]$

Section 6.2, Page 252

2.   We write the D.E. in the normalized form (6.3) of the text. This is

$$\frac{d^2y}{dx^2} + \frac{x-2}{x(x+1)} \frac{dy}{dx} + \frac{4}{x^2(x+1)} y = 0 \quad .$$

Here $P_1(x) = (x-2)/x(x+1)$ and $P_2(x) = 4/x^2(x+1)$. We see from this that the singular points are $x = 0$ and $x = -1$.

We consider $x = 0$ and form the functions defined by the products

$$xP_1(x) = (x-2)/(x+1) \quad \text{and} \quad x^2P_2(x) = 4/(x+1) \quad .$$

Both of the product functions thus defined are analytic at $x = 0$; so $x = 0$ is a <u>regular singular point</u>.

We now consider $x = -1$ and form the functions defined by the products

$$(x+1)P_1(x) = (x-2)/x \quad \text{and} \quad (x+1)^2P_2(x) = 4(x+1)/x^2 \quad .$$

Both of the product functions thus defined are analytic at $x = -1$; so $x = -1$ is also a <u>regular singular point</u>.

7.   We assume $y = \sum_{n=0}^{\infty} c_n x^{n+r}$, where $c_0 \neq 0$. Then

$$\frac{dy}{dx} = \sum_{n=0}^{\infty} (n+r)c_n x^{n+r-1} \quad , \quad \frac{d^2y}{dx^2} = \sum_{n=0}^{\infty} (n+r)(n+r-1)c_n x^{n+r-2}$$

Substituting these into the D.E., we obtain

$$\sum_{n=0}^{\infty} (n+r)(n+r-1)c_n x^{n+r} - \sum_{n=0}^{\infty} (n+r)c_n x^{n+r}$$

$$+ \sum_{n=0}^{\infty} c_n x^{n+r+2} + \sum_{n=0}^{\infty} \frac{8}{9} c_n x^{n+r} = 0 \quad .$$

Simplifying as in the solutions of Section 6.1, we write this as

$$\sum_{n=0}^{\infty} [(n+r)(n+r-1)-(n+r)+ \frac{8}{9}]c_n x^{n+r} + \sum_{n=2}^{\infty} c_{n-2} x^{n+r} = 0$$

or

$$[r(r-1)-r+ \frac{8}{9}]c_0 x^r + [(r+1)r-(r+1)+ \frac{8}{9}]c_1 x^{r+1}$$

$$+ \sum_{n=2}^{\infty} \left\{ [(n+r)(n+r-1)-(n+r)+ \frac{8}{9}]c_n + c_{n-2} \right\} x^{n+r} = 0. \quad (1)$$

Equating to zero the coefficient of the lowest power of x, we have the indicial equation $r(r-1)-r+(8/9) = 0$ or $r^2 - 2r + (8/9) = 0$, with roots $r_1 = 4/3$ and $r_2 = 2/3$. Since the difference between these roots is not zero or a positive integer, Conclusion 1 of Theorem 6.3 tells us that the D.E. has two linearly independent solutions of the assumed form, one corresponding to each of the roots r_1 and r_2. Equating to zero the coefficients of the higher powers of x in (1), we have

$$[(r+1)r-(r+1) + (8/9)]c_1 = 0 \qquad (2)$$

and

$$[(n+r)(n+r-1)-(n+r)+8/9]c_n + c_{n-2} = 0, \, n \geq 2. \qquad (3)$$

Letting $r = r_1 = 4/3$ in (2), we obtain $(5/3)c_1 = 0$; so $c_1 = 0$. Letting $r = r_1 = 4/3$ in (3), we obtain $n(n+2/3)c_n + c_{n-2} = 0$ or

$$c_n = - \frac{3c_{n-2}}{n(3n+2)}, \quad n \geq 2 .$$

From this, $c_2 = -3c_0/16$, $c_3 = -c_1/11 = 0$, $c_4 = -3c_2/56 = 9c_0/896$. Using these values, we obtain the solution corresponding to the larger root $r_1 = 4/3$:

$$y_1(x) = c_0 x^{4/3}(1 - 3x^2/16 + 9x^4/896 - \cdots) .$$

Letting $r = r_2 = 2/3$ in (2), we obtain $c_1/3 = 0$; so $c_1 = 0$. Letting $r = r_2 = 2/3$ in (3), we obtain $n(n - 2/3)c_n + c_{n-2} = 0$ or

$$c_n = - \frac{3c_{n-2}}{n(3n-2)}, \quad n \geq 2 .$$

From this, $c_2 = -3c_0/8$, $c_3 = -c_1/7 = 0$, $c_4 = -3c_2/40 = 9c_0/320$. Using these values, we obtain the solution corresponding to the smaller root $r_2 = 2/3$:

$$y_2(x) = c_0 x^{2/3}(1 - 3x^2/8 + 9x^4/320 - \cdots) .$$

The G.S. of the D.E. is $y = C_1 y_1(x) + C_2 y_2(x)$.

12. We assume $y = \sum_{n=0}^{\infty} c_n x^{n+r}$, where $c_0 \neq 0$. Then

$$\frac{dy}{dx} = \sum_{n=0}^{\infty} (n+r)c_n x^{n+r-1}, \quad \frac{d^2y}{dx^2} = \sum_{n=0}^{\infty} (n+r)(n+r-1)c_n x^{n+r-2}.$$

Substituting these into the D.E., we obtain

$$\sum_{n=0}^{\infty} (n+r)(n+r-1)c_n x^{n+r-1} + \sum_{n=0}^{\infty} 2(n+r)c_n x^{n+r-1}$$

$$+ \sum_{n=0}^{\infty} c_n x^{n+r+1} = 0 \ .$$

Simplifying, as in the solutions of Section 6.1, we write this as

$$\sum_{n=0}^{\infty} [(n+r)(n+r-1)+2(n+r)]c_n x^{n+r-1} + \sum_{n=2}^{\infty} c_{n-2} x^{n+r-1} = 0$$

or

$$[r(r-1)+2r]c_0 x^{r-1} + [(r+1)r+2(r+1)]c_1 x^r$$

$$+ \sum_{n=2}^{\infty} \left\{ [(n+r)(n+r-1)+2(n+r)]c_n + c_{n-2} \right\} x^{n+r-1} = 0. \quad (1)$$

Equating to zero the coefficient of the lowest power of x, we have the indicial equation $r^2 - r + 2r = 0$ or $r^2 + r = 0$, with roots $r_1 = 0$, $r_2 = -1$. Note that the difference between these roots is a positive integer. By Conclusion 2 of Theorem 6.3, the D.E. has a solution of the assumed form corresponding to the larger root $r_1 = 0$. Equating to zero the coefficients of the higher powers of x in (1), we have

$$[(r+1)r + 2(r+1)]c_1 = 0 \ , \qquad (2)$$

$$[(n+r)(n+r-1) + 2(n+r)]c_n + c_{n-2} = 0, \quad n \geq 2 \ . \qquad (3)$$

Letting $r = r_1 = 0$ in (2), we obtain $2c_1 = 0$; so $c_1 = 0$. Letting $r = r_1 = 0$ in (3), we obtain

$$n(n+1)c_n + c_{n-2} = 0 \qquad \text{or}$$

$$c_n = -\frac{c_{n-2}}{n(n+1)} \quad , \quad n \geq 2 \ .$$

From this, $c_2 = -c_0/3!$, $c_3 = -c_1/4! = 0$,

$c_4 = -c_2/(4)(5) = c_0/5!$, ... Using these values, we

obtain the solution corresponding to the larger root

$r_1 = 0$:

$$y_1(x) = c_0(1-x^2/3! + x^4/5! - \cdots)$$

$$= c_0 x^{-1}(x - x^3/3! + x^5/5! - \cdots)$$

or

$$y_1(x) = c_0 x^{-1} \sin x. \tag{4}$$

Letting $r = r_2 = -1$ in (2), we obtain $0c_1 = 0$; so

c_1 is arbitrary. Letting $r = r_2 = -1$ in (3), we obtain

$n(n-1)c_n + c_{n-2} = 0$ or

$$c_n = -\frac{c_{n-2}}{n(n-1)} \quad , \quad n \geq 2 \ .$$

From this, $c_2 = -c_0/2!$, $c_3 = -c_1/3!$,

$c_4 = -c_2/(4)(3) = c_0/4!$, $c_5 = -c_3/(5)(4) = c_1/5!$,

.... Using these values, we obtain the solution

corresponding to the smaller root $r_2 = -1$:

$$y_2(x) = c_0 x^{-1}(1 - x^2/2! + x^4/4! - \cdots)$$

$$+ c_1 x^{-1}(x - x^3/3! + x^5/5! - \cdots)$$

or $y_2(x) = c_0 x^{-1} \cos x + c_1 x^{-1} \sin x$. $\qquad$ (5)

The situation here is analogous to that of Example 6.13 of the text (in particular, see text, page 248). In like manner to the case of that example, we see from (4) and (5) that the G.S. of the given D.E. is of the form

$$y = C_1 x^{-1}(1 - x^2/2! + x^4/4! - \cdots) + C_2 x^{-1}(x - x^3/3!$$

$$+ x^5/5! - \cdots) \quad \text{or} \quad y = x^{-1}(C_1 \cos x + C_2 \sin x) \quad .$$

14. We assume $y = \sum\limits_{n=0}^{\infty} c_n x^{n+r}$, where $c_0 \neq 0$. Then

$$\frac{dy}{dx} = \sum_{n=0}^{\infty} (n+r)c_n x^{n+r-1} \quad , \quad \frac{d^2 y}{dx^2} = \sum_{n=0}^{\infty} (n+r)(n+r-1)c_n x^{n+r-2}$$

Substituting these into the D.E., we obtain

$$\sum_{n=0}^{\infty} (n+r)(n+r-1)c_n x^{n+r} + \sum_{n=0}^{\infty} (n+r)c_n x^{n+r+3}$$

$$+ \sum_{n=0}^{\infty} (n+r)c_n x^{n+r} - \sum_{n=0}^{\infty} c_n x^{n+r} = 0 \quad .$$

Simplifying, as in the solutions of Section 6.1, we write this as

$$\sum_{n=0}^{\infty} [(n+r)(n+r-1)+(n+r)-1]c_n x^{n+r} + \sum_{n=3}^{\infty} (n+r-3)c_{n-3} x^{n+r} = 0$$

or

$$[r(r-1)+r-1]c_0 x^r + [(r+1)r+r]c_1 x^{r+1}$$

$$+ [(r+2)(r+1)+r+1]c_2 x^{r+2}$$

$$+ \sum_{n=3}^{\infty} [(n+r+1)(n+r-1)c_n + (n+r-3)c_{n-3}]x^{n+r} = 0. \quad (1)$$

Equating to zero the coefficient of the lowest power
of x, we have the indicial equation $r(r-1)+r-1 = 0$ or
$r^2 - 1 = 0$, with roots $r_1 = 1$, $r_2 = -1$. Equating to zero
the coefficients of the higher powers of x in (1),
we have $\qquad (r^2 + 2r)c_1 = 0$, $\qquad\qquad$ (2)

$$(r^2 + 4r + 3)c_2 = 0 \quad ,\qquad\qquad (3)$$

$$(n+r+1)(n+r-1)c_n + (n+r-3)c_{n-3} = 0, \ n \geq 3. \qquad (4)$$

Letting $r = r_1 = 1$ in (2), we obtain $3c_1 = 0$, so
$c_1 = 0$. Letting $r = r_1 = 1$ in (3), we obtain $8c_2 = 0$,
so $c_2 = 0$. Letting $r = r_1 = 1$ in (4), we obtain
$(n+2)nc_n + (n-2)c_{n-3} = 0$ or

$$c_n = -\frac{(n-2)c_{n-3}}{n(n+2)} \quad , \quad n \geq 3 \ .$$

From this, $c_3 = -c_0/15$, $c_4 = -c_1/12 = 0$,

$c_5 = -3c_2/35 = 0$, $c_6 = -c_3/12 = c_0/180$, ... Using
these values we obtain the solution corresponding to the
larger root $r_1 = 1$:

$$y_1(x) = c_0 x(1 - x^3/15 + x^6/180 - \cdots) \ . \qquad (5)$$

Letting $r = r_2 = -1$ in (2), we obtain $-c_1 = 0$, so $c_1 = 0$.
Letting $r = r_2 = -1$ in (3), we obtain $0c_2 = 0$, so c_2 is
arbitrary. Letting $r = r_2 = -1$ in (4), we obtain

$n(n-2)c_n + (n-4)c_{n-3} = 0$ $\quad$ or
$$c_n = -\frac{(n-4)c_{n-3}}{n(n-2)} \quad , \quad n \geq 3 \ .$$

From this, $c_3 = c_0/3$, $c_4 = 0$, $c_5 = -c_2/15$, $c_6 = -c_3/12$
$= - c_0/36$, $c_7 = -3c_4/35 = 0$, $c_8 = -c_5/12 = c_2/180$, $\cdots$.
Using these values we obtain the solution corresponding
to the smaller root $r_2 = -1$:

$$y_2(x) = c_0 x^{-1}(1 + x^3/3 - x^6/36 + \cdots)$$
$$+ c_2 x^{-1}(x^2 - x^5/15 + x^8/180 - \cdots). \qquad (6)$$

The situation here is analogous to that of Example
6.13 of the text. In like manner to the case of that
example, we see from (5) and (6) that the G.S. of the
D.E. is of the form

$$y = C_1 x(1 - x^3/15 + x^6/180 - \cdots) + C_2 x^{-1}(1+x^3/3 - x^6/36$$
$$+ \cdots) \; .$$

17. We assume $y = \sum_{n=0}^{\infty} c_n x^{n+r}$, where $c_0 \neq 0$. Then

$$\frac{dy}{dx} = \sum_{n=0}^{\infty} (n+r)c_n x^{n+r-1}, \quad \frac{d^2y}{dx^2} = \sum_{n=0}^{\infty} (n+r)(n+r-1)c_n x^{n+r-2} \; .$$

Substituting these into the D.E., we obtain

$$\sum_{n=0}^{\infty} 2(n+r)(n+r-1)c_n x^{n+r} - \sum_{n=0}^{\infty} (n+r)(n+r-1)c_n x^{n+r-1}$$

$$+ \sum_{n=0}^{\infty} 2(n+r)c_n x^{n+r} - \sum_{n=0}^{\infty} 2(n+r)c_n x^{n+r-1}$$

$$- \sum_{n=0}^{\infty} 2c_n x^{n+r+2} + \sum_{n=0}^{\infty} 3c_n x^{n+r+1} - \sum_{n=0}^{\infty} 2c_n x^{n+r} = 0 \; .$$

Simplifying, as in the solutions of Section 6.1, we

write this successively as

$$\sum_{n=0}^{\infty} [2(n+r)(n+r-1)+2(n+r)-2]c_n x^{n+r}$$

$$- \sum_{n=-1}^{\infty} [(n+r+1)(n+r) + 2(n+r+1)]c_{n+1} x^{n+r}$$

$$- \sum_{n=2}^{\infty} 2c_{n-2} x^{n+r} + \sum_{n=1}^{\infty} 3c_{n-1} x^{n+r} = 0 \quad ,$$

$$[2r(r-1)+2r-2]c_0 x^r + [2(r+1)r+2(r+1)-2]c_1 x^{r+1}$$

$$+ \sum_{n=2}^{\infty} [2(n+r)(n+r-1)+2(n+r)-2]c_n x^{n+r}$$

$$-[r(r-1)+2r]c_0 x^{r-1} - [(r+1)r+2(r+1)]c_1 x^r$$

$$-[(r+2)(r+1)+2(r+2)]c_2 x^{r+1}$$

$$- \sum_{n=2}^{\infty} [(n+r+1)(n+r)+2(n+r+1)]c_{n+1} x^{n+r}$$

$$- \sum_{n=2}^{\infty} 2c_{n-2} x^{n+r} + 3c_0 x^{r+1} + \sum_{n=2}^{\infty} 3c_{n-1} x^{n+r} = 0 \quad ,$$

$$-[r(r-1)+2r]c_0 x^{r-1}$$

$$+ \left\{ [2r(r-1)+2r-2]c_0 - [(r+1)r+2(r+1)]c_1 \right\} x^r$$

$$+ \left\{ [2(r+1)r+2(r+1)-2]c_1 - [(r+2)(r+1)+2(r+2)]c_2 + 3c_0 \right\} x^{r+1}$$

$$+ \sum_{n=2}^{\infty} \left\{ [2(n+r)(n+r-1) + 2(n+r) -2]c_n \right.$$

$$\left. -[(n+r+1)(n+r)+2(n+r+1)]c_{n+1} - 2c_{n-2} + 3c_{n-1} \right\} x^{n+r} = 0 \quad .$$

Equating to zero the coefficient of the lowest power of x, we have the indicial equation $r(r-1)+2r = 0$ or $r^2 + r = 0$, with roots $r_1 = 0$, $r_2 = -1$. Equating to zero the coefficients of the higher powers of x, we have

$$[2r(r-1)+2r-2]c_0 - [(r+1)r+2(r+1)]c_1 = 0 \quad , \qquad (1)$$

$$[2(r+1)r+2(r+1)-2]c_1 - [(r+2)(r+1)+2(r+2)]c_2+3c_0=0 \quad (2)$$

$$[2(n+r)(n+r-1)+2(n+r)-2]c_n$$

$$-[(n+r+1)(n+r)+2(n+r+1)]c_{n+1} - 2c_{n-2} + 3c_{n-1} = 0 \quad ,$$

$$n \geq 2. \qquad (3)$$

Letting $r = r_1 = 0$ in (1), we find $-2c_0 - 2c_1 = 0$, so $c_1 = -c_0$. Letting $r = r_1 = 0$ in (2), we find $0c_1 - 6c_2 + 3c_0 = 0$, so $c_2 = c_0/2$. Letting $r = r_1 = 0$ in (3), we find successively

$$[2n(n-1)+2n-2]c_n - [(n+1)n+2(n+1)]c_{n+1} - 2c_{n-2} + 3c_{n-1} = 0,$$

$$(n^2+3n+2)c_{n+1} = (2n^2 - 2)c_n + 3c_{n-1} - 2c_{n-2} \quad ,$$

$$c_{n+1} = \frac{2(n^2-1)c_n + 3c_{n-1} - 2c_{n-2}}{(n+1)(n+2)} \quad , \quad n \geq 2 \ .$$

From this, $c_3 = (6c_2 + 3c_1 - 2c_0)/12 = -c_0/6$, ····.
Using these values, we obtain the solution corresponding to the larger root $r_1 = 0$:

$$y = c_0(1 - x + x^2/2 - x^3/6 + \cdots) \text{ or } y = c_0e^{-x}. \quad (4)$$

Letting $r = r_2 = -1$ in (1), we find $0c_0 - 0c_1 = 0$;
so c_1 is independent of c_0 and is a second arbitrary
constant. Letting $r = r_2 = -1$ in (2), we find
$-2c_1 - 2c_2 + 3c_0 = 0$, so $c_2 = (3/2)c_0 - c_1$. Letting
$r = r_2 = -1$ in (3), we find successively
$[2(n-1)(n-2)+2(n-1)-2]c_n - [n(n-1)+2n]c_{n+1}$

$$-2c_{n-2} + 3c_{n-1} = 0 \quad ,$$
$$(n^2 + n)c_{n+1} = (2n^2 - 4n)c_n + 3c_{n-1} - 2c_{n-2} \quad ,$$

$$c_{n+1} = \frac{2n(n-2)c_n + 3c_{n-1} - 2c_{n-2}}{n(n+1)} \quad , \quad n \geq 2 \quad .$$

From this $c_3 = (3c_1 - 2c_0)/6 = c_1/2 - c_0/3$, $\cdots$. Using
these values, we obtain the solution corresponding to
the smaller root $r_2 = -1$:

$$y = c_0 x^{-1}(1+3x^2/2-x^3/3+\cdots) + c_1 x^{-1}(x-x^2+x^3/2-\cdots)$$
or
$$y = c_0 x^{-1}(1+3x^2/2 - x^3/3 + \cdots) + c_1 e^x \quad . \qquad (5)$$

As in the case of Exercise 12, the situation here is
analogous to that of Example 6.13 of the text. In like
manner to the case of that example, we see from (4) and
(5) that the G.S. of the D.E. is of the form
$$y = C_1(1 - x + x^2/2 - x^3/6 + \cdots)$$
$$+ C_2 x^{-1}(1 + 3x^2/2 - x^3/3 + \cdots) \quad .$$

This can be written in the form

$y = C_1 e^{-x} + C_2(x^{-1}e^x - e^{-x})$ and hence

$y = Ce^{-x} + C_2 x^{-1}e^x$, where $C = C_1 - C_2$.

19. We assume $y = \sum_{n=0}^{\infty} c_n x^{n+r}$, where $c_0 \neq 0$. Then

$$\frac{dy}{dx} = \sum_{n=0}^{\infty} (n+r)c_n x^{n+r-1}, \frac{d^2 y}{dx^2} = \sum_{n=0}^{\infty} (n+r)(n+r-1)c_n x^{n+r-2}.$$

Substituting these into the D.E., we obtain

$$\sum_{n=0}^{\infty} (n+r)(n+r-1) c_n x^{n+r} + \sum_{n=0}^{\infty} (n+r)c_n x^{n+r}$$

$$+ \sum_{n=0}^{\infty} c_n x^{n+r+1} - \sum_{n=0}^{\infty} c_n x^{n+r} = 0 .$$

Simplifying, as in the solutions of Section 6.1, we write this as

$$\sum_{n=0}^{\infty} [(n+r)(n+r-1)+(n+r)-1]c_n x^{n+r} + \sum_{n=1}^{\infty} c_{n-1} x^{n+r} = 0$$

and hence as

$$[r(r-1)+r-1]c_0 + \sum_{n=1}^{\infty} [(n+r+1)(n+r-1)c_n + c_{n-1}]x^{n+r} = 0 .$$

Equating to zero the coefficient of the lowest power of x, we have the indicial equation $r(r-1)+r-1 = 0$ or $(r+1)(r-1) = 0$, with roots $r_1 = 1$, $r_2 = -1$. Equating to zero the coefficients of the higher powers of x, we have

$$(n+r+1)(n+r-1)c_n + c_{n-1} = 0, \quad n \geq 1. \tag{1}$$

Letting $r = r_1 = 1$ in (1), we find $(n+2)nc_n + c_{n-1} = 0$ and hence

$$c_n = -\frac{c_{n-1}}{n(n+2)} , \quad n \geq 1 .$$

From this, $c_1 = -c_0/(1)(3) = -2c_0/1!3!$, $c_2 = -c_1/(2)(4)$ $= 2c_0/2!4!$, $c_3 = -c_2/(3)(5) = -2c_0/3!5!$, and, in general, $\cdots c_n = (-1)^n 2c_0/n!(n+2)!$. Using these values, we obtain the solution corresponding to the larger root $r_1 = 1$ and given by

$$y = c_0 x[1 - 2x/1!3! + 2x^2/2!4! - 2x^3/3!5! + \cdots] .$$

Choosing, $c_0 = 1$, we obtain the particular solution denoted by $y_1(x)$ and defined by

$$y_1(x) = x[1 - 2x/1!3! + 2x^2/2!4! - 2x^3/3!5! + \cdots]$$

$$= x[1 + 2 \sum_{n=1}^{\infty} \frac{(-1)^n x^n}{n!(n+2)!}] . \tag{2}$$

Letting $r = r_2 = -1$ in (1), we find

$$n(n-2)c_n + c_{n-1} = 0 \tag{3}$$

and hence

$$c_n = -\frac{c_{n-1}}{n(n-2)} , \quad n \geq 1, n \neq 2 . \tag{4}$$

For $n=1$, (4) gives $c_1 = c_0$. For $n=2$, (4) does not apply and we must use (3). For $n=2$, (3) is $0c_2 + c_1 = 0$, and hence $c_1 = 0$. But then, since $c_1 = c_0$, we must have $c_0 = 0$. However, we assumed $c_0 \neq 0$ at the start. This

contradiction shows there is no solution of the assumed
form with $c_0 \neq 0$ corresponding to the smaller root
$r_2 = -1$. Moreover, use of (4) for $n \geq 3$ will only lead
us to the solution $y_1(x)$ already obtained and given by
(2) (the student should check that this is true).

We now seek a solution that is linearly independent
of the solution $y_1(x)$. From Theorem 6.3, we know that
it is of the form $\sum_{n=0}^{\infty} c_n{}^* x^{n-1} + Cy_1(x)\ln x$, where
$c_0{}^* \neq 0$ and $C \neq 0$. We shall use reduction of order to
find it. We let

$$y = y_1(x) \, v \quad .$$

From this we obtain

$$\frac{dy}{dx} = y_1(x) \frac{dv}{dx} + y_1'(x) \, v \qquad \text{and}$$

$$\frac{d^2y}{dx^2} = y_1(x) \frac{d^2v}{dx^2} + 2y_1'(x) \frac{dv}{dx} + y_1''(x)v.$$

Substituting these into the given D.E., after some
simplifications we obtain

$$x \, y_1(x) \frac{d^2v}{dx^2} + [2x \, y_1'(x) + y_1(x)] \frac{dv}{dx} = 0. \qquad (5)$$

Letting $w = dv/dx$, this reduces to

$$x \, y_1(x) \frac{dw}{dx} + [2xy_1'(x) + y_1(x)]w = 0 \quad .$$

From this, we have

$$\frac{dw}{w} = - [2 \frac{y_1'(x)}{y_1(x)} + \frac{1}{x}] \, dx \quad ,$$

and integrating we obtain the particular solution

$$w = 1/x[y_1(x)]^2 \quad .$$

Writing out the first three terms of $y_1(x)$ defined by (2), we have

$$y_1(x) = x - x^2/3 + x^3/24 + \cdots \quad . \tag{6}$$

From this, using basic multiplication and division, we find

$$[y_1(x)]^2 = x^2 - 2x^3/3 + 7x^4/36 + \cdots$$

and $1/[y_1(x)]^2 = 1/x^2 + 2/3x + 1/4 + \cdots$. Hence we obtain

$$w = 1/x[y_1(x)]^2 = 1/x^3 + 2/3x^2 + 1/4x + \cdots.$$

Integrating we obtain the particular solution of (5) given by

$$v = -1/2x^2 - 2/3x + \ln|x|/4 + \cdots \quad . \tag{7}$$

Multiplying (6) by (7), we obtain the desired linearly independent solution of the given D.E. We thus find

$$y = x^{-1}(-1/2 - x/2 + 29x^2/144 + \cdots) + y_1(x)\ln|x|/4. \tag{8}$$

The general solution is a linear combination of (2) and (8).

22. We assume $y = \sum_{n=0}^{\infty} c_n x^{n+r}$, where $c_0 \neq 0$. Then

$$\frac{dy}{dx} = \sum_{n=0}^{\infty} (n+r)c_n x^{n+r-1}, \frac{d^2y}{dx^2} = \sum_{n=0}^{\infty} (n+r)(n+r-1)c_n x^{n+r-2}.$$

Substituting these into the D.E., we obtain

$$\sum_{n=0}^{\infty} (n+r)(n+r-1)c_n x^{n+r} + \sum_{n=0}^{\infty} (n+r)c_n x^{n+r+1}$$

$$- \sum_{n=0}^{\infty} \frac{3}{4} c_n x^{n+r} = 0 .$$

Simplifying, as in the solutions of Section 6.1, we write this as

$$\sum_{n=0}^{\infty} [(n+r)(n+r-1) - \frac{3}{4}]c_n x^{n+r} + \sum_{n=1}^{\infty} (n+r-1)c_{n-1} x^{n+r} = 0$$

and hence as

$$[r(r-1) - \frac{3}{4}]c_0 x^r$$

$$+ \sum_{n=1}^{\infty} \left\{ [(n+r)(n+r-1) - \frac{3}{4}]c_n + (n+r-1)c_{n-1} \right\} x^{n+r} = 0 .$$

Equating to zero the coefficient of the lowest power of x, we have the individual equation $r^2 - r - 3/4 = 0$, with roots $r_1 = 3/2$, $r_2 = -1/2$. Equating to zero the the coefficients of the higher powers of x, we have

$$[(n+r)(n+r-1) - \frac{3}{4}]c_n + (n+r-1)c_{n-1} = 0, n \geq 1. \qquad (1)$$

Letting $r = r_1 = 3/2$ in (1), we obtain $(n^2 + 2n)c_n + (n + 1/2)c_{n-1} = 0$ and hence

$$c_n = - \frac{(2n+1)c_{n-1}}{2n(n+2)} , \qquad n \geq 1 .$$

From this, $c_1 = -c_0/2$, $c_2 = -5c_1/16 = 5c_0/32$,

$c_3 = -7c_2/30 = -7c_0/192$, $\cdots$. Observe that we can also write

$$c_1 = -\frac{3c_0}{2^0 1! \, 3!} \;, \quad c_2 = \frac{(3)(5)c_0}{2^1 2! 4!} \;, \quad c_3 = -\frac{(3)(5)(7)c_0}{2^2 \, 3! \, 5!}, \cdots,$$

and in general,

$$c_n = (-1)^n \frac{(3)(5)(7)\cdots(2n+1)c_0}{2^{n-1} \, n! \, (n+2)!} \;, \quad n \geq 1 \;.$$

Using these values we obtain the solution corresponding to the larger root $r_1 = 3/2$ and given by

$$y = c_0 x^{3/2}(1 - x/2 + 5x^2/32 - 7x^3/192 + \cdots) \;.$$

Choosing $c_0 = 1$, we obtain the particular solution denoted by $y_1(x)$ and defined by

$$y_1(x) = x^{3/2}(1 - x/2 + 5x^2/32 - 7x^3/192 + \cdots)$$

$$= x^{3/2} \left[1 + \sum_{n=1}^{\infty} \frac{(-1)^n[(3)(5)(7)\cdots(2n+1)]}{2^{n-1} \, n! \, (n+2)!} x^n \right]. \quad (2)$$

Letting $r = r_2 = -1/2$ in (1), we obtain $(n^2 - 2n)c_n + (n - 3/2)c_{n-1} = 0$, or

$$n(n-2)c_n + (2n - 3) \, c_{n-1}/2 = 0 \tag{3}$$

and hence

$$c_n = -\frac{(2n-3)c_{n-1}}{2n(n-2)} \;, \quad n \geq 1, \; n \neq 2 \;. \tag{4}$$

For n = 1, (4) gives $c_1 = -c_0/2$. For n=2, (4) does not apply and we must use (3). For n=2, (3) is $0c_2 + c_1/2 = 0$, and hence $c_1 = 0$. But then, since $c_1 = -c_0/2$, we must have $c_0 = 0$. However, we assumed $c_0 \neq 0$ at the start. This contradiction shows there is no solution of the assumed form with $c_0 \neq 0$ corresponding to the smaller root $r_2 = -1/2$. Moreover, use of (4) for $n \geq 3$ will only lead us to the solution $y_1(x)$ already obtained and given by (2) (the student should check that this is true).

We now seek a solution that is linearly independent of the solution $y_1(x)$. From Theorem 6.3, we know that it is of the form

$$\sum_{n=0}^{\infty} c_n^* \, x^{n-1/2} + Cy_1(x)\ln x, \text{ where } c_0^* \neq 0 \text{ and } C \neq 0.$$

We shall use reduction of order to find it. We let $y = y_1(x)v$. From this we obtain $\frac{dy}{dx} = y_1(x) \frac{dv}{dx} + y_1'(x)v$ and $\frac{d^2y}{dx^2} = y_1(x) \frac{d^2v}{dx^2} + 2y_1'(x) \frac{dv}{dx} + y_1''(x)v$. Substituting these into the given D.E., after some simplifications, we obtain

$$y_1(x) \frac{d^2v}{dx^2} + [2y_1'(x) + y_1(x)] \frac{dv}{dx} = 0. \qquad (5)$$

Letting w = dv/dx, this reduces to

$$y_1(x) \frac{dw}{dx} + [2y_1'(x) + y_1(x)]w = 0 \quad .$$

From this, we have $\frac{dw}{w} = - [\frac{2y_1'(x)}{y_1(x)} + 1] \, dx$, and

integrating we obtain the particular solution

$$w = e^{-x}/[y_1(x)]^2 \quad .$$

From (2), using basic multiplication and division, we find

$$[y_1(x)]^2 = x^3 - x^4 + 9x^5/16 - 11x^6/48 + \cdots$$

and

$$1/[y_1(x)]^2 = x^{-3} + x^{-2} + 7x^{-1}/16 + 5/48 + \cdots \quad .$$

Thus using $e^{-x} = 1 - x + x^2/2 - x^3/6 + \cdots$. we obtain

$$w = e^{-x}/[y_1(x)]^2 = x^{-3} - x^{-1}/16 + \cdots \quad .$$

Integrating we obtain the particular solution of (5) given by

$$v = -x^{-2}/2 - \ln|x|/16 + \cdots \, , \tag{6}$$

where the next nonzero term in this expansion is the term in x^2. Multiplying (2) by (6), we obtain the desired linearly independent solution of the given D.E. We thus find

$$y = x^{-1/2}(-1/2 + x/4 - 5x^2/64 + \cdots) - y_1(x)\ln|x|/16. \tag{7}$$

The general solution is a linear combination of (2) and (7).

25. We assume $y = \sum\limits_{n=0}^{\infty} c_n x^{n+r}$, where $c_0 \neq 0$. Then

$$\frac{dy}{dx} = \sum_{n=0}^{\infty} (n+r)c_n x^{n+r-1}, \quad \frac{d^2 y}{dx^2} = \sum_{n=0}^{\infty} (n+r)(n+r-1)c_n x^{n+r-2}.$$

Substituting these into the D.E., we obtain

$$\sum_{n=0}^{\infty} (n+r)(n+r-1)c_n x^{n+r} - \sum_{n=0}^{\infty} (n+r)c_n x^{n+r}$$

$$+ \sum_{n=0}^{\infty} c_n x^{n+r+2} + \sum_{n=0}^{\infty} c_n x^{n+r} = 0 \ .$$

Simplifying, as in the solutions of Section 6.1, we write this as

$$\sum_{n=0}^{\infty} [(n+r)(n+r-1)-(n+r)+1]c_n x^{n+r} + \sum_{n=2}^{\infty} c_{n-2} x^{n+r} = 0$$

and hence as

$$[r(r-1)-r+1]c_0 x^r + [(r+1)r-(r+1)+1]c_1 x^{r+1}$$

$$+ \sum_{n=2}^{\infty} \left\{ [(n+r)(n+r-1)-(n+r)+1]c_n + c_{n-2} \right\} x^{n+r} = 0 \ .$$

Equating to zero the coefficient of the lowest power of x, we have the indicial equation $r^2 - 2r + 1 = 0$, with the double root $r = 1$. Equating to zero the coefficients of the higher powers of x, we have

$$[(r+1)r - (r+1) + 1]c_1 = 0, \tag{1}$$

$$[(n+r)(n+r-1)-(n+r)+1]c_n + c_{n-2} = 0, \quad n \geq 2. \tag{2}$$

Letting $r = 1$ in (1), we have $c_1 = 0$. Letting $r=1$

in (2), we obtain $n^2 c_n + c_{n-2} = 0$ and hence

$$c_n = -\frac{c_{n-2}}{n^2} \ , \quad n \geq 2 \ .$$

From this, $c_2 = -c_0/4$, $c_3 = -c_1/9 = 0$, $c_4 = -c_2/16$ $= c_0/64$, $\cdots$. Note that all odd coefficients are zero and we can write the general even coefficients as

$$c_{2n} = \frac{(-1)^n}{[(2)(4)(6)\cdots(2n)]^2} \ , \quad n \geq 1.$$

Using these values we obtain the solution of the assumed form corresponding to the double root r=1 and given by

$$y = c_0 x(1 - x^2/4 + x^4/64 - x^6/2304 + \cdots) \ .$$

Choosing $c_0 = 1$, we obtain the particular solution denoted by $y_1(x)$ and defined by

$$y_1(x) = x(1 - x^2/4 + x^4/64 - x^6/2304 + \cdots)$$

$$= x[1 + \sum_{n=1}^{\infty} \frac{(-1)^n x^{2n}}{[(2)(4)(6)\cdots(2n)]^2}] \ . \tag{3}$$

Since the indicial equation has the double root r=1, by Conclusion 3 of Theorem 6.3, a linearly independent solution is of the form

$$\sum_{n=0}^{\infty} c_n^* x^{n+2} + y_1(x) \ln x.$$

We shall use reduction of order to find it. We let $y = y_1(x)v$. From this we obtain

$$\frac{dy}{dx} = y_1(x) \frac{dv}{dx} + y_1'(x)v \quad \text{and}$$

$$\frac{d^2y}{dx^2} = y_1(x) \frac{d^2v}{dx^2} + 2y_1'(x) \frac{dv}{dx} + y_1''(x)v.$$

Substituting these into the given D.E., after some simplifications, we obtain

$$xy_1(x) \frac{d^2v}{dx^2} + [2xy_1'(x) - y_1(x)] \frac{dv}{dx} = 0. \tag{4}$$

Letting $w = dv/dx$, this reduces to

$$xy_1(x) \frac{dw}{dx} + [2xy_1'(x) - y_1(x)]w = 0.$$

From this, we have

$$\frac{dw}{w} = [\frac{1}{x} - \frac{2y_1'(x)}{y_1(x)}] \, dx$$

and integrating, we obtain the particular solution

$$w = x/[y_1(x)]^2 .$$

From (3), using basic multiplication and division, we find

$$[y_1(x)]^2 = x^2 - x^4/2 + 3x^6/32 - 5x^8/576 + \cdots \quad \text{and}$$

$$1/[y_1(x)]^2 = x^{-2} + 1/2 + 5x^2/32 + 23x^4/576 + \cdots .$$

Thus we obtain

$$w = x^{-1} + x/2 + 5x^3/32 + 23x^5/576 + \cdots .$$

Integrating we obtain the particular solution of (4) given by

$$v = \ln|x| + x^2/4 + 5x^4/128 + 23x^6/3456 + \cdots . \tag{5}$$

Multiplying (3) by (5) we obtain the desired linearly
independent solution of the given D.E. We thus find

$$y = x^3/4 - 3x^5/128 + 11x^7/13824 + \cdots + y_1(x) \ln |x|. \quad (6)$$

The general solution is a linear combination of (3) and
(6).

CHAPTER 7

Section 7.1, Page 278

1. We introduce operator notation and write the system in the form

$$\begin{cases} (D - 2)x + (D - 4)y = e^t, \\ \\ Dx + (D - 1)y = e^{4t}. \end{cases} \tag{1}$$

We apply the operator $(D - 1)$ to the first equation and the operator $(D - 4)$ to the second, obtaining

$$\begin{cases} (D-1)(D-2)x + (D-1)(D-4)y = (D-1)e^t, \\ \\ (D-4)Dx + (D-4)(D-1)y = (D-4)e^{4t}. \end{cases}$$

Subtracting the second equation from the first, we obtain $[(D-1)(D-2) - (D-4)D]x = (D-1)e^t - (D-4)e^{4t}$ or $(D+2)x = 0$. The G.S. of this D.E. is

$$x = c\, e^{-2t}. \tag{2}$$

Now there are two ways to obtain y, and we give both here. First, we proceed by returning to the system (1), applying the operator D to the first equation and the

253

operator $(D - 2)$ to the second equation, obtaining

$$\begin{cases} D(D-2)x + D(D-4)y = De^t, \\ (D-2)Dx + (D-2)(D-1)y = (D-2)e^{4t}. \end{cases}$$

Subtracting the first equation from the second, we obtain
$[(D-2)(D-1) - D(D-4)]y = (D-2)e^{4t} - De^t$ or
$(D+2)y = 2e^{4t} - e^t$. Using undetermined coefficients, we
find the G.S. of this D.E. to be

$$y = ke^{-2t} + e^{4t}/3 - e^t/3 \ . \qquad\qquad (3)$$

Now the determinant of the operator "coefficients"
of x and y in (1) is

$$\begin{vmatrix} D - 2 & D - 4 \\ D & D - 1 \end{vmatrix} = D + 2 \ .$$

Since this is of order 1, only one of the two constants
c and k in (1) and (2) can be independent. To determine
the relation which must thus exist between c and k, we
substitute x given by (2) and y given by (3) into system
(1). Substituting into the second equation of (1), we
have

$$-2ce^{-2t} + [-2ke^{-2t} + 4e^{4t}/3 - e^t/3]$$
$$- [ke^{-2t} + e^{4t}/3 - e^t/3] = e^{4t}$$

or $(-2c - 3k)e^{-2t} = 0$. Thus $k = -2c/3$. Hence we obtain
the G.S. of (1) in the form

$$\begin{cases} x = ce^{-2t}, \\ y = -2ce^{-2t}/3 + e^{4t}/3 - e^{t}/3 \ . \end{cases} \qquad (4)$$

We now obtain y by the alternative procedure of the text, pages 276-277. We return to system (1) and sub-tract the second equation from the first, thereby elimi-nating Dy but not y, and thus obtaining $-2x - 3y = e^{t} - e^{4t}$. From this, $y = -2x/3 + e^{4t}/3 - e^{t}/3$. Substituting x into this from (2), we at once obtain $y = -2ce^{-2t}/3 + e^{4t}/3 - e^{t}/3$. Using this and (2), we again have the G.S., (4) of the original system.

5. We introduce operator notation and write the system in the form

$$\begin{cases} (2D-1)x + (D-1)y = e^{-t}, \\ (D+2)x + (D+1)y = e^{t}. \end{cases} \qquad (1)$$

We apply the operator (D + 1) to the first equation and the operator (D − 1) to the second, obtaining

$$\begin{cases} (D+1)(2D-1)x + (D+1)(D-1)y = (D+1)e^{-t}, \\ (D-1)(D+2)x + (D-1)(D+1)y = (D-1)e^{t} \ . \end{cases}$$

Subtracting the second from the first, we obtain
$[(D+1)(2D-1) - (D-1)(D+2)]x = (D+1)e^{-t} - (D-1)e^{t}$
or $(D^2 + 1)x = 0$.
The G.S. of this D.E. is

$$x = c_1 \sin t + c_2 \cos t \; . \tag{2}$$

We now find y using the alternate procedure of the text, pages 276-277. We subtract the second equation of (1) from the first, thereby eliminating Dy but not y, and thus obtaining $(D-3)x - 2y = e^{-t} - e^{t}$. From this, $y = (Dx - 3x - e^{-t} + e^{t})/2$. From (2), $Dx = c_1 \cos t - c_2 \sin t$. Substituting into the preceeding expression for y, we get $y = (c_1 \cos t - c_2 \sin t - 3c_1 \sin t - 3c_2 \cos t + e^{t} - e^{-t})/2$

or $y = -[(3c_1 + c_2)\sin t]/2 + [(c_1 - 3c_2) \cos t]/2$

$$+ e^{t}/2 - e^{-t}/2.$$

The G.S. of system (1) is thus:

$$\begin{cases} x = c_1 \sin t + c_2 \cos t, \\ y = - \dfrac{3c_1 + c_2}{2} \sin t + \dfrac{c_1 - 3c_2}{2} \cos t \\ \qquad\qquad\qquad + \dfrac{e^{t}}{2} - \dfrac{e^{-t}}{2} \; . \end{cases}$$

7. We introduce operator notation and write the system in the form

$$\begin{cases} (D-1)x + (D-6)y = e^{3t}, \\ (D-2)x + (2D-6)y = t. \end{cases} \tag{1}$$

We apply the operator $(2D-6)$ to the first equation and

the operator $(D - 6)$ to the second equation, obtaining

$$\begin{cases} (2D-6)(D-1)x + (2D-6)(D-6)y = (2D-6)e^{3t}, \\ (D-6)(D-2)x + (D-6)(2D-6)y = (D-6)t \quad . \end{cases}$$

Subtracting the second equation from the first, we

obtain

$$[(2D-6)(D-1) - (D-6)(D-2)]x = (2D-6)e^{3t} - (D-6)t$$

or $(D^2 - 6)x = 6t - 1 \quad .$

Using undetermined coefficients, the G.S. of this D.E.

is found to be

$$x = c_1 e^{\sqrt{6}t} + c_2 e^{-\sqrt{6}t} - t + 1/6. \tag{2}$$

We find y using the alternate procedure of the text,

pages 276-277. We multiply the first equation of (1) by

2, obtaining $(2D-2)x + (2D-12)y = 2e^{3t}$. We now subtract

the second equation of (1) from this, obtaining

$Dx - 6y = 2e^{3t} - t$, which involves y but not Dy. From

this,

$$y = (Dx - 2e^{3t} + t)/6. \tag{3}$$

From (2), $Dx = \sqrt{6}c_1 e^{\sqrt{6}t} - \sqrt{6}c_2 e^{-\sqrt{6}t} - 1$. Substituting

this into (3), we get

$$y = \sqrt{6}c_1 e^{\sqrt{6}t}/6 - \sqrt{6}c_2 e^{-\sqrt{6}t}/6 + t/6 - 1/6 - e^{3t}/3. \tag{4}$$

The pair (2) and (4) together constitute the G.S. of

system (1).

12. We introduce operator notation and write the system in the form

$$\begin{cases} (D-1)x + (D+5)y = t^2, \\ (D-2)x + (2D+4)y = 2t+1 . \end{cases} \tag{1}$$

We apply the operator $(2D+4)$ to the first equation and the operator $(D+5)$ to the second equation, obtaining

$$\begin{cases} (2D+4)(D-1)x + (2D+4)(D+5)y = (2D+4)t^2, \\ (D+5)(D-2)x + (D+5)(2D+4)y = (D+5)(2t+1). \end{cases}$$

Subtracting the second from the first, we obtain

$$[(2D+4)(D-1) - (D+5)(D-2)]x = (2D+4)t^2 - (D+5)(2t+1)$$

or $\quad (D^2 - D + 6)x = 4t + 4t^2 - 2 - 10t - 5$

or $\quad (D^2 - D + 6)x = 4t^2 - 6t - 7.$

The auxiliary equation of this D.E. is $m^2 - m + 6 = 0$, with roots $1/2 \pm \sqrt{23}i/2$; and the complementary function is

$$x = e^{t/2}[c_1 \sin(\sqrt{23}t/2) + c_2 \cos(\sqrt{23}t/2)] .$$

Using undetermined coefficients we find a particular integral $x_p = 2t^2/3 - 7t/9 - 41/27$. Thus

$$x = e^{t/2}[c_1 \sin(\sqrt{23}t/2) + c_2 \cos(\sqrt{23}t/2)]$$
$$+ 2t^2/3 - 7t/9 - 41/27 . \tag{2}$$

We find y using the alternate procedure of the text, pages 276-277. We multiply the first equation of (1) by 2, obtaining $(2D-2)x + (2D+10)y = 2t^2$. We now subtract the second equation of (1) from this, obtaining

$Dx + 6y = 2t^2 - 2t - 1$, which involves y but not Dy.

From this,

$$y = (-Dx + 2t^2 - 2t - 1)/6 \ . \qquad (3)$$

From (2),

$$Dx = e^{t/2}[(c_1/2 - \sqrt{23}c_2/2) \sin(\sqrt{23}t/2)$$

$$+ (\sqrt{23}c_1/2 + c_2/2) \cos (\sqrt{23}t/2)] + 4t/3 - 7/9.$$

Substituting this into (3), we get

$$y = e^{t/2}[(-c_1/12 + \sqrt{23}c_2/12) \sin(\sqrt{23}t/2)$$

$$+ (-\sqrt{23}c_1/12 - c_2/12) \cos (\sqrt{23}t/2)]$$

$$+ t^2/3 - 5t/9 - 1/27 \qquad (4)$$

The pair (2) and (4) together constitute the G.S. of system (1).

17. We introduce operator notation and write the system in the form

$$\begin{cases} (2D-1)x + (D-1)y = 1, \\ (D+2)x + (D-1)y = t \ . \end{cases} \qquad (1)$$

For this system we merely have to subtract the second equation from the first to eliminate both y and Dy. We have

$$[(2D-1) - (D+2)]x = 1 - t \quad \text{or} \quad (D-3)x = 1 - t \ .$$

The G.S. of this is

$$x = c_1 e^{3t} + t/3 - 2/9. \qquad (2)$$

We cannot apply the alternate procedure of pages 276-277 to find y here; for, as noted above, elimination of Dy from (1) also results in elimination of y itself. Thus we proceed to eliminate x and Dx from (1). We apply the operator $(D+2)$ to the first equation of (1) and the operator $(2D-1)$ to the second equation, obtaining

$$\begin{cases} (D+2)(2D-1)x + (D+2)(D-1)y = (D+2)1, \\ (2D-1)(D+2)x + (2D-1)(D-1)y = (2D-1)t \;. \end{cases}$$

Subtracting the first from the second, we obtain

$$[(2D-1)(D-1) - (D+2)(D-1)]y = (2D-1)t - (D+2)1$$

or $(D^2 - 4D + 3)y = -t$. Using undetermined coefficients, the G.S. of this D.E. is found to be

$$y = k_1 e^t + k_2 e^{3t} - t/3 - 4/9. \qquad (3)$$

The determinant of the operator "coefficients" of x and y in (1) is

$$\begin{vmatrix} 2D-1 & D-1 \\ D+2 & D-1 \end{vmatrix} = D^2 - 4D + 3 \;.$$

Since this is of order 2, only two of the three constants c_1, k_1, k_2 in (2) and (3) can be independent. To determine the relations which must thus exist among these three constants, we substitute x given by (2) and

y given by (3) into system (1). Substituting into the
first equation of (1), we have

$2(3c_1e^{3t} + 1/3) + (k_1e^t + 3k_2e^{3t} - 1/3)$

$\quad - (c_1e^{3t} + t/3 - 2/9) - (k_1e^t + k_2e^{3t} - t/3 - 4/9) = 1$

or $\qquad (5c_1 + 2k_2)e^{3t} + 0k_1e^t = 0$.

Thus $k_2 = -5c_1/2$ and k_1 is arbitrary. Thus k_1 is the
"second" arbitrary constant of the G.S., and we now
write it as c_2. Hence we have the G.S. in the form

$$\begin{cases} x = c_1e^{3t} + t/3 - 2/9 \ , \\ y = c_2e^t - 5c_1e^{3t}/2 - t/3 - 4/9 \ . \end{cases}$$

18. We introduce operator notation and write the
system in the form

$$\begin{cases} D^2x + Dy = e^{2t} \ , \\ (D-1)x + (D-1)y = 0 \ . \end{cases} \qquad (1)$$

We apply the operator $(D-1)$ to the first equation and
the operator D to the second equation, obtaining

$$\begin{cases} (D-1)D^2x + (D-1)Dy = (D-1)e^{2t}, \\ D(D-1)x + D(D-1)y = D0. \end{cases}$$

Subtracting the second equation from the first, we
obtain

$$[(D-1)D^2 - D(D-1)]x = (D-1)e^{2t} - 0$$

or
$$(D^3 - 2D^2 + D)x = e^{2t} \quad .$$

The auxiliary equation of this D.E. is $m^3 - 2m^2 + m = 0$
with roots $m = 0, 1, 1$ (double root); and the G.S.,
found by using undetermined coefficients, is

$$x = c_1 + (c_2 + c_3t)e^t + e^{2t}/2 \quad . \tag{2}$$

We find y using the alternate procedure of pages
276-277 of the text. We subtract the second equation of
(1) from the first, thereby eliminating Dy but not y,
and thus obtaining $D^2x - Dx + x + y = e^{2t}$. From this,

$$y = -x + Dx - D^2x + e^{2t}. \tag{3}$$

From (2), we find $Dx = (c_2 + c_3 + c_3t)e^t + e^{2t}$,
$D^2x = (c_2 + 2c_3 + c_3t)e^t + 2e^{2t}$. Substituting x from
(2) and these derivatives into (3), we get

$$y = -[c_1 + (c_2 + c_3t)e^t + e^{2t}/2]$$
$$+ [(c_2 + c_3 + c_3t)e^t + e^{2t}] - [(c_2 + 2c_3 + c_3t)e^t$$
$$+ 2e^{2t}] + e^{2t}$$

or

$$y = -c_1 - (c_2 + c_3 + c_3t)e^t - e^{2t}/2. \tag{4}$$

The pair (2) and (4) together constitute the G.S. of
system (1).

22. We introduce operator notation and write the
system in the form

$$\begin{cases} (D^2 + 1)x + (4D - 4)y = 0 \ , \\ (D-1)x + (D+9)y = e^{2t} \ . \end{cases} \qquad (1)$$

We apply the operator $(D+9)$ to the first equation and
the operator $(4D-4)$ to the second equation, obtaining

$$\begin{cases} (D+9)(D^2+1)x + (D+9)(4D-4)y = (D+9)0, \\ (4D-4)(D-1)x + (4D-4)(D+9)y = (4D-4)e^{2t} \ . \end{cases}$$

Subtracting the second equation from the first, we
obtain

$$[(D+9)(D^2+1) - (4D-4)(D-1)]x = 0 - (4D-4)e^{2t}$$

or $\qquad\qquad (D^3 + 5D^2 + 9D + 5)x = -4e^{2t}$.

The auxiliary equation of this D.E. is $m^3 + 5m^2 + 9m$
$+ 5 = 0$ or $(m+1)(m^2+4m+5) = 0$, with roots $m = -1$,
$m = -2 \pm i$. Using this information and undetermined
coefficients, the G.S. of this D.E. is found to be

$$x = c_1 e^{-t} + e^{-2t}(c_2 \sin t + c_3 \cos t) - 4e^{2t}/51.$$

$$(2)$$

We find y using the alternate procedure of pages
276-277 of the text. We multiply the second equation
of (1) by 4, obtaining $(4D-4)x + (4D+36)y = 4e^{2t}$. We
now subtract this from the first equation of (1),
obtaining $D^2 x - 4Dx + 5x - 40y = -4e^{2t}$. From this,

$$y = (D^2x - 4Dx + 5x + 4e^{2t})/40. \tag{3}$$

From (2), $Dx = -c_1 e^{-t} + e^{-2t}(-2c_2 - c_3) \sin t$

$$+ e^{-2t}(c_2 - 2c_3)\cos t - 8e^{2t}/51,$$

$D^2x = c_1 e^{-t} + e^{-2t}(3c_2 + 4c_3)\sin t$

$$+ e^{-2t}(-4c_2 + 3c_3)\cos t - 16e^{2t}/51.$$

Substituting x from (2) and these derivatives into (3), we get

$y = [c_1 e^{-t} + e^{-2t}(3c_2 + 4c_3)\sin t + e^{-2t}(-4c_2+3c_3)\cos t$

$$- 16e^{2t}/51]/40 - 4[-c_1 e^{-t} - e^{-2t}(-2c_2 - c_3)\sin t$$

$$+ e^{-2t}(c_2 - 2c_3)\cos t - 8e^{2t}/51]/40$$

$$+ 5[c_1 e^{-t} + e^{-2t}(c_2 \sin t + c_3 \cos t) - 4e^{2t}/51]/40$$

$$+ 4e^{2t}/40$$

or

$y = (c_1 + 4c_1 + 5c_1)e^{-t}/40 + e^{-2t}(3c_2 + 4c_3 + 8c_2 + 4c_3$

$$+ 5c_2)(\sin t)/40 + e^{-2t}(-4c_2 + 3c_3 - 4c_2 + 8c_3$$

$$+ 5c_3)(\cos t)/40 + (-16 + 32 - 20 + 204)e^{2t}/(40)(51)$$

or finally

$y = c_1 e^{-t}/4 + e^{-2t}[(2c_2/5 + c_3/5)\sin t$

$$+ (-c_2/5 + 2c_3/5)\cos t + 5e^{2t}/51. \tag{4}$$

The pair (2) and (4) constitute the G.S. of system (1).

26.   By text, page 268, equations (7.7), we let
$x_1 = x$, $x_2 = \frac{dx}{dt}$, $x_3 = \frac{d^2x}{dt^2}$, $x_4 = \frac{d^3x}{dt^3}$. From these

and the given fourth order D.E., we have

$$\frac{dx_1}{dt} = \frac{dx}{dt} = x_2, \quad \frac{dx_2}{dt} = \frac{d^2x}{dt^2} = x_3, \quad \frac{dx_3}{dt} = \frac{d^3x}{dt^3} = x_4 \quad ,$$

$$\frac{dx_4}{dt} = \frac{d^4x}{dt^4} = -2tx + t^2 \frac{d^2x}{dt^2} + \cos t = -2tx_1 + t^2x_3 + \cos t$$

Thus we have the system

$$\begin{cases} \dfrac{dx_1}{dt} = x_2, \quad \dfrac{dx_2}{dt} = x_3, \quad \dfrac{dx_3}{dt} = x_4, \\[4mm] \dfrac{dx_4}{dt} = -2tx_1 + t^2x_3 + \cos t. \end{cases}$$

Section 7.2, Page 290

3.   We apply Kirchhoff's voltage law (text, Section
5.6) to each of the three loops indicated in the Figure
7.6. For the left hand loop, the voltage drops are as
follows:

1.   across the resistor R_1: $20i_1$.
2.   across the inductor L_1: $0.01 \frac{di}{dt}$

Thus applying the voltage law to this loop, we have the
D.E.
$$0.01 \frac{di}{dt} + 20i_1 = 100. \tag{1}$$

For the right hand loop, the voltage drops are as
follows:

1. across the resistor R_1: $-20i_1$.

2. across the resistor R_2: $40i_2$.

3. across the inductor L_2: $0.02 \frac{di_2}{dt}$.

Thus applying the voltage law to this loop, we have the
D.E.

$$0.02 \frac{di_2}{dt} - 20i_1 + 40i_2 = 0. \tag{2}$$

For the outside loop, the voltage drops are

1. across the resistor R_2: $40i_2$.

2. across the inductor L_1: $0.01 \frac{di}{dt}$.

3. across the inductor L_2: $0.02 \frac{di_2}{dt}$.

Thus applying the voltage law to this loop, we have the
D.E.

$$0.02 \frac{di_2}{dt} + 0.01 \frac{di}{dt} + 40i_2 = 100. \tag{3}$$

The three equations are not all independent. Equation
(3) may be obtained by adding equations (1) and (2).
Hence we retain only equations (1) and (2).

We now apply Kirchhoff's current law to the upper
junction point in the figure. We have $i = i_1 + i_2$.
Hence we replace i by $i_1 + i_2$ in (1), retain (2) as is,
and obtain the linear system

$$\begin{cases} 0.01 \dfrac{di_1}{dt} + 0.01 \dfrac{di_2}{dt} + 20i_1 = 100, \\[4mm] 0.02 \dfrac{di_2}{dt} - 20i_1 + 40i_2 = 0. \end{cases} \qquad (4)$$

The initially zero currents give the I.C.'s

$$i_1(0) = 0, \qquad\qquad i_2(0) = 0.$$

We introduce operator notation and write (4) in the form

$$\begin{cases} (0.01D + 20)i_1 + 0.01Di_2 = 100, \\[3mm] -20i_1 \;\; + (0.02D + 40)i_2 = 0. \end{cases} \qquad (5)$$

We apply the operator $0.02D + 40$ to the first equation of (5) and the operator $0.01D$ to the second and subtract to obtain

$$[(0.01D + 20)(0.02D + 40) + 0.2D]i_1 = (0.02D + 40)100$$

or $\qquad\qquad (.0002D^2 + D + 800)i_1 = 4000$

or finally

$$(D^2 + 5000D + 4,000,000)i_1 = 20,000,000. \qquad (6)$$

The auxiliary equation of the corresponding homogeneous D.E. is $m^2 + 5000m + 4,000,000 = 0$, with roots $m = -1000, -4000$. Using this information and the method of undetermined coefficients, we see that the solution of (6) is

$$i_1 = c_1 e^{-1000t} + c_2 e^{-4000t} + 5. \qquad (7)$$

We use the alternative procedure of page 276 of the text to find i_2. Returning to (5), we multiply the first equation by 2 and then subtract the second equation from this, obtaining

$$.02Di_1 + 60i_1 - 40i_2 = 200 .$$

Note that this contains i_2 but not Di_2. From it, we have

$$i_2 = 3i_1/2 + .0005Di_1 - 5 . \qquad (8)$$

From (7), we find

$$Di_1 = -1000c_1e^{-1000t} - 4000c_2e^{-4000t} .$$

Now substituting i_1 from (7) and Di_1 first obtained into (8) and simplifying, we obtain

$$i_2 = c_1e^{-1000t} - c_2e^{-4000t}/2 + 5/2 . \qquad (9)$$

Finally, applying the I.C.'s to (7) and (9), we have $c_1 + c_2 + 5 = 0$ and $c_1 - c_2/2 + 5/2 = 0$, from which we find $c_1 = -10/3$, $c_2 = -5/3$. Thus we obtain

$$\begin{cases} i_1 = -10e^{-1000t}/3 - 5e^{-4000t}/3 + 5, \\ i_2 = -10e^{-1000t}/3 + 5e^{-4000t}/6 + 5/2. \end{cases}$$

6. Let x = the amount of salt in tank X at time t, and let y = the amount of salt in tank Y at time t, each measured in kilograms. Each tank always contains 30

liters of fluid, so the concentration of salt in tank X
is x/30 (kg/liter) and that in tank Y is y/30 (kg/liter).
Salt enters tank X two ways: (a) 1 kg. of salt per
liter of brine enters at the rate of 2 liters/min.
from outside the system, and (b) salt in the brine
pumped from tank Y back to tank X at the rate of 1
liter/min. By (a), salt enters tank X at the rate of 2
kg./min.; and by (b), it enters at the rate of y/30
kg./min. Salt only leaves tank X in the brine flowing
from tank X into tank Y at the rate of 4 liters/min.
Thus salt leaves tank X at the rate of 4x/30 kg./min.
Hence we obtain the D.E.

$$\frac{dx}{dt} = 2 + \frac{y}{30} - \frac{4x}{30} \tag{1}$$

for the amount of salt in tank X at time t. The D.E.
for the amount in tank Y is obtained similarly. It is

$$\frac{dy}{dt} = \frac{4x}{30} - \frac{4y}{30} \ . \tag{2}$$

Since initially there was 30 kg. of salt in tank X and
pure water in tank Y, the I.C.'s are x(0) = 30,
y(0) = 0.

We introduce operator notation and write the D.E.'s
(1) and (2) in the forms

$$\begin{cases} (D + \tfrac{2}{15})x - \tfrac{1}{30}y = 2 \ , \\[2mm] - \tfrac{2}{15}x + (D + \tfrac{2}{15})y = 0 \ . \end{cases} \tag{3}$$

We apply the operator $(D + 2/15)$ to the first equation of (3), multiply the second by $1/30$, and add to obtain

$$[(D + \tfrac{2}{15})^2 - \tfrac{1}{225}]x = 4/15$$

or $$(D^2 + \tfrac{4}{15}D + \tfrac{3}{225})x = 4/15 \ . \tag{4}$$

The auxiliary equation of the corresponding homogeneous D.E. is

$$m^2 + 4m/15 + 3/225 = 0$$

with roots $m = -1/5, -1/15$. Using this information and the method of undetermined coefficients, we see that the solution of (4) is

$$x = c_1 e^{-t/5} + c_2 e^{-t/15} + 20. \tag{5}$$

To find y, we return to (3), multiply the first equation by $2/15$, apply the operator $(D + 2/15)$ to the second, and add. After simplification, we obtain

$$(D^2 + \tfrac{4}{15}D + \tfrac{3}{225})y = 4/15.$$

Comparing this with (4) and its solution (5), we see that its solution is

$$y = c_3 e^{-t/5} + c_4 e^{-t/15} + 20 \ . \tag{6}$$

Only two of the four constants c_1, c_2, c_3, c_4 in (5)

and (6) are independent. To determine the relations
which must exist among these, we substitute (5) and (6)
into the second equation of (3). We find

$$(-2c_1 - c_3)e^{-t/5}/15 + (-2c_2 + c_4)e^{-t/15}/15 = 0 \quad .$$

Hence we must have $c_3 = -2c_1$, $c_4 = 2c_2$. Thus we have

$$\begin{cases} x = c_1 e^{-t/5} + c_2 e^{-t/15} + 20 \quad , \\ y = -2c_1 e^{-t/5} + 2c_2 e^{-t/15} + 20 \quad . \end{cases}$$

Finally we apply the I.C.'s. We have $c_1 + c_2 + 20 = 30$,
$-2c_1 + 2c_2 + 20 = 0$, from which we find $c_1 = 10$, $c_2 = 0$.
Thus we obtain the solution

$$\begin{cases} x = 10e^{-t/5} + 20, \\ y = -20e^{-t/5} + 20 \quad . \end{cases}$$

Section 7.3, Page 300

4. The determinant Δ is defined by

$$\Delta(t) = \begin{vmatrix} f_1(t) & f_2(t) \\ g_1(t) & g_2(t) \end{vmatrix} = f_1(t)g_2(t) - f_2(t)g_1(t).$$

Differentiating, we find $\quad \Delta'(t) = f_1(t)g_2'(t)$
$+ f_1'(t)g_2(t) - f_2(t)g_1'(t) - f_2'(t)g_1(t). \qquad (1)$

Since $x = f_1(t)$, $y = g_1(t)$ is a solution of system

(7.67), we have $f_1'(t) = a_{11}(t)f_1(t) + a_{12}(t)g_1(t)$ and

$g_1'(t) = a_{21}(t)f_1(t) + a_{22}(t)g_1(t)$. Since $x = f_2(t)$,

$y = g_2(t)$ is a solution of system (7.67), we have

$f_2'(t) = a_{11}(t)f_2(t) + a_{12}(t)g_2(t)$ and

$g_2'(t) = a_{21}(t)f_2(t) + a_{22}(t)g_2(t)$. Substituting these

derivatives into (1), we find

$$\Delta'(t) = f_1(t)[a_{21}(t)f_2(t) + a_{22}(t)g_2(t)]$$
$$+ g_2(t)[a_{11}(t)f_1(t) + a_{12}(t)g_1(t)]$$
$$- f_2(t)[a_{21}(t)f_1(t) + a_{22}(t)g_1(t)]$$
$$- g_1(t)[a_{11}(t)f_2(t) + a_{12}(t)g_2(t)]$$
$$= a_{11}(t)[f_1(t)g_2(t) - f_2(t)g_1(t)]$$
$$+ a_{22}(t)[f_1(t)g_2(t) - f_2(t)g_1(t)]$$
$$= [a_{11}(t) + a_{22}(t)][f_1(t)g_2(t) - f_2(t)g_1(t)] .$$

But

$$f_1(t)g_2(t) - f_2(t)g_1(t) = \begin{vmatrix} f_1(t) & f_2(t) \\ g_1(t) & g_2(t) \end{vmatrix} = \Delta(t),$$

so we have proved $\Delta'(t) = [a_{11}(t) + a_{22}(t)] \Delta(t)$,

as required.

Section 7.4, Page 310

1. We assume a solution of the form $x = Ae^{\lambda t}$,

$y = Be^{\lambda t}$. Substituting into the given system, we obtain

$$\begin{cases} A\lambda e^{\lambda t} = 5Ae^{\lambda t} - 2Be^{\lambda t} \ , \\ B\lambda e^{\lambda t} = 4Ae^{\lambda t} - Be^{\lambda t} \ , \end{cases}$$

which leads at once to the algebraic system

$$\begin{cases} (5-\lambda)A - 2B = 0 \ , \\ 4A + (-1-\lambda)B = 0. \end{cases} \tag{1}$$

For nontrivial solutions of this, we must have

$$\begin{vmatrix} 5-\lambda & -2 \\ 4 & -1-\lambda \end{vmatrix} = 0.$$ This leads to the characteristic

equation $\lambda^2 - 4\lambda + 3 = 0$, whose roots are $\lambda_1 = 1$, $\lambda_2 = 3$.

Setting $\lambda = \lambda_1 = 1$ in (1), we obtain $4A - 2B = 0$ (twice), a nontrivial solution of which is $A = 1$, $B = 2$. With these values of A, B, and λ, we find the non-trivial solution $x = e^t$, $y = 2e^t$. $\tag{2}$

Setting $\lambda = \lambda_2 = 3$ in (1), we obtain $2A - 2B = 0$ and $4A - 4B = 0$, a nontrivial solution of which is $A = B = 1$. With these values of A, B, and λ, we find the nontrivial solution $x = e^{3t}$, $y = e^{3t}$. $\tag{3}$

The G. S. of the given system is a linear combination of (2) and (3). Thus we find $x = c_1 e^t + c_2 e^{3t}$, $y = 2c_1 e^t + c_2 e^{3t}$.

6. We assume a solution of the form $x = Ae^{\lambda t}$,

$y = Be^{\lambda t}$. Substituting into the given system, we obtain

$$\begin{cases} A\lambda e^{\lambda t} = 6Ae^{\lambda t} - Be^{\lambda t}, \\ B\lambda e^{\lambda t} = 3Ae^{\lambda t} + 2Be^{\lambda t}, \end{cases}$$

which leads at once to the algebraic system

$$\begin{cases} (6 - \lambda)A - B = 0, \\ 3A + (2 - \lambda)B = 0. \end{cases} \qquad (1)$$

For nontrivial solutions of this, we must have

$$\begin{vmatrix} 6 - \lambda & -1 \\ 3 & 2 - \lambda \end{vmatrix} = 0.$$ This leads at once to the

characteristic equation $\lambda^2 - 8\lambda + 15 = 0$, whose roots

are $\lambda_1 = 5, \lambda_2 = 3$.

Setting $\lambda = \lambda_1 = 5$ in (1), we obtain $A - B = 0$,

$3A - 3B = 0$, a nontrivial solution of which is $A = 1$,

$B = 1$. With these values of A, B, and λ, we find the

nontrivial solution $x = e^{5t}$, $y = e^{5t}$.   (2)

Setting $\lambda = \lambda_2 = 3$ in (1), we obtain $3A - B = 0$

(twice), a nontrivial solution of which is $A = 1$,

$B = 3$. With these values of A, B, and λ, we find the

nontrivial solution $x = e^{3t}$, $y = 3e^{3t}$.   (3)

The G. S. of the given system is a linear combina-

tion of (2) and (3). Thus we find $x = c_1 e^{5t} + c_2 e^{3t}$,

$y = c_1 e^{5t} + 3c_2 e^{3t}$.

8. We assume a solution of the form $x = Ae^{\lambda t}$,
$y = Be^{\lambda t}$. Substituting into the given system, we
obtain

$$\begin{cases} A\lambda e^{\lambda t} = 2Ae^{\lambda t} - Be^{\lambda t}, \\ B\lambda e^{\lambda t} = 9Ae^{\lambda t} + 2Be^{\lambda t}, \end{cases}$$

which leads at once to the algebraic system

$$\begin{cases} (2 - \lambda)A - B = 0, \\ 9A + (2 - \lambda)B = 0. \end{cases} \tag{1}$$

For nontrivial solutions of this, we must have

$$\begin{vmatrix} 2 - \lambda & -1 \\ 9 & 2 - \lambda \end{vmatrix} = 0.$$ This leads to the characteristic

equation $\lambda^2 - 4\lambda + 13 = 0$, whose roots are
$\lambda = 2 \pm 3i$.

Letting $\lambda = 2 + 3i$ in (1), we obtain $-3iA - B = 0$,
$9A - 3iB = 0$, a nontrivial solution of which is $A = 1$,
$B = -3i$. With these values of A, B, and λ, we find
the complex solution

$$\begin{cases} x = e^{(2+3i)t}, \\ y = -3ie^{(2+3i)t}. \end{cases}$$

Using Euler's formula this takes the form

$$\begin{cases} x = e^{2t}(\cos 3t + i \sin 3t), \\ y = e^{2t}(3 \sin 3t - 3i \cos 3t). \end{cases}$$

Both the real and imaginary parts of this are solutions.
Thus we obtain the two linearly independent solutions

$$
\begin{cases} x = e^{2t} \cos 3t \\ y = 3e^{2t} \sin 3t \end{cases} \quad \text{and} \quad \begin{cases} x = e^{2t} \sin 3t \\ y = -3e^{2t} \cos 3t \ . \end{cases}
$$

The G.S. of the system is thus

$$
\begin{cases} x = e^{2t}(c_1 \cos 3t + c_2 \sin 3t), \\ y = e^{2t}(3c_1 \sin 3t - 3c_2 \cos 3t) \ . \end{cases}
$$

11. We assume a solution of the form $x = Ae^{\lambda t}$,
$y = Be^{\lambda t}$. Substituting into the given system, we
obtain

$$
\begin{cases} A\lambda e^{\lambda t} = 3Ae^{\lambda t} - Be^{\lambda t} \ , \\ B\lambda e^{\lambda t} = 4Ae^{\lambda t} - Be^{\lambda t} \ , \end{cases}
$$

which leads at once to the algebraic system

$$
\begin{cases} (3 - \lambda)A - B = 0 \ , \\ 4A + (-1 - \lambda)B = 0 \ . \end{cases} \tag{1}
$$

For nontrivial solutions of this, we must have

$$
\begin{vmatrix} 3 - \lambda & -1 \\ 4 & -1 - \lambda \end{vmatrix} = 0.
$$ This leads to the characteristic equation $\lambda^2 - 2\lambda + 1 = 0$, whose roots are 1, 1
(real and equal).

Setting $\lambda = 1$ in (1), we obtain $2A - B = 0$,
$4A - 2B = 0$, a nontrivial solution of which is $A = 1$,

B = 2. With these values of A, B, and λ , we find the
nontrivial solution x = e^t, y = $2e^t$.   (2)

We now seek a second solution of the form
x = $(A_1 t + A_2)e^t$, y = $(B_1 t + B_2)e^t$. Substituting
these into the given system, we obtain

$$\begin{cases} (A_1 t + A_2 + A_1)e^t = 3(A_1 t + A_2)e^t - (B_1 t + B_2)e^t , \\ (B_1 t + B_2 + B_1)e^t = 4(A_1 t + A_2)e^t - (B_1 t + B_2)e^t . \end{cases}$$

These equations reduce to

$$\begin{cases} (2A_1 - B_1)t + (2A_2 - A_1 - B_2) = 0 , \\ (4A_1 - 2B_1)t + (4A_2 - B_1 - 2B_2) = 0 . \end{cases}$$

Thus we must have

$$\begin{cases} 2A_1 - B_1 = 0, \quad 2A_2 - A_1 - B_2 = 0, \\ 4A_1 - 2B_1 = 0, \quad 4A_2 - B_1 - 2B_2 = 0 . \end{cases}$$

A simple nontrivial solution of these is $A_1 = 1$, $B_1 = 2$,
$A_2 = 0$, $B_2 = -1$. This leads to the solution
x = te^t, y = $(2t - 1)e^t$.   (3)

The G. S. of the given system is a linear combination of (2) and (3). Thus we find x = $c_1 e^t + c_2 te^t$,
y = $2c_1 e^t + c_2(2t - 1)e^t$.

14. We assume a solution of the form x = $Ae^{\lambda t}$,

$y = Be^{\lambda t}$. Substituting into the given system, we obtain

$$\begin{cases} A\lambda e^{\lambda t} = Ae^{\lambda t} - 2Be^{\lambda t}, \\ B\lambda e^{\lambda t} = 2Ae^{\lambda t} - 3Be^{\lambda t}, \end{cases}$$

which leads at once to

the algebraic system $\begin{cases} (1 - \lambda)A - 2B = 0, \\ 2A + (-3 - \lambda)B = 0. \end{cases}$ (1)

For nontrivial solutions of this, we must have

$$\begin{vmatrix} 1 - \lambda & -2 \\ 2 & -3 - \lambda \end{vmatrix} = 0.$$ This leads to the characteris-

tic equation $\lambda^2 + 2\lambda + 1 = 0$ whose roots are -1, -1 (real and equal).

Setting $\lambda = -1$ in (1), we obtain $2A - 2B = 0$ (twice), a nontrivial solution of which is $A = 1$, $B = 1$. With these values of A, B, and λ , we find the nontrivial solution $x = e^{-t}$, $y = e^{-t}$.   (2)

We now seek a second solution of the form $x = (A_1 t + A_2)e^{-t}$, $y = (B_1 t + B_2)e^{-t}$. Substituting these into the given system, we obtain

$$\begin{cases} (-A_1 t - A_2 + A_1)e^{-t} = (A_1 t + A_2)e^{-t} - 2(B_1 t + B_2)e^{-t}, \\ (-B_1 t - B_2 + B_1)e^{-t} = 2(A_1 t + A_2)e^{-t} - 3(B_1 t + B_2)e^{-t}. \end{cases}$$

These equations reduce to

$$\begin{cases} (2A_1 - 2B_1)t + (2A_2 - 2B_2 - A_1) = 0 , \\ (2A_1 - 2B_1)t + (2A_2 - 2B_2 - B_1) = 0 . \end{cases}$$

Thus we must have

$$\begin{cases} 2A_1 - 2B_1 = 0, & 2A_2 - 2B_2 - A_1 = 0, \\ 2A_1 - 2B_1 = 0, & 2A_2 - 2B_2 - B_1 = 0. \end{cases}$$

A simple nontrivial solution of these is $A_1 = 2$, $B_1 = 2$, $A_2 = 1$, $B_2 = 0$. This leads to the solution

$$x = (2t + 1)e^{-t}, \quad y = 2te^{-t}. \qquad (3)$$

The G.S. of the given system is a linear combination of (2) and (3). Thus we find

$$x = c_1 e^{-t} + c_2(2t + 1)e^{-t}, \quad y = c_1 e^{-t} + 2c_2 te^{-t}.$$

15. We assume a solution of the form $x = Ae^{\lambda t}$, $y = Be^{\lambda t}$. Substituting into the given system, we obtain

$$A\lambda e^{\lambda t} = Ae^{\lambda t} - 4Be^{\lambda t},$$
$$B\lambda e^{\lambda t} = Ae^{\lambda t} + Be^{\lambda t}, \text{ which leads at}$$

once to the algebraic system

$$\begin{cases} (1 - \lambda)A - 4B = 0, \\ A + (1 - \lambda)B = 0. \end{cases} \qquad (1)$$

For nontrivial solutions of this, we must have

$$\begin{vmatrix} 1 - \lambda & -4 \\ 1 & 1 - \lambda \end{vmatrix} = 0.$$ This leads to the characteris-

tic equation $\lambda^2 - 2\lambda + 5 = 0$, whose roots are $1 \pm 2i$.

Letting $\lambda = 1 + 2i$ in (1), we obtain $-2iA - 4B = 0$, $A - 2iB = 0$, a nontrivial solution of which is $A = 2i$,

$B = 1$. With these values of A, B, and λ , we find
the complex solution

$$
\begin{cases}
x = 2ie^{(1 + 2i)t} , \\
y = e^{(1 + 2i)t} .
\end{cases}
$$

Using Euler's formula this takes the form

$$
\begin{cases}
x = 2ie^t (\cos 2t + i \sin 2t), \\
y = e^t(\cos 2t + i \sin 2t).
\end{cases}
$$

Both the real and imaginary parts of this are solutions.
Thus we obtain the two linearly independent solutions

$$
\begin{cases}
x = -2e^t \sin 2t, \\
y = e^t \cos 2t.
\end{cases}
\quad \text{and} \quad
\begin{cases}
x = 2e^t \cos 2t, \\
y = e^t \sin 2t.
\end{cases}
$$

The G.S. of the system is thus

$$
\begin{cases}
x = 2e^t(-c_1 \sin 2t + c_2 \cos 2t), \\
y = e^t(c_1 \cos 2t + c_2 \sin 2t).
\end{cases}
$$

22. We assume a solution of the form $x = Ae^{\lambda t}$,
$y = Be^{\lambda t}$. Substituting into the given system, we
obtain

$$
\begin{cases}
A\lambda e^{\lambda t} = 6Ae^{\lambda t} - 5Be^{\lambda t}, \\
B\lambda e^{\lambda t} = Ae^{\lambda t} + 2Be^{\lambda t},
\end{cases}
$$

which leads at once to the algebraic system

$$\begin{cases} (6 - \lambda)A - 5B = 0, \\ A + (2 - \lambda)B = 0. \end{cases} \quad (1)$$

For nontrivial solutions of this, we must have

$\begin{vmatrix} 6 - \lambda & -5 \\ 1 & 2 - \lambda \end{vmatrix} = 0.$ This leads to the characteristic

equation $\lambda^2 - 8\lambda + 17 = 0$, whose roots are

$\lambda = 4 \pm i.$

Letting $\lambda = 4 + i$ in (1), we obtain $(2-i)A - 5B = 0$, $A - (2+i)B = 0$, a nontrivial solution of which is $A = -5$, $B = i - 2$. With these values of A, B, and λ , we find the complex solution

$$\begin{cases} x = -5e^{(4+i)t} \\ y = (i-2)e^{(4+i)t}. \end{cases}$$

Using Euler's formula this takes the form

$$\begin{cases} x = -5e^{4t}(\cos t + i \sin t), \\ y = (i-2)e^{4t}(\cos t + i \sin t). \end{cases}$$

Both real and imaginary parts of this are solutions.
Thus we obtain the two linearly independent solutions

$$\begin{cases} x = -5e^{4t} \cos t, \\ y = e^{4t}(-2 \cos t - \sin t), \end{cases} \text{ and } \begin{cases} x = -5e^{4t} \sin t, \\ y = e^{4t}(\cos t - 2 \sin t) \end{cases}$$

The G.S. of the system is thus

$$\begin{cases} x = -5e^{4t}(c_1 \cos t + c_2 \sin t), \\ y = e^{4t}[(c_2 - 2c_1)\cos t - (c_1 + 2c_2)\sin t]. \end{cases}$$

23. We assume a solution of the form $x = Ae^{\lambda t}$, $y = Be^{\lambda t}$. Substituting into the given system, we obtain

$$\begin{cases} A\lambda e^{\lambda t} = -2Ae^{\lambda t} + 7Be^{\lambda t}, \\ B\lambda e^{\lambda t} = 3Ae^{\lambda t} + 2Be^{\lambda t}, \end{cases}$$

which leads at once to the algebraic system

$$\begin{cases} (-2 - \lambda)A + 7B = 0, \\ 3A + (2 - \lambda)B = 0. \end{cases} \quad (1)$$

For nontrivial solutions of this, we must have

$$\begin{vmatrix} -2 - \lambda & 7 \\ 3 & 2 - \lambda \end{vmatrix} = 0.$$ This leads to the characteristic equation $\lambda^2 - 25 = 0$, whose roots are $\lambda_1 = 5$, $\lambda_2 = -5$.

Setting $\lambda = \lambda_1 = 5$ in (1), we obtain $-7A + 7B = 0$, $3A - 3B = 0$, a nontrivial solution of which is $A = 1$, $B = 1$. With these values of A, B, and λ, we find the nontrivial solution $x = e^{5t}$, $y = e^{5t}$.   (2)

Setting $\lambda = \lambda_2 = -5$ in (1), we obtain $3A + 7B = 0$ (twice), a nontrivial solution of which is $A = 7$, $B = -3$. With these values of A, B, and λ, we find the non trivial solution $x = 7e^{-5t}$, $y = -3e^{-5t}$.   (3)

The G.S. of the given system is a linear combination of (2) and (3). Thus we find the G.S. in the form

$$x = c_1 e^{5t} + 7c_2 e^{-5t}, \qquad y = c_1 e^{5t} - 3c_2 e^{-5t}.$$ (4)

We apply the I.C.'s $x(0) = 9$, $y(0) = -1$ to (4), obtaining $c_1 + 7c_2 = 9$, $c_1 - 3c_2 = -1$. The solution of this is $c_1 = 2$, $c_2 = 1$. Substituting these values back into (4), we get the desired particular solution

$$x = 2e^{5t} + 7e^{-5t}, \qquad y = 2e^{5t} - 3e^{-5t} .$$

25. We assume a solution of the form $x = Ae^{\lambda t}$, $y = Be^{\lambda t}$. Substituting into the given system, we obtain

$$\begin{cases} A \lambda e^{\lambda t} = 6Ae^{\lambda t} - 4Be^{\lambda t}, \\ B \lambda e^{\lambda t} = Ae^{\lambda t} + 2Be^{\lambda t}, \end{cases}$$

which leads at once to the algebraic system

$$\begin{cases} (6 - \lambda)A - 4B = 0, \\ A + (2 - \lambda)B = 0 . \end{cases}$$ (1)

For nontrivial solutions of this, we must have

$$\begin{vmatrix} 6 - \lambda & -4 \\ 1 & 2 - \lambda \end{vmatrix} = 0.$$ This leads to the characteris-

tic equation $\lambda^2 - 8\lambda + 16 = 0$, whose roots are 4, 4 (real and equal).

Setting $\lambda = 4$ in (1), we obtain $2A - 4B = 0$, $A - 2B = 0$, a nontrivial solution of which is $A = 2$, $B = 1$. With these values of A, B, and λ, we find the nontrivial solution $x = 2e^{4t}$, $y = e^{4t}$. (2)

We now seek a second solution of the form $x = (A_1 t + A_2)e^{4t}$, $y = (B_1 t + B_2)e^{4t}$. Substituting these into the given system, we obtain

$$\begin{cases} (4A_1 t + 4A_2 + A_1)e^{4t} = 6(A_1 t + A_2)e^{4t} - 4(B_1 t + B_2)e^{4t}, \\ (4B_1 t + 4B_2 + B_1)e^{4t} = (A_1 t + A_2)e^{4t} + 2(B_1 t + B_2)e^{4t}. \end{cases}$$

These equations reduce to

$$\begin{cases} (2A_1 - 4B_1)t + (2A_2 - 4B_2 - A_1) = 0, \\ (A_1 - 2B_1)t + (A_2 - 2B_2 - B_1) = 0. \end{cases}$$

Thus we must have

$$\begin{cases} 2A_1 - 4B_1 = 0, \\ A_1 - 2B_1 = 0, \end{cases} \qquad \begin{cases} 2A_2 - 4B_2 - A_1 = 0, \\ A_2 - 2B_2 - B_1 = 0. \end{cases}$$

A simple nontrivial solution of this is $A_1 = 2$, $B_1 = 1$, $A_2 = 1$, $B_2 = 0$. This leads to the solution

$$x = (2t + 1)e^{4t}, \qquad y = te^{4t}. \tag{3}$$

The G.S. of the given system is a linear combination

of (2) and (3). Thus we find the G.S. in the form

$$x = 2c_1 e^{4t} + c_2(2t + 1)e^{4t}, \quad y = c_1 e^{4t} + c_2 te^{4t}. \quad (4)$$

We apply the I.C.'s $x(0) = 2$, $y(0) = 3$ to (4), obtaining

$2c_1 + c_2 = 2$, $c_1 + 0 = 3$. Thus $c_1 = 3$, $c_2 = -4$.

Substituting these values back into (4) we get the

desired particular solution $x = 6e^{4t} - 4(2t + 1)e^{4t}$,

$y = 3e^{4t} - 4te^{4t}$; or $x = 2e^{4t} - 8te^{4t}$, $y = 3e^{4t} - 4te^{4t}$.

27. We assume a solution of the form $x = Ae^{\lambda t}$,

$y = Be^{\lambda t}$. Substituting into the given system, we

obtain

$$\begin{cases} A\lambda e^{\lambda t} = 2Ae^{\lambda t} - 8Be^{\lambda t}, \\ B\lambda e^{\lambda t} = Ae^{\lambda t} + 6Be^{\lambda t}, \end{cases}$$

which leads at once to the algebraic system

$$\begin{cases} (2 - \lambda)A - 8B = 0, \\ A + (6 - \lambda)B = 0. \end{cases} \quad (1)$$

For nontrivial solutions of this, we must have

$$\begin{vmatrix} 2 - \lambda & -8 \\ 1 & 6 - \lambda \end{vmatrix} = 0.$$ This leads to the characteristic

equation $\lambda^2 - 8\lambda + 20 = 0$, whose roots are $4 \pm 2i$.

Letting $\lambda = 4 + 2i$ in (1), we obtain

$(-2 - 2i)A - 8B = 0$, $A + (2 - 2i)B = 0$, a nontrivial

solution of which is $A = 2(i - 1)$, $B = 1$. With these

values of A, B, and λ, we find the complex solution

$$\begin{cases} x = 2(i - 1)e^{(4 + 2i)t} \\ y = e^{(4 + 2i)t} \end{cases}.$$

Using Euler's formula this takes the form

$$\begin{cases} x = 2(i - 1)e^{4t}(\cos 2t + i \sin 2t), \\ y = e^{4t}(\cos 2t + i \sin 2t). \end{cases}$$

Both real and imaginary parts of this are solutions.
Thus we obtain the two linearly independent solutions

$$\begin{cases} x = -2e^{4t}(\cos 2t + \sin 2t), \\ y = e^{4t}\cos 2t, \end{cases} \quad \text{and} \quad \begin{cases} x = 2e^{4t}(\cos 2t - \sin 2t), \\ y = e^{4t}\sin 2t. \end{cases}$$

The G.S. of the system is thus

$$\begin{cases} x = 2e^{4t}[(-c_1 + c_2)\cos 2t - (c_1 + c_2)\sin 2t], \\ y = e^{4t}(c_1 \cos 2t + c_2 \sin 2t). \end{cases} \quad (2)$$

We apply the I.C.'s $x(0) = 4$, $y(0) = 1$ to (2), obtaining
$2(-c_1 + c_2) = 4$, $c_1 = 1$. Thus $c_1 = 1$, $c_2 = 3$.
Substituting back into (2), we get the desired particular solution

$$\begin{cases} x = 4e^{4t}(\cos 2t - 2 \sin 2t), \\ y = e^{4t}(\cos 2t + 3 \sin 2t). \end{cases}$$

30. We let $t = e^w$, where we assume $t > 0$. Then
$w = \ln t$, and $dx/dt = (dx/dw)(dw/dt) = (1/t)dx/dw$. Thus

t dx/dt = dx/dw; and similarly t dy/dt = dy/dw.
Substituting into the original system, we obtain the
constant coefficient system

$$\begin{cases} \dfrac{dx}{dw} = x + y, \\[2mm] \dfrac{dy}{dw} = -3x + 5y. \end{cases} \tag{1}$$

We assume a solution of (1) of the form $x = Ae^{\lambda w}$,
$y = Be^{\lambda w}$. Substituting this into (1), we obtain

$$\begin{cases} A\lambda e^{\lambda w} = Ae^{\lambda w} + Be^{\lambda w}, \\[2mm] B\lambda e^{\lambda w} = -3Ae^{\lambda w} + 5Be^{\lambda w}, \end{cases}$$

which leads at once to the algebraic system

$$\begin{cases} (1 - \lambda)A + B = 0, \\ -3A + (5 - \lambda)B = 0. \end{cases} \tag{2}$$

For nontrivial solutions of this, we must have

$\begin{vmatrix} 1 - \lambda & 1 \\ -3 & 5 - \lambda \end{vmatrix} = 0.$ This leads to the characteris-

tic equation $\lambda^2 - 6\lambda + 8 = 0$, whose roots are
$\lambda_1 = 2$, $\lambda_2 = 4$.

Setting $\lambda = \lambda_1 = 2$ in (2), we obtain
$-A + B = 0$, $-3A + 3B = 0$, a nontrivial solution of
which is $A = 1$, $B = 1$. With these values of A, B, and
λ, we find the nontrivial solution of (1), $x = e^{2w}$,
$y = e^{2w}$.

Setting $\lambda = \lambda_2 = 4$ in (2), we obtain $-3A + B = 0$
(twice), a nontrivial solution of which is $A = 1$, $B = 3$.
With these values of A, B, and λ , we find the nontrivial solution of (1), $x = e^{4w}$, $y = 3e^{4w}$.

The G.S. of (1) is thus $x = c_1 e^{2w} + c_2 e^{4w}$,
$y = c_1 e^{2w} + 3c_2 e^{4w}$. We must now return to the original
variable t. Since $t = e^{w}$, we see that the G.S. of the
original system is $x = c_1 t^2 + c_2 t^4$, $y = c_1 t^2 + 3c_2 t^4$.

Section 7.5A, Page 321

3. c. We have

$$-\bar{x}_1 + 5\bar{x}_2 - 2\bar{x}_3 + 3\bar{x}_4 = -\begin{pmatrix} 3 \\ 2 \\ -1 \\ 4 \end{pmatrix} + 5\begin{pmatrix} -1 \\ 3 \\ 5 \\ -2 \end{pmatrix} - 2\begin{pmatrix} 2 \\ 4 \\ 0 \\ 6 \end{pmatrix} + 3\begin{pmatrix} -1 \\ 2 \\ -3 \\ 5 \end{pmatrix}$$

$$= \begin{pmatrix} -3 \\ -2 \\ 1 \\ -4 \end{pmatrix} + \begin{pmatrix} -5 \\ 15 \\ 25 \\ -10 \end{pmatrix} + \begin{pmatrix} -4 \\ -8 \\ 0 \\ -12 \end{pmatrix} + \begin{pmatrix} -3 \\ 6 \\ -9 \\ 15 \end{pmatrix} = \begin{pmatrix} -15 \\ 11 \\ 17 \\ -11 \end{pmatrix} .$$

4. c. We have

$$\begin{pmatrix} 1 & 0 & -3 \\ 2 & -5 & 4 \\ -3 & 1 & 2 \end{pmatrix} \begin{pmatrix} x_1 + x_2 \\ x_1 + 2x_2 \\ x_2 - x_3 \end{pmatrix} = \begin{pmatrix} (x_1 + x_2) + 0 - 3(x_2 - x_3) \\ 2(x_1 + x_2) - 5(x_1 + 2x_2) + 4(x_2 - x_3) \\ -3(x_1 + x_2) + (x_1 + 2x_2) + 2(x_2 - x_3) \end{pmatrix}$$

$$= \begin{pmatrix} x_1 - 2x_2 + 3x_3 \\ -3x_1 - 4x_2 - 4x_3 \\ -2x_1 + x_2 - 2x_3 \end{pmatrix} .$$

Section 7.5B, Page 331

Here and following, we shall indicate vectors and matrices by placing a bar over the corresponding letter. Thus, for example, we write $\overline{A}$, $\overline{v}$, etc.

1. $\overline{A}\,\overline{B} = \begin{pmatrix} 3 & 5 \\ 1 & 7 \end{pmatrix} \begin{pmatrix} 4 & 6 \\ 2 & 1 \end{pmatrix}$

$$= \begin{pmatrix} (3)(4) + (5)(2) & (3)(6) + (5)(1) \\ (1)(4) + (7)(2) & (1)(6) + (7)(1) \end{pmatrix} = \begin{pmatrix} 22 & 23 \\ 18 & 13 \end{pmatrix}.$$

$$\overline{B}\,\overline{A} = \begin{pmatrix} 4 & 6 \\ 2 & 1 \end{pmatrix} \begin{pmatrix} 3 & 5 \\ 1 & 7 \end{pmatrix} = \begin{pmatrix} (4)(3) + (6)(1) & (4)(5) + (6)(7) \\ (2)(3) + (1)(1) & (2)(5) + (1)(7) \end{pmatrix}$$

$$= \begin{pmatrix} 18 & 62 \\ 7 & 17 \end{pmatrix} .$$

6. $\bar{A}\,\bar{B} = \begin{pmatrix} 3 & 2 & 1 \\ 0 & 1 & 2 \\ 5 & 4 & 3 \end{pmatrix} \begin{pmatrix} 2 & -1 & -3 \\ -6 & 0 & 1 \\ 1 & -3 & 4 \end{pmatrix}$

$= \begin{pmatrix} (3)(2)+(2)(-6)+(1)(1) & (3)(-1)+(2)(0)+(1)(-3) \\ (0)(2)+(1)(-6)+(2)(1) & (0)(-1)+(1)(0)+(2)(-3) \\ (5)(2)+(4)(-6)+(3)(1) & (5)(-1)+(4)(0)+(3)(-3) \end{pmatrix}$

$\begin{matrix} (3)(-3)+(2)(1)+(1)(4) \\ (0)(-3)+(1)(1)+(2)(4) \\ (5)(-3)+(4)(1)+(3)(4) \end{matrix}\Big) = \begin{pmatrix} -5 & -6 & -3 \\ -4 & -6 & 9 \\ -11 & -14 & 1 \end{pmatrix}$.

$\bar{B}\,\bar{A} = \begin{pmatrix} 2 & -1 & -3 \\ -6 & 0 & 1 \\ 1 & -3 & 4 \end{pmatrix} \begin{pmatrix} 3 & 2 & 1 \\ 0 & 1 & 2 \\ 5 & 4 & 3 \end{pmatrix}$

$= \begin{pmatrix} (2)(3)+(-1)(0)+(-3)(5) & (2)(2)+(-1)(1)+(-3)(4) \\ (-6)(3)+(0)(0)+(1)(5) & (-6)(2)+(0)(1)+(1)(4) \\ (1)(3)+(-3)(0)+(4)(5) & (1)(2)+(-3)(1)+(4)(4) \end{pmatrix}$

$\begin{matrix} (2)(1)+(-1)(2)+(-3)(3) \\ (-6)(1)+(0)(2)+(1)(3) \\ (1)(1)+(-3)(2)+(4)(3) \end{matrix}\Big) = \begin{pmatrix} -9 & -9 & -9 \\ -13 & -8 & -3 \\ 23 & 15 & 7 \end{pmatrix}$.

9. $\bar{A}\,\bar{B} = \begin{pmatrix} 2 & 1 & 0 & 1 \\ 0 & 2 & 3 & -1 \\ 1 & -2 & 1 & 0 \end{pmatrix} \begin{pmatrix} 1 & 2 \\ 0 & 3 \\ -1 & 0 \\ 1 & -2 \end{pmatrix}$

$$= \begin{pmatrix} (2)(1)+(\ 1)(0)+(0)(-1)+(1)(1) \\ (0)(1)+(\ 2)(0)+(3)(-1)+(-1)(1) \\ (1)(1)+(-2)(0)+(1)(-1)+(0)(1) \end{pmatrix}$$

$$\begin{pmatrix} (2)(2)+(\ 1)(3)+(0)(0)+(\ 1)(-2) \\ (0)(2)+(\ 2)(3)+(3)(0)+(-1)(-2) \\ (1)(2)+(-2)(3)+(1)(0)+(0)(-2) \end{pmatrix} = \begin{pmatrix} 3 & 5 \\ -4 & 8 \\ 0 & -4 \end{pmatrix}.$$

Since the number of columns in $\overline{B}$, 2, is unequal to the number of rows in $\overline{A}$, 3, the product $\overline{B}\,\overline{A}$ is not defined.

12. We have

$$\overline{A}^2 = \overline{A}\,\overline{A} = \begin{pmatrix} 2 & 3 & 3 \\ 1 & -2 & 1 \\ -3 & -1 & 0 \end{pmatrix}\begin{pmatrix} 2 & 3 & 3 \\ 1 & -2 & 1 \\ -3 & -1 & 0 \end{pmatrix}$$

$$= \begin{pmatrix} (2)(2)+(\ 3)(1)+(3)(-3) & (2)(3)+(\ 3)(-2)+(3)(-1) \\ (1)(2)+(-2)(1)+(1)(-3) & (1)(3)+(-2)(-2)+(1)(-1) \\ (-3)(2)+(-1)(1)+(0)(-3) & (-3)(3)+(-1)(-2)+(0)(-1) \end{pmatrix}$$

$$\begin{pmatrix} (2)(3)+(\ 3)(1)+(3)(0) \\ (1)(3)+(-2)(1)+(1)(0) \\ (-3)(3)+(-1)(1)+(0)(0) \end{pmatrix} = \begin{pmatrix} -2 & -3 & 9 \\ -3 & 6 & 1 \\ -7 & -7 & -10 \end{pmatrix}.$$

Then

$$\overline{A}^2 + 3\overline{A} + 2\overline{I} = \begin{pmatrix} -2 & -3 & 9 \\ -3 & 6 & 1 \\ -7 & -7 & -10 \end{pmatrix} + \begin{pmatrix} 6 & 9 & 9 \\ 3 & -6 & 3 \\ -9 & -3 & 0 \end{pmatrix} + \begin{pmatrix} 2 & 0 & 0 \\ 0 & 2 & 0 \\ 0 & 0 & 2 \end{pmatrix}$$

$$= \begin{pmatrix} 6 & 6 & 18 \\ 0 & 2 & 4 \\ -16 & -10 & -8 \end{pmatrix} .$$

13. $\bar{A} = \begin{pmatrix} 1 & 3 \\ 2 & 5 \end{pmatrix}$. We have $\text{cof } \bar{A} = \begin{pmatrix} 5 & -2 \\ -3 & 1 \end{pmatrix}$;

$\text{adj } \bar{A} = (\text{cof } \bar{A})^T = \begin{pmatrix} 5 & -3 \\ -2 & 1 \end{pmatrix}$; and $|\bar{A}| = \begin{vmatrix} 1 & 3 \\ 2 & 5 \end{vmatrix} = -1.$

Thus
$$\bar{A}^{-1} = \frac{1}{|\bar{A}|} (\text{adj } \bar{A}) = \begin{pmatrix} -5 & 3 \\ 2 & -1 \end{pmatrix} .$$

17. We find

$$\text{cof } \bar{A} = \begin{pmatrix} \begin{vmatrix} 3 & 2 \\ 1 & 1 \end{vmatrix} & -\begin{vmatrix} 3 & 2 \\ -1 & 1 \end{vmatrix} & \begin{vmatrix} 3 & 3 \\ -1 & 1 \end{vmatrix} \\ -\begin{vmatrix} 3 & 1 \\ 1 & 1 \end{vmatrix} & \begin{vmatrix} 4 & 1 \\ -1 & 1 \end{vmatrix} & -\begin{vmatrix} 4 & 3 \\ -1 & 1 \end{vmatrix} \\ \begin{vmatrix} 3 & 1 \\ 3 & 2 \end{vmatrix} & -\begin{vmatrix} 4 & 1 \\ 3 & 2 \end{vmatrix} & \begin{vmatrix} 4 & 3 \\ 3 & 3 \end{vmatrix} \end{pmatrix}$$

$$= \begin{pmatrix} 1 & -5 & 6 \\ -2 & 5 & -7 \\ 3 & -5 & 3 \end{pmatrix} ; \quad \text{adj } \bar{A} = (\text{cof } \bar{A})^T$$

$$= \begin{pmatrix} 1 & -2 & 3 \\ -5 & 5 & -5 \\ 6 & -7 & 3 \end{pmatrix} ; \text{ and } |\bar{A}| = \begin{vmatrix} 4 & 3 & 1 \\ 3 & 3 & 2 \\ -1 & 1 & 1 \end{vmatrix} = -5 .$$

Thus

$$\bar{A}^{-1} = \frac{1}{|\bar{A}|} (\text{adj } \bar{A}) = \begin{pmatrix} -1/5 & 2/5 & -3/5 \\ 1 & -1 & 1 \\ -6/5 & 7/5 & -3/5 \end{pmatrix} \quad .$$

    22. We have

$$\text{cof } \bar{A} = \begin{pmatrix} \begin{vmatrix} 7 & 1 \\ 3 & 3 \end{vmatrix} & -\begin{vmatrix} 3 & 1 \\ 1 & 3 \end{vmatrix} & \begin{vmatrix} 3 & 7 \\ 1 & 3 \end{vmatrix} \\ -\begin{vmatrix} 2 & 0 \\ 3 & 3 \end{vmatrix} & \begin{vmatrix} 1 & 0 \\ 1 & 3 \end{vmatrix} & -\begin{vmatrix} 1 & 2 \\ 1 & 3 \end{vmatrix} \\ \begin{vmatrix} 2 & 0 \\ 7 & 1 \end{vmatrix} & -\begin{vmatrix} 1 & 0 \\ 3 & 1 \end{vmatrix} & \begin{vmatrix} 1 & 2 \\ 3 & 7 \end{vmatrix} \end{pmatrix}$$

$$= \begin{pmatrix} 18 & -8 & 2 \\ -6 & 3 & -1 \\ 2 & -1 & 1 \end{pmatrix} \; ; \; \text{adj } \bar{A} = (\text{cof } \bar{A})^T = \begin{pmatrix} 18 & -6 & 2 \\ -8 & 3 & -1 \\ 2 & -1 & 1 \end{pmatrix}$$

and $|\bar{A}| = \begin{vmatrix} 1 & 2 & 0 \\ 3 & 7 & 1 \\ 1 & 3 & 3 \end{vmatrix} = 2.$

Thus $\bar{A}^{-1} = \frac{1}{|\bar{A}|} (\text{adj } \bar{A}) = \begin{pmatrix} 9 & -3 & 1 \\ -4 & 3/2 & -1/2 \\ 1 & -1/2 & 1/2 \end{pmatrix} \quad .$

Section 7.5C, Page 336

    1. a. We must show there exist numbers c_1, c_2, c_3, not all zero, such that $c_1\bar{v}_1 + c_2\bar{v}_2 + c_3\bar{v}_3 = \bar{0}$. This is

$$c_1 \begin{pmatrix} 3 \\ -1 \\ 2 \end{pmatrix} + c_2 \begin{pmatrix} 13 \\ 5 \\ -4 \end{pmatrix} + c_3 \begin{pmatrix} 2 \\ 4 \\ -5 \end{pmatrix} = \begin{pmatrix} 0 \\ 0 \\ 0 \end{pmatrix} \ .$$

This is equivalent to the homogeneous linear system

$$\begin{cases} 3c_1 + 13c_2 + 2c_3 = 0 \ , \\ -c_1 + 5c_2 + 4c_3 = 0 \ , \\ 2c_1 - 4c_2 - 5c_3 = 0 \ , \end{cases} \tag{1}$$

the determinant of coefficients of which is

$$\begin{vmatrix} 3 & 13 & 2 \\ -1 & 5 & 4 \\ 2 & -4 & -5 \end{vmatrix} = 0 \ .$$

Hence, by Theorem A (text, page 332), the system (1) has a nontrivial solution for c_1, c_2, c_3. For example, one solution is $c_1 = 3$, $c_2 = -1$, $c_3 = 2$. Thus $\bar{v}_1$, $\bar{v}_2$, $\bar{v}_3$ are linearly dependent.

2. a. Here we must show that if $c_1\bar{v}_1 + c_2\bar{v}_2 + c_3\bar{v}_3 = \bar{0}$, then $c_1 = c_2 = c_3 = 0$. Thus, we suppose

$$c_1 \begin{pmatrix} 2 \\ 1 \\ 0 \end{pmatrix} + c_2 \begin{pmatrix} 1 \\ 0 \\ 3 \end{pmatrix} + c_3 \begin{pmatrix} 0 \\ -1 \\ 1 \end{pmatrix} = \begin{pmatrix} 0 \\ 0 \\ 0 \end{pmatrix} .$$

This is equivalent to the homogeneous linear system

$$\begin{cases} 2c_1 + c_2 = 0 \ , \\ c_1 - c_3 = 0 \ , \\ 3c_2 + c_3 = 0 \ , \end{cases} \tag{1}$$

the determinant of which is

$$\begin{vmatrix} 2 & 1 & 0 \\ 1 & 0 & -1 \\ 0 & 3 & 1 \end{vmatrix} = 5 \neq 0 \ .$$

Thus by Theorem A (text, page 332), the system (1) has only the trivial solution $c_1 = c_2 = c_3 = 0$, and so $\bar{v}_1$, $\bar{v}_2$, $\bar{v}_3$ are linearly independent.

  3. a. The given vectors are linearly dependent if and only if there exist numbers c_1, c_2, c_3, not zero, such that $c_1\bar{v}_1 + c_2\bar{v}_2 + c_3\bar{v}_3 = \bar{0}$.   (1)
This is

$$c_1 \begin{pmatrix} k \\ 2 \\ -1 \end{pmatrix} + c_2 \begin{pmatrix} 1 \\ -1 \\ 2 \end{pmatrix} + c_3 \begin{pmatrix} 3 \\ 7 \\ -8 \end{pmatrix} = \begin{pmatrix} 0 \\ 0 \\ 0 \end{pmatrix} \ .$$

This is equivalent to the homogeneous linear system

$$\begin{cases} kc_1 + c_2 + 3c_3 = 0 \ , \\ 2c_1 - c_2 + 7c_3 = 0 \ , \\ -c_1 + 2c_2 - 8c_3 = 0 \ , \end{cases} \tag{2}$$

the determinant of coefficients of which is

$$\begin{vmatrix} k & 1 & 3 \\ 2 & -1 & 7 \\ -1 & 2 & -8 \end{vmatrix} = -6k + 18. \qquad (3)$$

By Theorem A (text, page 332), the system (2) has a
nontrivial solution for c_1, c_2, c_3, and hence there
exist c_1, c_2, c_3, not all zero, such that (1) holds,
if and only if the determinant (3) is zero. Thus we
have $-6k + 18 = 0$, and hence $k = 3$.

   4. b. Note that

$$4\overline{\phi}_1(t) - 2\overline{\phi}_2(t) - \overline{\phi}_3(t)$$

$$= \begin{pmatrix} 4 \sin t + 4 \cos t \\ 8 \sin t \\ -4 \cos t \end{pmatrix} + \begin{pmatrix} -4 \sin t \\ -8 \sin t + 2 \cos t \\ 2 \sin t \end{pmatrix}$$

$$+ \begin{pmatrix} -4 \cos t \\ -2 \cos t \\ -2 \sin t + 4 \cos t \end{pmatrix} = \begin{pmatrix} 0 \\ 0 \\ 0 \end{pmatrix}.$$

Thus there
exists the set of three numbers 4, -2, -1, none of which
are zero, such that $4\overline{\phi}_1(t) + (-2)\overline{\phi}_2(t) + (-1)\overline{\phi}_3(t) = \overline{0}$
for all t such that $a \leq t \leq b$, for any a and b. There-
fore $\overline{\phi}_1$, $\overline{\phi}_2$, and $\overline{\phi}_3$ are linearly dependent on
$a \leq t \leq b$.

   5. b. Suppose there exist numbers c_1 and c_2 such

that $c_1\overline{\phi}_1(t) + c_2\overline{\phi}_2(t) = \overline{0}$ for all t on a $\leq$ t $\leq$ b.   Then

$$
\begin{cases}
2c_1 e^{2t} + c_2 e^{-t} = 0, \\
- c_1 e^{2t} + 3c_2 e^{-t} = 0, \text{ for all t on a } \leq t \leq b. \text{ From}
\end{cases}
$$

this, we have both

$$
\begin{cases}
2c_1 + c_2 e^{-3t} = 0, \\
-2c_1 + 6c_2 e^{-3t} = 0,
\end{cases}
\text{ and }
\begin{cases}
6c_1 e^{3t} + 3c_2 = 0, \\
c_1 e^{3t} - 3c_2 = 0.
\end{cases}
\quad (1)
$$

From the former of (1), $7c_2 e^{-3t} = 0$ on a $\leq$ t $\leq$ b, so
$c_2 = 0$; and from the latter of (1), $7c_1 e^{3t} = 0$ on
a $\leq$ t $\leq$ b, so $c_1 = 0$.   Thus if $c_1\overline{\phi}_1(t) + c_2\overline{\phi}_2(t) = \overline{0}$
for all t on a $\leq$ t $\leq$ b, we must have $c_1 = c_2 = 0$, and
hence $\overline{\phi}_1$ and $\overline{\phi}_2$ are linearly independent on a $\leq$ t $\leq$ b,
for any a and b.

Section 7.5D, Page 344

1.   The characteristic equation is $\begin{vmatrix} 1-\lambda & 2 \\ 3 & 2-\lambda \end{vmatrix} = 0$

or $\lambda^2 - 3\lambda - 4 = 0$ or $(\lambda + 1)(\lambda - 4) = 0$.
Thus the characteristic values are $\lambda = -1$ and 4.

The characteristic vectors corresponding to $\lambda = -1$
have components x_1 and x_2 such that

$\begin{pmatrix} 1 & 2 \\ 3 & 2 \end{pmatrix} \begin{pmatrix} x_1 \\ x_2 \end{pmatrix} = (-1) \begin{pmatrix} x_1 \\ x_2 \end{pmatrix}$.   Thus x_1 and x_2 must satisfy the system

$$\begin{cases} x_1 + 2x_2 = -x_1 \\ 3x_1 + 2x_2 = -x_2 \end{cases} \quad \text{or} \quad \begin{cases} 2x_1 + 2x_2 = 0 \\ 3x_1 + 3x_2 = 0 \end{cases}.$$

We find $x_1 = k$, $x_2 = -k$ for every real k. Hence the
characteristic vectors corresponding to $\lambda = -1$ are
$\begin{pmatrix} k \\ -k \end{pmatrix}$ for every nonzero real k.

The characteristic vectors corresponding to
$\lambda = 4$ have components x_1 and x_2 such that

$$\begin{pmatrix} 1 & 2 \\ 3 & 2 \end{pmatrix}\begin{pmatrix} x_1 \\ x_2 \end{pmatrix} = 4 \begin{pmatrix} x_1 \\ x_2 \end{pmatrix}.$$ Thus x_1 and x_2 must

satisfy the system

$$\begin{cases} x_1 + 2x_2 = 4x_1 \\ 3x_1 + 2x_2 = 4x_2 \end{cases} \quad \text{or} \quad \begin{cases} -3x_1 + 2x_2 = 0 \\ 3x_2 - 2x_2 = 0 \end{cases}.$$

We find $x_1 = 2k$, $x_2 = 3k$ for every real k. Hence the
characteristic vectors corresponding to $\lambda = 4$ are
$\begin{pmatrix} 2k \\ 3k \end{pmatrix}$ for every nonzero real k.

7. The characteristic equation is

$$\begin{vmatrix} 1-\lambda & 1 & -1 \\ 2 & 3-\lambda & -4 \\ 4 & 1 & -4-\lambda \end{vmatrix} = 0,$$

which reduces to

$$(1 - \lambda) \begin{vmatrix} 3 - \lambda & -4 \\ 1 & 4 - \lambda \end{vmatrix} - \begin{vmatrix} 2 & -4 \\ 4 & -4 - \lambda \end{vmatrix} - \begin{vmatrix} 2 & 3 - \lambda \\ 4 & 1 \end{vmatrix} = 0$$

or $\lambda^3 - 7\lambda + 6 = 0$ or $(\lambda - 1)(\lambda - 2)(\lambda + 3) = 0$.
Thus the characteristic values are $\lambda = 1, 2, -3$.

The characteristic vectors corresponding to $\lambda = 1$
have components x_1, x_2, x_3 such that

$$\begin{pmatrix} 1 & 1 & -1 \\ 2 & 3 & -4 \\ 4 & 1 & -4 \end{pmatrix} \begin{pmatrix} x_1 \\ x_2 \\ x_3 \end{pmatrix} = 1 \begin{pmatrix} x_1 \\ x_2 \\ x_3 \end{pmatrix}$$. Thus x_1, x_2, x_3

must satisfy the system

$$\begin{cases} x_1 + x_2 - x_3 = x_1 \\ 2x_1 + 3x_2 - 4x_3 = x_2, \\ 4x_1 + x_2 - 4x_3 = x_3, \end{cases} \quad \text{or} \quad \begin{cases} x_2 - x_3 = 0, \\ 2x_1 + 2x_2 - 4x_3 = 0, \\ 4x_1 + x_2 - 5x_3 = 0 . \end{cases}$$

From the first of these, $x_3 = x_2$; and then the second
and third become

$$\begin{cases} 2x_1 - 2x_2 = 0, \\ 4x_1 - 4x_2 = 0, \end{cases}$$

a solution of which is $x_1 = x_2 = k$.   Thus we find
$x_1 = k$, $x_2 = k$, $x_3 = k$ for every real k.   Hence the
characteristic vectors corresponding to $\lambda = 1$ are
$$\begin{pmatrix} k \\ k \\ k \end{pmatrix}$$ for every nonzero real k.

The characteristic vectors corresponding to $\lambda = 2$ have components x_1, x_2, x_3 such that

$$\begin{pmatrix} 1 & 1 & -1 \\ 2 & 3 & -4 \\ 4 & 1 & -4 \end{pmatrix} \begin{pmatrix} x_1 \\ x_2 \\ x_3 \end{pmatrix} = 2 \begin{pmatrix} x_1 \\ x_2 \\ x_3 \end{pmatrix}.$$ Thus x_1, x_2, x_3 must

satisfy the system

$$\begin{cases} x_1 + x_2 - x_3 = 2x_1, \\ 2x_1 + 3x_2 - 4x_3 = 2x_2, \\ 4x_1 + x_2 - 4x_3 = 2x_3 \end{cases} \text{ or } \begin{cases} -x_1 + x_2 - x_3 = 0, \\ 2x_1 + x_2 - 4x_3 = 0, \\ 4x_1 + x_2 - 6x_3 = 0. \end{cases}$$

From the first and second, $3x_1 - 3x_3 = 0$, so $x_3 = x_1$, and $3x_2 - 6x_3 = 0$, so $x_2 = 2x_3$. Thus we find $x_1 = k$, $x_2 = 2k$, $x_3 = k$ for every real k. Hence the characteristic vectors corresponding to $\lambda = 2$ are $\begin{pmatrix} k \\ 2k \\ k \end{pmatrix}$ for every nonzero real k.

The characteristic vectors corresponding to $\lambda = -3$ have components x_1, x_2, x_3 such that

$$\begin{pmatrix} 1 & 1 & -1 \\ 2 & 3 & -4 \\ 4 & 1 & -4 \end{pmatrix} \begin{pmatrix} x_1 \\ x_2 \\ x_3 \end{pmatrix} = -3 \begin{pmatrix} x_1 \\ x_2 \\ x_3 \end{pmatrix}.$$ Thus x_1, x_2, x_3 must

satisfy the system

$$\begin{cases} x_1 + x_2 - x_3 = -3x_1, \\ 2x_1 + 3x_2 - 4x_3 = -3x_2, \\ 4x_1 + x_2 - 4x_3 = -3x_3, \end{cases} \text{ or } \begin{cases} 4x_1 + x_2 - x_3 = 0, \\ 2x_1 + 6x_2 - 4x_3 = 0, \\ 4x_1 + x_2 - x_3 = 0 . \end{cases}$$

Regarding the first and second equations as two equations
in x_2 and x_3, we find $x_2 = 7x_1$, $x_3 = 11x_1$. Thus we
find $x_1 = k$, $x_2 = 7k$, $x_3 = 11k$ for every real k. Hence
the characteristic vectors corresponding to $\lambda = -3$ are
$\begin{pmatrix} k \\ 7k \\ 11k \end{pmatrix}$ for every nonzero real k.

10. The characteristic equation is

$$\begin{vmatrix} 1-\lambda & 1 & 0 \\ 1 & -\lambda & 1 \\ 0 & 1 & 1-\lambda \end{vmatrix} = 0 \text{ which reduces to}$$

$$(1-\lambda) \begin{vmatrix} -\lambda & 1 \\ 1 & 1-\lambda \end{vmatrix} - \begin{vmatrix} 1 & 1 \\ 0 & 1-\lambda \end{vmatrix} = 0$$

or $\lambda^3 - 2\lambda^2 - \lambda + 2 = 0$ or

$(\lambda - 1)(\lambda - 2)(\lambda + 1) = 0$. Thus the characteristic values are $\lambda = 1, 2, -1$.

The characteristic vectors corresponding to $\lambda = 1$
have components x_1, x_2, x_3 such that

$$\begin{pmatrix} 1 & 1 & 0 \\ 1 & 0 & 1 \\ 0 & 1 & 1 \end{pmatrix} \begin{pmatrix} x_1 \\ x_2 \\ x_3 \end{pmatrix} = 1 \begin{pmatrix} x_1 \\ x_2 \\ x_3 \end{pmatrix}. \text{ Thus } x_1, x_2, x_3 \text{ must}$$

satisfy the system

$$\begin{cases} x_1 + x_2 = x_1, \\ x_1 + x_3 = x_2, \\ x_2 + x_3 = x_3, \end{cases} \quad \text{or} \quad \begin{cases} x_2 = 0, \\ x_1 - x_2 + x_3 = 0, \\ x_2 = 0. \end{cases}$$

From these, we see at once that $x_2 = 0$ and then $x_3 = -x_1$.
Thus we find $x_1 = k$, $x_2 = 0$, $x_3 = -k$ for every real k.
Hence the characteristic vectors corresponding to
$\lambda = 1$ are $\begin{pmatrix} k \\ 0 \\ -k \end{pmatrix}$ for every nonzero real k.

The characteristic vectors corresponding to $\lambda = 2$
have components x_1, x_2, x_3 such that

$$\begin{pmatrix} 1 & 1 & 0 \\ 1 & 0 & 1 \\ 0 & 1 & 1 \end{pmatrix} \begin{pmatrix} x_1 \\ x_2 \\ x_3 \end{pmatrix} = 2 \begin{pmatrix} x_1 \\ x_2 \\ x_3 \end{pmatrix} \quad . \quad \text{Thus } x_1, x_2, x_3$$

must satisfy the system

$$\begin{cases} x_1 + x_2 = 2x_1, \\ x_1 + x_3 = 2x_2, \\ x_2 + x_3 = 2x_3, \end{cases} \quad \text{or} \quad \begin{cases} -x_1 + x_2 = 0 \\ x_1 - 2x_2 + x_3 = 0, \\ x_2 - x_3 = 0. \end{cases}$$

We find $x_1 = k$, $x_2 = k$, $x_3 = k$ for every real k. Hence
the characteristic vectors corresponding to $\lambda = 2$ are
$\begin{pmatrix} k \\ k \\ k \end{pmatrix}$ for every nonzero real k.

The characteristic vectors corresponding to $\lambda = -1$
have components x_1, x_2, x_3 such that

$$\begin{pmatrix} 1 & 1 & 0 \\ 1 & 0 & 1 \\ 0 & 1 & 1 \end{pmatrix} \begin{pmatrix} x_1 \\ x_2 \\ x_3 \end{pmatrix} = (-1) \begin{pmatrix} x_1 \\ x_2 \\ x_3 \end{pmatrix} \quad . \quad \text{Thus } x_1, x_2, x_3$$

must satisfy the system

$$\begin{cases} x_1 + x_2 = -x_1, \\ x_1 + x_3 = -x_2, \\ x_2 + x_3 = -x_3, \end{cases} \quad \text{or} \quad \begin{cases} 2x_1 + x_2 = 0, \\ x_1 + x_2 + x_3 = 0, \\ x_2 + 2x_3 = 0 \end{cases}$$

We find $x_1 = k$, $x_2 = -2k$, $x_3 = k$ for every real k.
Hence the characteristic vectors corresponding to
$\lambda = -1$ are $\begin{pmatrix} k \\ -2k \\ k \end{pmatrix}$ for every nonzero real k.

14.   The characteristic equation is

$$\begin{vmatrix} -2-\lambda & 6 & -18 \\ 12 & -23-\lambda & 66 \\ 5 & -10 & 29-\lambda \end{vmatrix} = 0$$

or

$$(-2-\lambda)\begin{vmatrix} -23-\lambda & 66 \\ -10 & 29-\lambda \end{vmatrix} - 6\begin{vmatrix} 12 & 66 \\ 5 & 29-\lambda \end{vmatrix}$$

$$- 18 \begin{vmatrix} 12 & -23-\lambda \\ 5 & -10 \end{vmatrix} = 0$$

or $\lambda^3 - 4\lambda^2 - \lambda + 4 = 0$ or $(\lambda - 1)(\lambda + 1)(\lambda - 4) = 0$.
Thus the characteristic values are $\lambda = 1, -1, 4$.

The characteristic vectors corresponding to $\lambda = 1$
have components x_1, x_2, x_3 such that

$$\begin{pmatrix} -2 & 6 & -18 \\ 12 & -23 & 66 \\ 5 & -10 & 29 \end{pmatrix} \begin{pmatrix} x_1 \\ x_2 \\ x_3 \end{pmatrix} = 1 \begin{pmatrix} x_1 \\ x_2 \\ x_3 \end{pmatrix}$$. Thus x_1, x_2,

x_3 must satisfy the system

$$\begin{cases} - 2x_1 + 6x_2 - 18x_3 = x_1 \ , \\ 12x_1 - 23x_2 + 66x_3 = x_2, \\ 5x_1 - 10x_2 + 29x_3 = x_3 \ , \end{cases}$$

or

$$\begin{cases} -3x_1 + 6x_2 - 18x_3 = 0, \\ 12x_1 - 24x_2 + 66x_3 = 0, \\ 5x_1 - 10x_2 + 28x_3 = 0. \end{cases}$$

Adding four times the first to the second, we find
$-6x_3 = 0$, and so $x_3 = 0$. Then the first reduces to
$-3x_1 + 6x_2 = 0$, from which $x_1 = 2x_2$. Thus we find
$x_1 = 2k$, $x_2 = k$, $x_3 = 0$ for every real k. Hence the
characteristic vectors corresponding to $\lambda = 1$ are
$\begin{pmatrix} 2k \\ k \\ 0 \end{pmatrix}$ for every nonzero real k.

The characteristic vectors corresponding to $\lambda = -1$
have components x_1, x_2, x_3 such that

$$\begin{pmatrix} -2 & 6 & -18 \\ 12 & -23 & 66 \\ 5 & -10 & 29 \end{pmatrix} \begin{pmatrix} x_1 \\ x_2 \\ x_3 \end{pmatrix} = (-1) \begin{pmatrix} x_1 \\ x_2 \\ x_3 \end{pmatrix}. \qquad \text{Thus } x_1,$$

x_2, x_3 must satisfy the system

$$\begin{cases} -2x_1 + 6x_2 - 18x_3 = -x_1 , \\ 12x_1 - 23x_2 + 66x_3 = -x_2, \\ 5x_1 - 10x_2 + 29x_3 = -x_3 , \end{cases} \qquad \text{or}$$

$$\begin{cases} -x_1 + 6x_2 - 18x_3 = 0, \\ 12x_1 - 22x_2 + 66x_3 = 0, \\ 5x_1 - 10x_2 + 30x_3 = 0. \end{cases}$$

Adding 3/5 the third equation to the first gives $2x_1 = 0$, so $x_1 = 0$. Then the first equation becomes $6x_2 - 18x_3 = 0$, from which $x_2 = 3x_3$. Thus we find $x_1 = 0$, $x_2 = 3k$, $x_3 = k$ for every real k. Hence the characteristic vectors corresponding to $\lambda = -1$ are $\begin{pmatrix} 0 \\ 3k \\ k \end{pmatrix}$ for every nonzero real k.

The characteristic vectors corresponding to $\lambda = 4$ have components x_1, x_2, x_3 such that

$$\begin{pmatrix} -2 & 6 & -18 \\ 12 & -23 & 66 \\ 5 & -10 & 29 \end{pmatrix} \begin{pmatrix} x_1 \\ x_2 \\ x_3 \end{pmatrix} = 4 \begin{pmatrix} x_1 \\ x_2 \\ x_3 \end{pmatrix}. \qquad \text{Thus } x_1, x_2, x_3$$

must satisfy the system

$$\begin{cases} -2x_1 + 6x_2 - 18x_3 = 4x_1, \\ 12x_1 - 23x_2 + 66x_3 = 4x_2, \text{ or } \\ 5x_1 - 10x_2 + 29x_3 = 4x_3, \end{cases} \begin{cases} -6x_1 + 6x_2 - 18x_3 = 0, \\ 12x_1 - 27x_2 + 66x_3 = 0, \\ 5x_1 - 10x_2 + 25x_3 = 0. \end{cases}$$

The first and third reduce to $\begin{cases} -x_1 + x_2 - 3x_3 = 0, \\ x_1 - 2x_2 + 5x_3 = 0. \end{cases}$

Regarding these as two equations in x_1 and x_2, we find $x_1 = -x_3$, $x_2 = 2x_3$. Letting $x_3 = -k$, we find $x_1 = k$, $x_2 = -2k$, $x_3 = -k$ for every real k. Hence the characteristic vectors corresponding to $\lambda = 4$ are $\begin{pmatrix} k \\ -2k \\ -k \end{pmatrix}$ for every nonzero real k.

Section 7.6B, Page 367

1. The three vectors defined by the individual columns of the matrix

$$\bar{\bar{\Phi}}(t) = \begin{pmatrix} e^t & e^{2t} & e^{-3t} \\ e^t & 2e^{2t} & 7e^{-3t} \\ e^t & e^{2t} & 11e^{-3t} \end{pmatrix}$$

are each solutions of the stated system. For example, the vector

$$\bar{x} = \begin{pmatrix} e^{-3t} \\ 7e^{-3t} \\ 11e^{-3t} \end{pmatrix}$$

defined by the third column is a solution, since

$$
\begin{pmatrix} -3e^{-3t} \\ -21e^{-3t} \\ -33e^{-3t} \end{pmatrix} = \begin{pmatrix} 1 & 1 & -1 \\ 2 & 3 & -4 \\ 4 & 1 & -4 \end{pmatrix} \begin{pmatrix} e^{-3t} \\ 7e^{-3t} \\ 11e^{-3t} \end{pmatrix} .
$$

Now note that

$$
|\overline{\overline{\Phi}}(t)| = \begin{vmatrix} e^t & e^{2t} & e^{-3t} \\ e^t & 2e^{2t} & 7e^{-3t} \\ e^t & e^{2t} & 11e^{-3t} \end{vmatrix} = 10 \neq 0
$$

for all real t. Thus by Theorem 7.15, the solutions defined by the columns of $\overline{\overline{\Phi}}(t)$ are linearly independent on every real interval. Thus, by definition, $\overline{\overline{\Phi}}(t)$ is a fundamental matrix of the stated system.

8. a. By direct substitution one verifies that each column of $\overline{\overline{\Phi}}(t)$ is a solution of the stated system. Then, since

$$
|\overline{\overline{\Phi}}(t)| = \begin{vmatrix} e^{5t} & e^{3t} \\ e^{5t} & 3e^{3t} \end{vmatrix} = 2e^{8t} \neq 0
$$

on every real interval, by Theorem 7.15, $\overline{\overline{\Phi}}(t)$ is a fundamental matrix of the system.

b. We find

$$
\overline{\overline{\Phi}}^{-1}(t) = \begin{pmatrix} 3e^{-5t}/2 & -e^{-5t}/2 \\ -e^{-3t}/2 & e^{-3t}/2 \end{pmatrix} ,
$$

and hence $\overline{\overline{\Phi}}^{-1}(t_0) = \overline{\overline{\Phi}}^{-1}(0) = \begin{pmatrix} 3/2 & -1/2 \\ -1/2 & 1/2 \end{pmatrix}$.

Then by Formula (7.138) of Theorem 7.19,

$$\overline{\phi}(t) = \begin{pmatrix} e^{5t} & e^{3t} \\ e^{5t} & 3e^{3t} \end{pmatrix} \begin{pmatrix} 3/2 & -1/2 \\ -1/2 & 1/2 \end{pmatrix} \begin{pmatrix} 2 \\ 0 \end{pmatrix}$$

$$= \begin{pmatrix} e^{5t} & e^{3t} \\ e^{5t} & 3e^{3t} \end{pmatrix} \begin{pmatrix} 3 \\ -1 \end{pmatrix} = \begin{pmatrix} 3e^{5t} - e^{3t} \\ 3e^{5t} - 3e^{3t} \end{pmatrix}$$

Thus the desired solution is given by

$$x_1 = 3e^{5t} - e^{3t} \quad , \quad x_2 = 3e^{5t} - 3e^{3t} \quad .$$

c. We have

$$\begin{pmatrix} 5e^{5t} & 3e^{3t} \\ 5e^{5t} & 9e^{3t} \end{pmatrix} = \begin{pmatrix} 6 & -1 \\ 3 & 2 \end{pmatrix} \begin{pmatrix} e^{5t} & e^{3t} \\ e^{5t} & 3e^{3t} \end{pmatrix} \quad .$$

11. a. By direct substitution one verifies that each column of $\overline{\overline{\Phi}}(t)$ is a solution of the stated system. Then, since

$$|\overline{\overline{\Phi}}(t)| = \begin{vmatrix} 2e^t & e^{2t} & 0 \\ 2e^t & 2e^{2t} & 3e^{5t} \\ e^t & e^{2t} & e^{5t} \end{vmatrix} = -e^{8t} \neq 0$$

on every real interval, by Theorem 7.15, $\overline{\overline{\Phi}}(t)$ is a fundamental matrix of the system.

    b. We find

$$\overline{\Phi}^{-1}(t) = \begin{pmatrix} e^{-t} & e^{-t} & -3e^{-t} \\ -e^{-2t} & -2e^{-2t} & 6e^{-2t} \\ 0 & e^{-5t} & -2e^{-5t} \end{pmatrix}$$

and hence

$$\overline{\Phi}^{-1}(t_0) = \overline{\Phi}^{-1}(0) = \begin{pmatrix} 1 & 1 & -3 \\ -1 & -2 & 6 \\ 0 & 1 & -2 \end{pmatrix} .$$

Then by Formula (7.138) of Theorem 7.19,

$$\overline{\Phi}(t) = \begin{pmatrix} 2e^t & e^{2t} & 0 \\ 2e^t & 2e^{2t} & 3e^{5t} \\ e^t & e^{2t} & e^{5t} \end{pmatrix} \begin{pmatrix} 1 & 1 & -3 \\ -1 & -2 & 6 \\ 0 & 1 & -2 \end{pmatrix} \begin{pmatrix} 6 \\ 3 \\ 2 \end{pmatrix}$$

$$= \begin{pmatrix} 2e^t & e^{2t} & 0 \\ 2e^t & 2e^{2t} & 3e^{5t} \\ e^t & e^{2t} & e^{5t} \end{pmatrix} \begin{pmatrix} 3 \\ 0 \\ -1 \end{pmatrix} = \begin{pmatrix} 6e^t \\ 6e^t - 3e^{5t} \\ 3e^t - e^{5t} \end{pmatrix} .$$

Thus the desired solution is given by

$$x_1 = 6e^t, \quad x_2 = 6e^t - 3e^{5t}, \quad x_3 = 3e^t - e^{5t} .$$

    c. We have

$$\begin{pmatrix} 2e^t & 2e^{2t} & 0 \\ 2e^t & 4e^{2t} & 15e^{5t} \\ e^t & 2e^{2t} & 5e^{5t} \end{pmatrix} = \begin{pmatrix} 0 & -2 & 6 \\ -2 & 9 & -12 \\ -1 & 2 & -1 \end{pmatrix} \begin{pmatrix} 2e^t & e^{2t} & 0 \\ 2e^t & 2e^{2t} & 3e^{5t} \\ e^t & e^{2t} & e^{5t} \end{pmatrix}.$$

13.   a.   By hypothesis, $\bar{\Phi}'(t) = \bar{A}(t)\,\bar{\Phi}(t)$, and also $|\bar{\Phi}(t)| \neq 0$, $|\bar{C}| \neq 0$, on [a,b]. Then, using Result E of Section 7.5, $[\bar{\Phi}(t)\,\bar{C}]' = \bar{\Phi}'(t)\,\bar{C} + \bar{\Phi}(t)\,\bar{C}'$. But since $\bar{C}$ is a constant matrix, $\bar{C}' = \bar{0}$; so $[\bar{\Phi}(t)\,\bar{C}]' = \bar{\Phi}'(t)\,\bar{C}$. Then we have $[\bar{\Phi}(t)\,\bar{C}]' = \bar{\Phi}'(t)\,\bar{C} = [\bar{A}(t)\,\bar{\Phi}(t)]\bar{C} = \bar{A}(t)[\bar{\Phi}(t)\,\bar{C}]$. Thus $\bar{\Phi}(t)\,\bar{C}$ satisfies $d\bar{x}/dt = \bar{A}(t)\bar{x}$. Also, $|\bar{\Phi}(t)\,\bar{C}| = |\bar{\Phi}(t)||\bar{C}| \neq 0$ on [a,b]. Hence $\bar{\Phi}\,\bar{C}$ is a fundamental matrix of $d\bar{x}/dt = \bar{A}(t)\bar{x}$ on [a,b].

b.   For $\bar{C}\bar{\Phi}$ to be a fundamental matrix on [a.b], we must have $[\bar{C}\,\bar{\Phi}(t)]' = \bar{A}(t)[\bar{C}\,\bar{\Phi}(t)]$ on [a,b]. But $[\bar{C}\,\bar{\Phi}(t)] = \bar{C}'\,\bar{\Phi}(t) + \bar{C}\,\bar{\Phi}'(t) = \bar{C}\,\bar{\Phi}'(t) = \bar{C}[\bar{A}(t)\,\bar{\Phi}(t)] = [\bar{C}\bar{A}(t)]\,\bar{\Phi}(t) \neq [A(t)\bar{C}]\,\bar{\Phi}(t) = \bar{A}(t)[\bar{C}\,\bar{\Phi}(t)]$ in general.

c.   We seek $\bar{B}(t)$ such that $[\bar{C}\,\bar{\Phi}(t)]' = \bar{B}(t)[\bar{C}\,\bar{\Phi}(t)]$ on [a,b]. Since $[\bar{C}\,\bar{\Phi}(t)]' = \bar{C}'\,\bar{\Phi}(t) + \bar{C}\,\bar{\Phi}'(t) = C\,\bar{\Phi}'(t)$, this may be written as $\bar{C}\,\bar{\Phi}'(t) = \bar{B}(t)\bar{C}\,\bar{\Phi}(t)$. Since by hypothesis $\bar{\Phi}'(t) = \bar{A}(t)\,\bar{\Phi}(t)$, this becomes $\bar{C}A(t)\,\bar{\Phi}(t) = \bar{B}(t)\bar{C}\,\bar{\Phi}(t)$. Then $\bar{B}(t)\bar{C} = \bar{C}A(t)$, and so $\bar{B}(t) = \bar{C}A(t)\bar{C}^{-1}$. Hence

$\overline{B} = \overline{CAC}^{-1}$.

Section 7.6C, Page 378

    1. a. The corresponding homogeneous system is

$$\frac{d\overline{x}}{dt} = \begin{pmatrix} 6 & -3 \\ 2 & 1 \end{pmatrix} \overline{x}, \text{ that is, } \begin{cases} dx_1/dt = 6x_1 - 3x_2, \\ dx_2/dt = 2x_1 + x_2 . \end{cases}$$

As in Section 7.4, we seek a solution of this of the form $x_1 = Ae^{\lambda t}$, $x_2 = Be^{\lambda t}$. Substituting into this system, we find, after simplification,

$$\begin{cases} (6 - \lambda)A - 3B = 0, \\ 2A + (1 - \lambda)B = 0. \end{cases} \tag{1}$$

For a nontrivial solution for A and B, we must have

$\begin{vmatrix} 6 - \lambda & -3 \\ 2 & 1 - \lambda \end{vmatrix} = 0$. This leads to the characteristic

equation $\lambda^2 - 7\lambda + 12 = 0$, whose roots are $\lambda = 3, 4$.

    Letting $\lambda = 3$ in (1), we obtain $3A - 3B = 0$, $2A - 2B = 0$, a nontrivial solution of which is $A = B = 1$. With these values of A, B, and λ , we have the solution $x_1 = e^{3t}$, $x_2 = e^{3t}$, that is, $\begin{pmatrix} e^{3t} \\ e^{3t} \end{pmatrix}$.

    Letting $\lambda = 4$ in (1), we obtain $2A - 3B = 0$ (twice), a nontrivial solution of which is $A = 3$, $B = 2$. With these values of A, B, and λ , we have the solution $x_1 = 3e^{4t}$, $x_2 = 2e^{4t}$, that is, $\begin{pmatrix} 3e^{4t} \\ 2e^{4t} \end{pmatrix}$.

Since $\begin{vmatrix} e^{3t} & 3e^{4t} \\ e^{3t} & 2e^{4t} \end{vmatrix} = -e^{7t} \neq 0$, we see that the

solutions $\begin{pmatrix} e^{3t} \\ e^{3t} \end{pmatrix}$ and $\begin{pmatrix} 3e^{4t} \\ 2e^{4t} \end{pmatrix}$ are linearly

independent. Hence $\overline{\overline{\Phi}}(t) = \begin{pmatrix} e^{3t} & 3e^{4t} \\ e^{3t} & 2e^{4t} \end{pmatrix}$ is a

fundamental matrix of the homogeneous system.

   b. We use Formula (7.150) of Theorem 7.22 with

$\overline{\overline{\Phi}}(t)$ found in part (a), $\overline{F}(u) = \begin{pmatrix} e^{2u} \\ -e^{2u} \end{pmatrix}$, and $t_0 = 0$.

We find

$$\overline{\overline{\Phi}}^{-1}(u) = \begin{pmatrix} -2e^{-3u} & 3e^{-3u} \\ e^{-4u} & -e^{-4u} \end{pmatrix} .$$

Then

$$\overline{\phi}_0(t) = \begin{pmatrix} e^{3t} & 3e^{4t} \\ e^{3t} & 2e^{4t} \end{pmatrix} \int_0^t \begin{pmatrix} -2e^{-3u} & 3e^{-3u} \\ e^{-4u} & -e^{-4u} \end{pmatrix} \begin{pmatrix} e^{2u} \\ -e^{2u} \end{pmatrix} du$$

$$= \begin{pmatrix} e^{3t} & 3e^{4t} \\ e^{3t} & 2e^{4t} \end{pmatrix} \int_0^t \begin{pmatrix} -5e^{-u} \\ 2e^{-2u} \end{pmatrix} du$$

$$= \begin{pmatrix} e^{3t} & 3e^{4t} \\ e^{3t} & 2e^{4t} \end{pmatrix} \begin{pmatrix} 5e^{-t} - 5 \\ -e^{-2t} + 1 \end{pmatrix}$$

$$= \begin{pmatrix} 2e^{2t} - 5e^{3t} + 3e^{4t} \\ 3e^{2t} - 5e^{3t} + 2e^{4t} \end{pmatrix} .$$

Note that this can be written in the form

$$\overline{\phi}(t) = -5\overline{\phi}_1(t) + \overline{\phi}_2(t) + \overline{\phi}_0^{*}(t),$$

where $\overline{\phi}_1(t) = \begin{pmatrix} e^{3t} \\ e^{3t} \end{pmatrix}$ and $\overline{\phi}_2(t) = \begin{pmatrix} 3e^{4t} \\ 2e^{4t} \end{pmatrix}$ are the

fundamental set of the homogeneous system found in part
(a) and $\overline{\phi}_0^{*}(t) = \begin{pmatrix} 2e^{2t} \\ 3e^{2t} \end{pmatrix}$ is a particular solution of

the given nonhomogeneous system. Thus, in particular,
$\begin{pmatrix} 2e^{2t} \\ 3e^{2t} \end{pmatrix}$ is a solution of the given system.

4. a. The corresponding homogeneous system is
$$\frac{d\overline{x}}{dt} = \begin{pmatrix} 3 & -1 \\ 4 & -1 \end{pmatrix} \overline{x}, \text{ that is } \begin{cases} dx_1/dt = 3x_1 - x_2, \\ dx_2/dt = 4x_1 - x_2. \end{cases}$$

As in Section 7.4, we seek a solution of this of the
form $x_1 = Ae^{\lambda t}$, $x_2 = Be^{\lambda t}$. Substituting into this
system, we find, after simplification,

$$\begin{cases} (3 - \lambda)A - B = 0 \\ 4A + (-1 - \lambda)B = 0. \end{cases} \tag{1}$$

For a nontrivial solution for A and B, we must have

$\begin{vmatrix} 3-\lambda & -1 \\ 4 & -1-\lambda \end{vmatrix} = 0$. This leads to the characteristic

equation $\lambda^2 - 2\lambda + 1 = 0$, whose roots are $\lambda = 1, 1$
(double root).

Letting $\lambda = 1$ in (1), we obtain $2A - B = 0$,
$4A - 2B = 0$, a nontrivial solution of which is $A = 1$,
$B = 2$. With these values of A, B, and λ, we have the
solution $x_1 = e^t$, $x_2 = 2e^t$.

We now seek a solution of the form

$$\begin{cases} x_1 = (A_1 t + A_2)e^t, \\ x_2 = (B_1 t + B_2)e^t. \end{cases}$$ Substituting this into the

homogeneous system and simplifying, we find

$$\begin{cases} (2A_1 - B_1)t + (2A_2 - B_2 - A_1) = 0, \\ (4A_1 - 2B_1)t + (4A_2 - 2B_2 - B_1) = 0 \end{cases}.$$

For these to hold, we must have

$$\begin{cases} 2A_1 - B_1 = 0, \\ 4A_1 - 2B_1 = 0, \end{cases} \quad \begin{cases} 2A_2 - B_2 - A_1 = 0, \\ 4A_2 - 2B_2 - B_1 = 0. \end{cases}$$

A nontrivial solution of these is

$$A_1 = 1, B_1 = 2, A_2 = 1, B_2 = 1.$$

With these values, we obtain the solution

$$x_1 = (t + 1)e^t, \quad x_2 = (2t + 1)e^t.$$

Since $\begin{vmatrix} e^t & (t+1)e^t \\ 2e^t & (2t+1)e^t \end{vmatrix} = -e^{2t} \neq 0$,

we see that the solutions $\begin{pmatrix} e^t \\ 2e^t \end{pmatrix}$ and $\begin{pmatrix} (t+1)e^t \\ (2t+1)e^t \end{pmatrix}$ are

linearly independent. Hence

$$\overline{\Phi}(t) = \begin{pmatrix} e^t & (t+1)e^t \\ 2e^t & (2t+1)e^t \end{pmatrix}$$

is a fundamental matrix of the homogeneous system.

  b. We use Formula (7.150) of Theorem 7.22 with $\overline{\Phi}(t)$ found in part (a), $\overline{F}(u) = \begin{pmatrix} 2e^{2u} \\ -2 \end{pmatrix}$, and $t_0 = 0$.
We find

$$\overline{\Phi}^{-1}(u) = \begin{pmatrix} -(2u+1)e^{-u} & (u+1)e^{-u} \\ 2e^{-u} & -e^{-u} \end{pmatrix} .$$

Then

$$\Phi_0(t) = \begin{pmatrix} e^t & (t+1)e^t \\ 2e^t & (2t+1)e^t \end{pmatrix} \int_0^t \begin{pmatrix} -(2u+1)e^{-u} & (u+1)e^{-u} \\ 2e^{-u} & -e^{-u} \end{pmatrix} \begin{pmatrix} 2e^{2u} \\ -2 \end{pmatrix} du$$

$$= \begin{pmatrix} e^t & (t+1)e^t \\ 2e^t & (2t+1)e^t \end{pmatrix} \int_0^t \begin{pmatrix} -(4u+2)e^u - (2u+2)e^{-u} \\ 4e^u + 2e^{-u} \end{pmatrix} du$$

$$= \begin{pmatrix} e^t & (t+1)e^t \\ 2e^t & (2t+1)e^t \end{pmatrix} \begin{pmatrix} -4te^t + 2e^t + 2te^{-t} + 4e^{-t} - 6 \\ 4e^t - 2e^{-t} - 2 \end{pmatrix}$$

$$= \begin{pmatrix} -8^t - 2te^t + 6e^{2t} + 2 \\ -14e^t - 4te^t + 8e^{2t} + 6 \end{pmatrix}.$$

Note that this can be written in the form

$$\bar{\phi}_0(t) = -6\bar{\phi}_1(t) - 2\bar{\phi}_2(t) + \bar{\phi}_0^* (t),$$

where $\bar{\phi}_1(t) = \begin{pmatrix} e^t \\ 2e^t \end{pmatrix}$ and $\bar{\phi}_2(t) = \begin{pmatrix} (t+1)e^t \\ (2t+1)e^t \end{pmatrix}$ are

the fundamental set of the homogeneous system found in

part (a) and $\bar{\phi}_0^* (t) = \begin{pmatrix} 6e^{2t} + 2 \\ 8e^{2t} + 6 \end{pmatrix}$ is a particular

solution of the given nonhomogeneous system. Thus, in

particular, $\begin{pmatrix} 6e^{2t} + 2 \\ 8e^{2t} + 6 \end{pmatrix}$ is a solution of the given

system.

7. The corresponding homogeneous system is

$$\frac{d\bar{x}}{dt} = \begin{pmatrix} -10 & 6 \\ -12 & 7 \end{pmatrix} \bar{x}, \text{ that is } \begin{cases} dx_1/dt = -10x_1 + 6x_2, \\ dx_2/dt = -12x_1 + 7x_2. \end{cases} \quad (1)$$

We need a fundamental matrix of this in order to use

Theorem 7.23 to find the desired nonhomogeneous solu-

tion. We proceed as in Section 7.4, assuming a solu-

tion of (1) of the form $x_1 = Ae^{\lambda t}$, $x_2 = Be^{\lambda t}$.

Substituting into (1) and simplifying, we obtain

$$\begin{cases} (-10 - \lambda)A + 6B = 0, \\ -12A + (7 - \lambda)B = 0. \end{cases} \quad (2)$$

For a nontrivial solution of this, we must have

$$\begin{vmatrix} -10 - \lambda & 6 \\ -12 & 7 - \lambda \end{vmatrix} = 0.$$

This leads to the characteristic equation

$\lambda^2 + 3\lambda + 2 = 0$, whose roots are $\lambda = -1$, $\lambda = -2$.

Letting $\lambda = -1$ in (2), we obtain $-9A + 6B = 0$, $-12A + 8B = 0$, a nontrivial solution of which is $A = 2$, $B = 3$. With these values of A, B, and λ , we obtain the solution $x_1 = 2e^{-t}$, $x_2 = 3e^{-t}$.

Letting $\lambda = -2$ in (2), we obtain $-8A + 6B = 0$, $-12A + 9B = 0$, a nontrivial solution of which is $A = 3$, $B = 4$. With these values of A, B, and λ , we obtain the solution $x_1 = 3e^{-2t}$, $x_2 = 4e^{-2t}$.

Since $\begin{vmatrix} 2e^{-t} & 3e^{-2t} \\ 3e^{-t} & 4e^{-2t} \end{vmatrix} = -e^{-3t} \neq 0$, we see that the

solutions are linearly independent. Hence

$$\overline{\Phi}(t) = \begin{pmatrix} 2e^{-t} & 3e^{-2t} \\ 3e^{-t} & 4e^{-2t} \end{pmatrix}$$

is a fundamental matrix of the homogeneous system.

We can now apply Formula (7.160) of Theorem 7.23, with $\overline{\Phi}(t)$ just given, $\overline{F}(u) = \begin{pmatrix} 10e^{-3u} \\ 18e^{-3u} \end{pmatrix}$, $t_0 = 0$, and $\overline{x}_0 = \begin{pmatrix} 1 \\ -2 \end{pmatrix}$. We find $\overline{\Phi}^{-1}(u) = \begin{pmatrix} -4e^{u} & 3e^{u} \\ 3e^{2u} & -2e^{2u} \end{pmatrix}$ and $\overline{\Phi}^{-1}(0) = \begin{pmatrix} -4 & 3 \\ 3 & -2 \end{pmatrix}$.

Then the desired solution is given by

$$\bar{\phi}(t) = \begin{pmatrix} 2e^{-t} & 3e^{-2t} \\ 3e^{-t} & 4e^{-2t} \end{pmatrix} \begin{pmatrix} -4 & 3 \\ 3 & -2 \end{pmatrix} \begin{pmatrix} 1 \\ -2 \end{pmatrix}$$

$$+ \begin{pmatrix} 2e^{-t} & 3e^{-2t} \\ 3e^{-t} & 4e^{-2t} \end{pmatrix} \int_0^t \begin{pmatrix} -4e^{u} & 3e^{u} \\ 3e^{2u} & -2e^{2u} \end{pmatrix} \begin{pmatrix} 10e^{-3u} \\ 18e^{-3u} \end{pmatrix} du$$

$$= \begin{pmatrix} 2e^{-t} & 3e^{-2t} \\ 3e^{-t} & 4e^{-2t} \end{pmatrix} \begin{pmatrix} -10 \\ 7 \end{pmatrix} + \begin{pmatrix} 2e^{-t} & 3e^{-2t} \\ 3e^{-t} & 4e^{-2t} \end{pmatrix} \int_0^t \begin{pmatrix} 14e^{-2u} \\ -6e^{-u} \end{pmatrix} du$$

$$= \begin{pmatrix} -20e^{-t} + 21e^{-2t} \\ -30e^{-t} + 28e^{-2t} \end{pmatrix} + \begin{pmatrix} 4e^{-3t} + 14e^{-t} - 18e^{-2t} \\ 3e^{-3t} + 21e^{-t} - 24e^{-2t} \end{pmatrix}$$

$$= \begin{pmatrix} -6e^{-t} + 3e^{-2t} + 4e^{-3t} \\ -9e^{-t} + 4e^{-2t} + 3e^{-3t} \end{pmatrix} .$$

Section 7.7, Page 390

1. We assume a solution of the form $\bar{x} = \bar{\alpha} e^{\lambda t}$, that is, $x_1 = \alpha_1 e^{\lambda t}$, $x_2 = \alpha_2 e^{\lambda t}$, $x_3 = \alpha_3 e^{\lambda t}$. Substituting into the given system and dividing through by $e^{\lambda t} \neq 0$, we obtain

$$\begin{cases} \alpha_1 \lambda = \alpha_1 + \alpha_2 - \alpha_3, \\ \alpha_2 \lambda = 2\alpha_1 + 3\alpha_2 - 4\alpha_3, \\ \alpha_3 \lambda = 4\alpha_1 + \alpha_2 - 4\alpha_3, \end{cases}$$

or
$$\begin{cases} (1 - \lambda)\alpha_1 + \alpha_2 - \alpha_3 = 0, \\ 2\alpha_1 + (3 - \lambda)\alpha_2 - 4\alpha_3 = 0, \\ 4\alpha_1 + \alpha_2 + (-4 - \lambda)\alpha_3 = 0. \end{cases}$$

This has a nontrivial solution for α_1, α_2, α_3 if and only if

$$\begin{vmatrix} 1 - \lambda & 1 & -1 \\ 2 & 3 - \lambda & -4 \\ 4 & 1 & -4 - \lambda \end{vmatrix} = 0 . \tag{1}$$

This is the characteristic equation of the coefficient matrix

$$\overline{A} = \begin{pmatrix} 1 & 1 & -1 \\ 2 & 3 & -4 \\ 4 & 1 & -4 \end{pmatrix}$$

of the given system.

This matrix $\overline{A}$ is the matrix of Exercise 7 of Section 7.5D (text, page 344). On page 299 of this manual, we found that the characteristic equation (1) of $\overline{A}$ reduces to $\lambda^3 - 7\lambda + 6 = 0$. We also found that the roots of this equation, the characteristic values of $\overline{A}$, are 1, 2, and -3. Finally, we found that characteristic vectors corresponding to these respective characteristic values are $\begin{pmatrix} k \\ k \\ k \end{pmatrix}$, $\begin{pmatrix} k \\ 2k \\ k \end{pmatrix}$, and $\begin{pmatrix} k \\ 7k \\ 11k \end{pmatrix}$.

Choosing k = 1, we see that particular characteristic vectors corresponding to the respective characteristic values 1, 2, and -3 are

$$\begin{pmatrix} 1 \\ 1 \\ 1 \end{pmatrix} \ , \ \begin{pmatrix} 1 \\ 2 \\ 1 \end{pmatrix} \ \text{and} \ \begin{pmatrix} 1 \\ 7 \\ 11 \end{pmatrix} \ .$$

By Theorem 7.24, a fundamental set of solutions of the given system is

$$\begin{pmatrix} 1 \\ 1 \\ 1 \end{pmatrix} e^t \ , \ \begin{pmatrix} 1 \\ 2 \\ 1 \end{pmatrix} e^{2t} \ , \ \text{and} \ \begin{pmatrix} 1 \\ 7 \\ 11 \end{pmatrix} e^{-3t}$$

or

$$\begin{pmatrix} e^t \\ e^t \\ e^t \end{pmatrix} \ , \ \begin{pmatrix} e^{2t} \\ 2e^{2t} \\ e^{2t} \end{pmatrix} \ , \ \text{and} \ \begin{pmatrix} e^{-3t} \\ 7e^{-3t} \\ 11e^{-3t} \end{pmatrix} \ .$$

Thus a general solution is

$$\begin{cases} x_1 = c_1 e^t + c_2 e^{2t} + c_3 e^{-3t} \ , \\ x_2 = c_1 e^t + 2c_2 e^{2t} + 7c_3 e^{-3t} \ , \\ x_3 = c_1 e^t + c_2 e^{2t} + 11c_3 e^{-3t} \ . \end{cases}$$

6. We assume a solution of the form $\bar{x} = \bar{a}e^{\lambda t}$, that is, $x_1 = \alpha_1 e^{\lambda t}$, $x_2 = \alpha_2 e^{\lambda t}$, $x_3 = \alpha_3 e^{\lambda t}$. Substituting into the given system and dividing through by $e^{\lambda t} \neq 0$, we obtain

$$\begin{cases} \alpha_1 \lambda = \alpha_1 + \alpha_2, \\ \alpha_2 \lambda = \alpha_1 + \alpha_3, \\ \alpha_3 \lambda = \alpha_2 + \alpha_3, \end{cases} \quad \text{or} \quad \begin{cases} (1-\lambda)\alpha_1 + \alpha_2 = 0 \ , \\ \alpha_1 - \lambda\alpha_2 + \alpha_3 = 0, \\ \alpha_2 + (1-\lambda)\alpha_3 = 0 \ . \end{cases}$$

This has a nontrivial solution for α_1, α_2, α_3 if and only if

$$\begin{vmatrix} 1-\lambda & 1 & 0 \\ 1 & -\lambda & 1 \\ 0 & 1 & 1-\lambda \end{vmatrix} = 0 \ .$$

This is the characteristic equation of the coefficient matrix of the system. It reduces to $(\lambda - 1)(\lambda + 1)(\lambda - 2) = 0$; and its roots, the characteristic values, are 1, 2, -1.

A characteristic vector corresponding to $\lambda = 1$ has components α_1, α_2, α_3 such that

$$\begin{pmatrix} 1 & 1 & 0 \\ 1 & 0 & 1 \\ 0 & 1 & 1 \end{pmatrix} \begin{pmatrix} \alpha_1 \\ \alpha_2 \\ \alpha_3 \end{pmatrix} = 1 \begin{pmatrix} \alpha_1 \\ \alpha_2 \\ \alpha_3 \end{pmatrix} \ .$$

Thus α_1, α_2, α_3 must satisfy the system

$$\begin{cases} \alpha_2 = 0, \\ \alpha_1 - \alpha_2 + \alpha_3 = 0, \\ \alpha_2 = 0, \end{cases}$$

a nontrivial solution of which is $\alpha_1 = 1$, $\alpha_2 = 0$, $\alpha_3 = -1$. Thus we have the characteristic vector $\begin{pmatrix} 1 \\ 0 \\ -1 \end{pmatrix}$ corresponding to $\lambda = 1$.

A characteristic vector corresponding to $\lambda = 2$ has components α_1, α_2, α_3 such that

$$\begin{pmatrix} 1 & 1 & 0 \\ 1 & 0 & 1 \\ 0 & 1 & 1 \end{pmatrix} \begin{pmatrix} \alpha_1 \\ \alpha_2 \\ \alpha_3 \end{pmatrix} = 2 \begin{pmatrix} \alpha_1 \\ \alpha_2 \\ \alpha_3 \end{pmatrix} \quad .$$

Thus α_1, α_2, α_3 must satisfy the system

$$\begin{cases} -\alpha_1 + \alpha_2 = 0 \\ \alpha_1 - 2\alpha_2 + \alpha_3 = 0 \\ \alpha_2 - \alpha_3 = 0, \end{cases} \qquad \text{a nontrivial solution of}$$

which is $\alpha_1 = 1$, $\alpha_2 = 1$, $\alpha_3 = 1$. Thus we have the

characteristic vector $\begin{pmatrix} 1 \\ 1 \\ 1 \end{pmatrix}$ corresponding to $\lambda = 2$.

A characteristic vector corresponding to $\lambda = -1$

has components α_1, α_2, α_3 such that

$$\begin{pmatrix} 1 & 1 & 0 \\ 1 & 0 & 1 \\ 0 & 1 & 1 \end{pmatrix} \begin{pmatrix} \alpha_1 \\ \alpha_2 \\ \alpha_3 \end{pmatrix} = (-1) \begin{pmatrix} \alpha_1 \\ \alpha_2 \\ \alpha_3 \end{pmatrix} \quad . \text{ Thus } \alpha_1, \alpha_2,$$

α_3 must satisfy the system

$$\begin{cases} 2\alpha_1 + \alpha_2 = 0, \\ \alpha_1 + \alpha_2 + \alpha_3 = 0, \\ \alpha_2 + 2\alpha_3 = 0, \end{cases}$$

a nontrivial solution of which is $\alpha_1 = 1$, $\alpha_2 = -2$,

$\alpha_3 = 1$. Thus we have the characteristic vector $\begin{pmatrix} 1 \\ -2 \\ 1 \end{pmatrix}$

corresponding to $\lambda = -1$.

By Theorem 7.24, a fundamental set of solutions of the given system is

$$\begin{pmatrix} 1 \\ 0 \\ -1 \end{pmatrix} e^t, \quad \begin{pmatrix} 1 \\ 1 \\ 1 \end{pmatrix} e^{2t}, \quad \text{and} \quad \begin{pmatrix} 1 \\ -2 \\ 1 \end{pmatrix} e^{-t},$$

or

$$\begin{pmatrix} e^t \\ 0 \\ -e^t \end{pmatrix}, \quad \begin{pmatrix} e^{2t} \\ e^{2t} \\ e^{2t} \end{pmatrix}, \quad \text{and} \quad \begin{pmatrix} e^{-t} \\ -2e^{-t} \\ e^{-t} \end{pmatrix}.$$

Thus a general solution is

$$\begin{cases} x_1 = c_1 e^t + c_2 e^{2t} + c_3 e^{-t}, \\ x_2 = c_2 e^{2t} - 2c_3 e^{-t}, \\ x_3 = -c_1 e^t + c_2 e^{2t} + c_3 e^{-t}. \end{cases}$$

7. We assume a solution of the form $\bar{x} = \bar{\alpha} e^{\lambda t}$, that is, $x_1 = \alpha_1 e^{\lambda t}$, $x_2 = \alpha_2 e^{\lambda t}$, $x_3 = \alpha_3 e^{\lambda t}$. Substituting into the given system and dividing through by $e^{\lambda t} \neq 0$, we obtain

$$\begin{cases} \alpha_1 \lambda = \alpha_1 - 2\alpha_2, \\ \alpha_2 \lambda = -2\alpha_1 + 3\alpha_2, \\ \alpha_3 \lambda = 2\alpha_3, \end{cases}$$

or

$$\begin{cases} (1 - \lambda)\alpha_1 - 2\alpha_2 = 0, \\ -2\alpha_1 + (3 - \lambda)\alpha_2 = 0, \\ (2 - \lambda)\alpha_3 = 0. \end{cases}$$

This has nontrivial solutions for α_1, α_2, α_3 if and only if

$$
\begin{vmatrix}
1 - \lambda & -2 & 0 \\
-2 & 3 - \lambda & 0 \\
0 & 0 & 2 - \lambda
\end{vmatrix} = 0.
$$

This is the characteristic equation of the coefficient matrix of the system. It reduces to

$(\lambda - 2)(\lambda^2 - 4\lambda - 1) = 0$; and its roots, the characteristic values, are $\lambda = 2, \ 2 \pm \sqrt{5}$.

The characteristic vector corresponding to $\lambda = 2 + \sqrt{5}$ has components α_1, α_2, α_3 such that

$$
\begin{pmatrix}
1 & -2 & 0 \\
-2 & 3 & 0 \\
0 & 0 & 2
\end{pmatrix}
\begin{pmatrix}
\alpha_1 \\
\alpha_2 \\
\alpha_3
\end{pmatrix}
= (2 + \sqrt{5})
\begin{pmatrix}
\alpha_1 \\
\alpha_2 \\
\alpha_3
\end{pmatrix}.
$$

Thus α_1, α_2, α_3 must satisfy the system

$$
\begin{cases}
(-1 - \sqrt{5})\alpha_1 - 2\alpha_2 = 0, \\
-2\alpha_1 + (1 - \sqrt{5})\alpha_2 = 0, \\
-\sqrt{5}\alpha_3 = 0,
\end{cases}
$$

a solution of which is $\alpha_1 = -2$, $\alpha_2 = 1 + \sqrt{5}$, $\alpha_3 = 0$.

Thus we obtain the characteristic vector $\begin{pmatrix} -2 \\ 1 + \sqrt{5} \\ 0 \end{pmatrix}$

corresponding to $\lambda = 2 + \sqrt{5}$.

The characteristic vector corresponding to $\lambda = 2 - \sqrt{5}$ has components α_1, α_2, α_3 such that

$$
\begin{pmatrix} 1 & -2 & 0 \\ -2 & 3 & 0 \\ 0 & 0 & 2 \end{pmatrix} \begin{pmatrix} \alpha_1 \\ \alpha_2 \\ \alpha_3 \end{pmatrix} = (2 - \sqrt{5}) \begin{pmatrix} \alpha_1 \\ \alpha_2 \\ \alpha_3 \end{pmatrix} \quad .
$$

Thus α_1, α_2, α_3 must satisfy the system

$$
\begin{cases} (-1 + \sqrt{5})\alpha_1 - 2\alpha_2 = 0, \\ -2\alpha_1 + (1 + \sqrt{5})\alpha_2 = 0, \\ \sqrt{5}\alpha_3 = 0, \end{cases}
$$

a solution of which is $\alpha_1 = 2$, $\alpha_2 = -1 + \sqrt{5}$, $\alpha_3 = 0$.
Thus we obtain the characteristic vector

$$
\begin{pmatrix} 2 \\ -1 + \sqrt{5} \\ 0 \end{pmatrix} \quad \text{corresponding to } \lambda = 2 - \sqrt{5}.
$$

The characteristic vector corresponding to $\lambda = 2$
has components α_1, α_2, α_3 such that

$$
\begin{pmatrix} 1 & -2 & 0 \\ -2 & 3 & 0 \\ 0 & 0 & 2 \end{pmatrix} \begin{pmatrix} \alpha_1 \\ \alpha_2 \\ \alpha_3 \end{pmatrix} = 2 \begin{pmatrix} \alpha_1 \\ \alpha_2 \\ \alpha_3 \end{pmatrix} \quad .
$$

Thus α_1, α_2, α_3 must satisfy the system

$$
\begin{cases} -\alpha_1 - 2\alpha_2 = 0, \\ -2\alpha_1 + \alpha_2 = 0, \\ 0\alpha_3 = 0. \end{cases}
$$

From the first two equations, we must have $\alpha_1 = \alpha_2 = 0$;
from the third equation, α_3 is an arbitrary nonzero

number.   Choosing $\alpha_3 = 1$, we have the characteristic

vector $\begin{pmatrix} 0 \\ 0 \\ 1 \end{pmatrix}$ corresponding to $\lambda = 2$.

By Theorem 7.24, a fundamental set of solutions of the given system is

$$\begin{pmatrix} -2 \\ 1 + \sqrt{5} \\ 0 \end{pmatrix} e^{(2+\sqrt{5})t} , \quad \begin{pmatrix} 2 \\ -1+\sqrt{5} \\ 0 \end{pmatrix} e^{(2-\sqrt{5})t} , \text{ and } \begin{pmatrix} 0 \\ 0 \\ 1 \end{pmatrix} e^{2t} ,$$

or

$$\begin{pmatrix} -2e^{(2+\sqrt{5})t} \\ (1+\sqrt{5})e^{(2+\sqrt{5})t} \\ 0 \end{pmatrix} , \quad \begin{pmatrix} 2e^{(2-\sqrt{5})t} \\ (-1+\sqrt{5})e^{(2-\sqrt{5})t} \\ 0 \end{pmatrix} \text{ and } \begin{pmatrix} 0 \\ 0 \\ e^{2t} \end{pmatrix} .$$

Thus a general solution is

$$\begin{cases} x_1 = -2c_1 e^{(2+\sqrt{5})t} + 2c_2 e^{(2-\sqrt{5})t} \\ x_2 = (1+\sqrt{5})c_1 e^{(2+\sqrt{5})t} + (-1+\sqrt{5})c_2 e^{(2-\sqrt{5})t} \\ x_3 = c_3 e^{2t} \end{cases}$$

11. We assume a solution of the form $\bar{x} = \bar{\alpha}e^{\lambda t}$, that is, $x_1 = \alpha_1 e^{\lambda t}$, $x_2 = \alpha_2 e^{\lambda t}$, $x_3 = \alpha_3 e^{\lambda t}$. Substituting into the given system and dividing through by $e^{\lambda t} \neq 0$, we obtain

$$\begin{cases} \alpha_1 \lambda = 11\alpha_1 + 6\alpha_2 + 18\alpha_3, \\ \alpha_2 \lambda = 9\alpha_1 + 8\alpha_2 + 18\alpha_3, \\ \alpha_3 \lambda = -9\alpha_1 - 6\alpha_2 - 16\alpha_3, \end{cases}$$

or

$$\begin{cases} (11 - \lambda)\alpha_1 + 6\alpha_2 + 18\alpha_3 = 0, \\ 9\alpha_1 + (8 - \lambda)\alpha_2 + 18\alpha_3 = 0, \\ -9\alpha_1 - 6\alpha_2 + (-16 - \lambda)\alpha_3 = 0. \end{cases}$$

This has a nontrivial solution for α_1, α_2, α_3 if and only if

$$\begin{vmatrix} 11 - \lambda & 6 & 18 \\ 9 & 8 - \lambda & 18 \\ -9 & -6 & -16 - \lambda \end{vmatrix} = 0.$$

This is the characteristic equation of the coefficient matrix of the system. It reduces to $(\lambda - 2)^2(\lambda + 1) = 0$; and its roots, the characteristic values, are -1, 2, 2 (double root).

A characteristic vector corresponding to $\lambda = -1$ has components α_1, α_2, α_3 such that

$$\begin{pmatrix} 11 & 6 & 18 \\ 9 & 8 & 18 \\ -9 & -6 & -16 \end{pmatrix} \begin{pmatrix} \alpha_1 \\ \alpha_2 \\ \alpha_3 \end{pmatrix} = (-1) \begin{pmatrix} \alpha_1 \\ \alpha_2 \\ \alpha_3 \end{pmatrix} .$$

Thus α_1, α_2, α_3 must satisfy the system

$$\begin{cases} 12\alpha_1 + 6\alpha_2 + 18\alpha_3 = 0, \\ 9\alpha_1 + 9\alpha_2 + 18\alpha_3 = 0, \\ -9\alpha_1 - 6\alpha_2 - 15\alpha_3 = 0. \end{cases} \quad \text{These simplify to}$$

$$\begin{cases} 2\alpha_1 + \alpha_2 + 3\alpha_3 = 0, \\ \alpha_1 + \alpha_2 + 2\alpha_3 = 0, \\ 3\alpha_1 + 2\alpha_2 + 5\alpha_3 = 0, \text{ and we readily find that} \end{cases}$$

$\alpha_1 = 1$, $\alpha_2 = 1$, $\alpha_3 = -1$ is a nontrivial solution of this. Thus we have the characteristic vector

$$\begin{pmatrix} 1 \\ 1 \\ -1 \end{pmatrix} \qquad \text{corresponding to } \lambda = -1.$$

We now consider the repeated characteristic value $\lambda = 2$. A corresponding characteristic vector has components α_1, α_2, α_3 such that

$$\begin{pmatrix} 11 & 6 & 18 \\ 9 & 8 & 18 \\ -9 & -6 & -16 \end{pmatrix} \begin{pmatrix} \alpha_1 \\ \alpha_2 \\ \alpha_3 \end{pmatrix} = 2 \begin{pmatrix} \alpha_1 \\ \alpha_2 \\ \alpha_3 \end{pmatrix} .$$

Thus α_1, α_2, α_3 must satisfy the system

$$\begin{cases} 9\alpha_1 + 6\alpha_2 + 18\alpha_3 = 0, \\ 9\alpha_1 + 6\alpha_2 + 18\alpha_3 = 0, \\ -9\alpha_1 - 6\alpha_2 - 18\alpha_3 = 0. \end{cases}$$

Each of these three relations is equivalent to the other two, and each is equivalent to

$$3\alpha_1 + 2\alpha_2 + 6\alpha_3 = 0.$$

This is the only relationship which α_1, α_2, α_3 must satisfy here. Two linearly independent solutions of it are $\alpha_1 = 2$, $\alpha_2 = 0$, $\alpha_3 = -1$, and $\alpha_1 = 0$, $\alpha_2 = 3$, $\alpha_3 = -1$.

That is, corresponding to the double characteristic value $\lambda = 2$, we have the two linearly independent character-istic vectors $\begin{pmatrix} 2 \\ 0 \\ -1 \end{pmatrix}$ and $\begin{pmatrix} 0 \\ 3 \\ -1 \end{pmatrix}$.

From the discussion of subcase (1), text, page 386, a fundamental set of solutions of the given system is

$$\begin{pmatrix} 1 \\ 1 \\ -1 \end{pmatrix} e^{-t} , \quad \begin{pmatrix} 2 \\ 0 \\ -1 \end{pmatrix} e^{2t} , \quad \text{and} \begin{pmatrix} 0 \\ 3 \\ -1 \end{pmatrix} e^{2t} ,$$

or

$$\begin{pmatrix} e^{-t} \\ e^{-t} \\ -e^{-t} \end{pmatrix} , \quad \begin{pmatrix} 2e^{2t} \\ 0 \\ -e^{2t} \end{pmatrix} , \quad \text{and} \begin{pmatrix} 0 \\ 3e^{2t} \\ -e^{2t} \end{pmatrix} .$$

Thus a general solution is

$$\begin{cases} x_1 = c_1 e^{-t} + 2c_2 e^{2t}, \\ x_2 = c_1 e^{-t} + 3c_3 e^{2t}, \\ x_3 = -c_1 e^{-t} - c_2 e^{2t} - c_3 e^{2t} . \end{cases}$$

12. We assume a solution of the form $\bar{x} = \bar{\alpha} e^{\lambda t}$, that is, $x_1 = \alpha_1 e^{\lambda t}$, $x_2 = \alpha_2 e^{\lambda t}$, $x_3 = \alpha_3 e^{\lambda t}$.

Substituting into the given system and dividing through by $e^{\lambda t} \neq 0$, we obtain

$$\begin{cases} \alpha_1 \lambda = \alpha_1 + 9\alpha_2 + 9\alpha_3 \, , \\ \alpha_2 \lambda = 19\alpha_2 + 18\alpha_3 \, , \\ \alpha_3 \lambda = 9\alpha_2 + 10\alpha_3, \end{cases}$$

or

$$\begin{cases} (1 - \lambda)\alpha_1 + 9\alpha_2 + 9\alpha_3 = 0, \\ (19 - \lambda)\alpha_2 + 18\alpha_3 = 0, \\ 9\alpha_2 + (10 - \lambda)\alpha_3 = 0 \, . \end{cases}$$

This has a nontrivial solution for α_1, α_2, α_3 if and only if

$$\begin{vmatrix} 1 - \lambda & 9 & 9 \\ 0 & 19 - \lambda & 18 \\ 0 & 9 & 10 - \lambda \end{vmatrix} = 0.$$

This is the characteristic equation of the coefficient matrix of the system. It reduces to

$(\lambda - 1)(\lambda^2 - 29\lambda + 28) = 0$ or $(\lambda - 1)^2(\lambda - 28) = 0$;

and its roots, the characteristic values, are 28, 1, 1 (double root).

A characteristic vector corresponding to $\lambda = 28$ has components α_1, α_2, α_3 such that

$$\begin{pmatrix} 1 & 9 & 9 \\ 0 & 19 & 18 \\ 0 & 9 & 10 \end{pmatrix} \begin{pmatrix} \alpha_1 \\ \alpha_2 \\ \alpha_3 \end{pmatrix} = 28 \begin{pmatrix} \alpha_1 \\ \alpha_2 \\ \alpha_3 \end{pmatrix} \, .$$ Thus α_1, α_2, α_3

must satisfy the system

$$\begin{cases} -27\alpha_1 + 9\alpha_2 + 9\alpha_3 = 0, \\ -9\alpha_2 + 18\alpha_3 = 0, \\ 9\alpha_2 - 18\alpha_3 = 0, \end{cases}$$

a nontrivial solution of which is $\alpha_1 = 1$, $\alpha_2 = 2$, $\alpha_3 = 1$.
Thus we have the characteristic vector

$$\begin{pmatrix} 1 \\ 2 \\ 1 \end{pmatrix} \quad \text{corresponding to} \quad \lambda = 28.$$

We now consider the repeated characteristic value
$\lambda = 1$. A corresponding characteristic vector has
components α_1, α_2, α_3 such that

$$\begin{pmatrix} 1 & 9 & 9 \\ 0 & 19 & 18 \\ 0 & 9 & 10 \end{pmatrix} \begin{pmatrix} \alpha_1 \\ \alpha_2 \\ \alpha_3 \end{pmatrix} = 1 \begin{pmatrix} \alpha_1 \\ \alpha_2 \\ \alpha_3 \end{pmatrix}. \quad \text{Thus } \alpha_1, \alpha_2,$$

α_3 must satisfy the system

$$\begin{cases} 9\alpha_2 + 9\alpha_3 = 0, \\ 18\alpha_2 + 18\alpha_3 = 0, \\ 9\alpha_2 + 9\alpha_3 = 0, \end{cases}$$

Each of these three relations is equivalent to the
other two, and each is equivalent to

$$\alpha_2 + \alpha_3 = 0.$$

This is the only relationship which α_1, α_2, α_3 must
satisfy; and we note that it allows α_1 to be arbitrary.
Two linearly independent solutions of it are

$\alpha_1 = 1$, $\alpha_2 = \alpha_3 = 0$ and $\alpha_1 = 0$, $\alpha_2 = 1$, $\alpha_3 = -1$.

That is, corresponding to the double characteristic value $\lambda = 1$, we have the two linearly independent characteristic vectors $\begin{pmatrix} 1 \\ 0 \\ 0 \end{pmatrix}$ and $\begin{pmatrix} 0 \\ 1 \\ -1 \end{pmatrix}$.

From the discussion of subcase (1), text, page 386, a fundamental set of solutions of the given system is

$$\begin{pmatrix} 1 \\ 2 \\ 1 \end{pmatrix} e^{28t}, \quad \begin{pmatrix} 1 \\ 0 \\ 0 \end{pmatrix} e^t \quad \text{and} \quad \begin{pmatrix} 0 \\ 1 \\ -1 \end{pmatrix} e^t,$$

or

$$\begin{pmatrix} e^{28t} \\ 2e^{28t} \\ e^{28t} \end{pmatrix}, \quad \begin{pmatrix} e^t \\ 0 \\ 0 \end{pmatrix}, \quad \text{and} \quad \begin{pmatrix} 0 \\ e^t \\ -e^t \end{pmatrix} .$$

Thus a general solution is

$$\begin{cases} x_1 = c_1 e^{28t} + c_2 e^t, \\ x_2 = 2c_1 e^{28t} + c_3 e^t, \\ x_3 = c_1 e^{28t} - c_3 e^t. \end{cases}$$

CHAPTER 8

Section 8.1, Page 399

3. The isoclines of the D.E. are $y/x^2 = c$, that is,
the parabolas $y = cx^2$ for $c \neq 0$, and the x-axis $y = 0$
for $c = 0$. In the figure these appear for $c = 0, \pm 1/2,$
± 1, and ± 2; and several line elements having the
appropriate inclination are drawn along each. For example, for $c = 1$, the corresponding isocline is the para-
bola $y = x^2$; and along this the line elements have
inclination arctan $1 = 45°$. Several approximate integral curves are shown, drawn heavily.

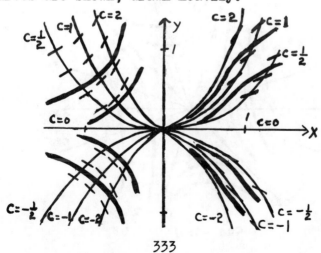

5.   The isoclines of the D.E. are $(3x - y)/(x+1) = c$.
Solving for y, we find that they are the straight lines

$$y = \frac{(3 - c)x}{c + 1} \quad \text{for } c \neq -1,$$

and the y-axis $x = 0$ for $c = -1$. In the figure these
appear for   each of the values   -3, -1, 0, 1, 2,
3, and 5; and several line elements having the appropriate inclincation are drawn along each. For example,
for $c = 2$, the corresponding isocline is the straight
line $y = x/3$; and along this the line elements have
inclination arctan $2 \approx 63°$. Several approximate integral curves are shown, drawn heavily. Note that the
isoclines $y = x$ (for $c = 1$) and $y = -3x$ (for $c = -3$) are
in fact special integral curves.

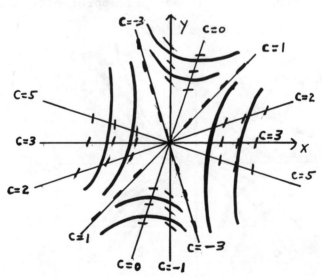

Section 8.2, Page 405

3. a. We have

$$y(x) = y(0) + y'(0)x + y''(0)x^2/2! + y'''(0)x^3/3!$$

$$+ y^{iv}(0)x^4/4! + \cdots. \tag{1}$$

The I.C. states $y(0) = 2$, and the D.E. gives $y'(0) = 1 + (0)(2)^2 = 1$. Differentiating the D.E., we obtain

$$\frac{d^2y}{dx^2} = 2xy \frac{dy}{dx} + y^2, \tag{2}$$

$$\frac{d^3y}{dx^3} = 2xy \frac{d^2y}{dx^2} + 2x(\frac{dy}{dx})^2 + 4y \frac{dy}{dx}, \tag{3}$$

$$\frac{d^4y}{dx^4} = 2xy \frac{d^3y}{dx^3} + 6x \frac{dy}{dx} \frac{d^2y}{dx^2} + 6y \frac{dy^2}{dx^2} + 6(\frac{dy}{dx})^2. \tag{4}$$

Substituting $x = 0$, $y = 2$, $\frac{dy}{dx} = 1$ into (2), we obtain

$$y''(0) = 2(0)(2)(1) + (2)^2 = 4.$$

Substituting $x = 0$, $y = 2$, $\frac{dy}{dx} = 1$, $\frac{d^2y}{dx^2} = 4$ into (3),

we obtain

$$y'''(0) = (2)(0)(2)(4) + (2)(0)(1)^2 + (4)(2)(1) = 8.$$

Substituting $x = 0$, $y = 2$, $\frac{dy}{dx} = 1$, $\frac{d^2y}{dx^2} = 4$, $\frac{d^3y}{dx^3} = 8$

into (4), we obtain

$$y^{iv}(0) = (2)(0)(1)(8) + (6)(0)(1)(4) + (6)(2)(4) + (6)(1)^2$$

$$= 54.$$

Now substituting the values of $y(0)$, $y'(0)$, $y''(0)$, $y'''(0)$, and $y^{iv}(0)$ so determined into (1), we obtain

$$y = 2 + 1x + 4x^2/2! + 8x^3/3! + 54x^4/4! + \cdots$$
$$= 2 + x + 2x^2 + 4x^3/3 + 9x^4/4 + \cdots.$$

  b. We assume

$$y = c_0 + c_1x + c_2x^2 + c_3x^3 + c_4x^4 + \cdots. \quad \text{To}$$

satisfy the I.C. $y(0) = 2$, we must have $c_0 = 2$, and hence

$$y = 2 + c_1x + c_2x^2 + c_3x^3 + c_4x^4 + \cdots. \qquad (5)$$

Differentiating this, we find

$$\frac{dy}{dx} = c_1 + 2c_2x + 3c_3x^2 + 4c_4x^3 + \cdots. \qquad (6)$$

  Since the initial y value is 2, we must express y^2 in the D.E. in powers of $y - 2$. Then the D.E. takes the form

$$\frac{dy}{dx} = 1 + x[(y-2)^2 + 4(y-2) + 4].$$

Now substituting (5) and (6) into this, we obtain

$$c_1 + 2c_2x + 3c_3x^2 + 4c_4x^3 + \cdots$$
$$= 1 + x[(c_1x + c_2x^2 + \cdots)^2 + 4(c_1x + c_2x^2 + c_3x^3 + \cdots)$$
$$+ 4]$$

or

$$c_1 + 2c_2x + 3c_3x^2 + 4c_4x^3 + \cdots$$
$$= 1 + 4x + 4c_1x^2 + (c_1^2 + 4c_2)x^3 + \cdots.$$

From this, $c_1 = 1$, $2c_2 = 4$, $3c_3 = 4c_1$, $4c_4 = c_1^2 + 4c_2$;

and from these, $c_1 = 1$, $c_2 = 2$, $c_3 = 4c_1/3 = 4/3$,

$c_4 = (c_1^2 + 4c_2)/4 = 9/4$. Substituting these into (5)

we again obtain

$$y = 2 + x + 2x^2 + 4x^3/3 + 9x^4/4 + \cdots.$$

4. a. We have

$$y(x) = y(0) + y'(0)x + y''(0)x^2/2! + y'''(0)x^3/3!$$

$$+ y^{iv}(0) \, x^4/4! + \cdots. \tag{1}$$

The I.C. states that $y(0) = 3$, and the D.E. gives

$y'(0) = (0)^3 + (3)^3 = 27$. Differentiating the D.E.,

we obtain

$$\frac{d^2y}{dx^2} = 3x^2 + 3y^2 \frac{dy}{dx}, \tag{2}$$

$$\frac{d^3y}{dx^3} = 6x + 3y^2 \frac{d^2y}{dx^2} + 6y\left(\frac{dy}{dx}\right)^2, \tag{3}$$

$$\frac{d^4y}{dx^4} = 6 + 3y^2 \frac{d^3y}{dx^3} + 18y \frac{dy}{dx} \frac{d^2y}{dx^2} + 6\left(\frac{dy}{dx}\right)^3. \tag{4}$$

Substituting $x = 0$, $y = 3$, $\frac{dy}{dx} = 27$ into (1), we obtain

$$y''(0) = 3(0)^2 + 3(3)^2(27) = 729.$$

Substituting $x = 0$, $y = 3$, $\frac{dy}{dx} = 27$, $\frac{d^2y}{dx^2} = 729$ into (3),

we obtain

$$y'''(0) = 6(0) + 3(3)^2(729) + 6(3)(27)^2 = 32,805.$$

Substituting $y = 3$, $\frac{dy}{dx} = 27$, $\frac{d^2y}{dx^2} = 729$, $\frac{d^3y}{dx^3} = 32{,}805$

into (4), we obtain

$$y^{iv}(0) = 6 + 3(3)^2(32{,}805) + (18)(3)(27)(729) + 6(27)^3$$

$$= 2{,}066{,}721.$$

Now substituting the values of $y(0)$, $y'(0)$, $y''(0)$,
$y'''(0)$ and $y^{iv}(0)$ so determined into (1), we obtain

$$y = 3 + 27x + 729x^2/2 + 32{,}805x^3/6 + 2{,}066{,}721x^4/24 + \cdots$$

$$= 3 + 27x + 729x^2/2 + 10{,}935x^3/2 + 688{,}907x^4/8 + \cdots .$$

   b. We assume $c_0 + c_1 x + c_2 x^2 + c_3 x^3 + c_4 x^4 + \cdots$.
To satisfy the I.C. $y(0) = 3$, we must have $c_0 = 3$, and
hence
$$y = 3 + c_1 x + c_2 x^2 + c_3 x^3 + c_4 x^4 + \cdots . \qquad (5)$$
Differentiating this, we find
$$\frac{dy}{dx} = c_1 + 2c_2 x + 3c_3 x^2 + 4c_4 x^3 + \cdots . \qquad (6)$$

   Since the initial y value is 3, we must express y^3
in the D.E. in powers of $y - 3$. Then the D.E. takes
the form
$$\frac{dy}{dx} = x^3 + (y - 3)^3 + 9(y - 3)^2 + 27(y - 3) + 27.$$

Now substituting (5) and (6) into this, we obtain

$$c_1 + 2c_2x + 3c_3x^2 + 4c_4x^3 + \cdots$$

$$= x^3 + (c_1x + c_2x^2 + \cdots)^3 + 9(c_1x + c_2x^2 + \cdots)^2$$

$$+ 27(c_1x + c_2x^2 + c_3x^3 + \cdots) + 27$$

or

$$c_1 + 2c_2x + 3c_3x^2 + 4c_4x^3 + \cdots$$

$$= 27 + 27c_1x + (9c_1^2 + 27c_2)x^2$$

$$+ (1 + c_1^3 + 18c_1c_2 + 27c_3)x^3 + \cdots.$$

From this, $c_1 = 27$, $2c_2 = 27c_1$, $3c_3 = 9c_1^2 + 27c_2$, $4c_4 = 1 + c_1^3 + 18c_1c_2 + 27c_3$; and from these, $c_1 = 27$, $c_2 = 27c_1/2 = 729/2$, $c_3 = 3c_1^2 + 9c_2 = 3(27)^2 + 9(729)/2 = 10,935/2$, $c_4 = (1 + c_1^3 + 18c_1c_2 + 27c_3)/4 = (1 + 19,683 + 177,147 + 295,245/2)/4 = 688,907/8$. Substituting these into (5), we again obtain

$$y = 3 + 27x + 729x^2/2 + 10,935x^3/2 + 688,907x^4/8 + \cdots .$$

11.   a.   We have

$$y(x) = y(1) + y'(1)(x-1) + y''(1)(x-1)^2/2!$$

$$+ y'''(1)(x-1)^3/3! + \cdots$$

$$= \sum_{n=0}^{\infty} y^{(n)}(1)(x-1)^n/n! \,. \tag{1}$$

The I.C. states that $y(1) = \pi$, and the D.E. gives

$y'(1) = 1+ \cos \pi = 0$. Differentiating the D.E., we obtain

$$\frac{d^2y}{dx^2} = 1 - (\sin y) \frac{dy}{dx} , \qquad (2)$$

$$\frac{d^3y}{dx^3} = - (\sin y) \frac{d^2y}{dx^2} - (\cos y)(\frac{dy}{dx})^2, \qquad (3)$$

$$\frac{d^4y}{dx^4} = - (\sin y) \frac{d^3y}{dx^3} - 3(\cos y) \frac{d^2y}{dx^2} \frac{dy}{dx}$$

$$+ (\sin y) (\frac{dy}{dx})^2, \qquad (4)$$

$$\frac{d^5y}{dx^5} = - (\sin y) \frac{d^4y}{dx^4} - 4(\cos y) \frac{d^3y}{dx^3} \frac{dy}{dx}$$

$$- 3(\cos y)(\frac{d^2y}{dx^2})^2 + 3(\sin y)(\frac{d^2y}{dx^2})(\frac{dy}{dx})^2$$

$$+ 2(\sin y)(\frac{dy}{dx})(\frac{d^2y}{dx^2}) + (\cos y)(\frac{dy}{dx})^3. \qquad (5)$$

Substituting $x = 1$, $y = \pi$, $\frac{dy}{dx} = 0$ into (2), we obtain $y''(1) = 1 - (\sin \pi)(0) = 1$. Substituting $y = \pi$, $\frac{dy}{dx} = 0$, $\frac{d^2y}{dx^2} = 1$ into (3), we obtain $y'''(1) = -(\sin \pi)(1)$ $- (\cos \pi)(0)^2 = 0$. Substituting $y = \pi$, $\frac{dy}{dx} = 0$, $\frac{d^2y}{dx^2} = 1$, $\frac{d^3y}{dx^3} = 0$ into (4), we obtain

$$y^{iv}(1) = -(\sin \pi)(0) - (3 \cos \pi)(1)(0)$$
$$+ (\sin \pi)(0)^2 = 0.$$

Substituting $y = \pi$, $\frac{dy}{dx} = 0$, $\frac{d^2y}{dx^2} = 1$, $\frac{d^3y}{dx^3} = 0$, $\frac{d^4y}{dx^4} = 0$

into (5), we obtain

$y^V(1) = -(\sin \pi)(0) - 4(\cos \pi)(0)(0) - 3(\cos \pi)(1)^2$

$\qquad + 3(\sin \pi)(1)(0)^2 + 2(\sin \pi)(0)(1) + (\cos \pi)(0)^3 = 3.$

Now substituting the values of $y(1)$, $y'(1)$, $\cdots$, $y^V(1)$

so determined into (1), we obtain

$y = \pi + (x-1)^2/2! + 3(x-1)^5/5! + \cdots$

$\qquad = \pi + (x-1)^2/2 + (x-1)^5/40 + \cdots$.

     b. We assume

$y = c_0 + c_1(x-1) + c_2(x-1)^2 + c_3(x-1)^3 + c_4(x-1)^4$

$\qquad\qquad + c_5(x-1)^5 + \cdots$.

To satisfy the I.C. $y(1) = \pi$, we must have $c_0 = \pi$, and

hence

$y = \pi + c_1(x-1) + c_2(x-1)^2 + c_3(x-1)^3 + c_4(x-1)^4$

$\qquad\qquad + c_5(x-1)^5 + \cdots$.
$\hfill$ (6)

Differentiating this, we find

$\frac{dy}{dx} = c_1 + 2c_2(x-1) + 3c_3(x-1)^2 + 4c_4(x-1)^3 + 5c_5(x-1)^4 + \cdots$.
$\hfill$ (7)

     Since the initial y value is π, we must express $\cos y$

in the D.E. in powers of $y - \pi$. By Taylor's Theorem,

$$\phi(y) = \sum_{n=0}^{\infty} \frac{\phi^{(n)}(\pi)}{n!} (y - \pi)^n,$$

with $\phi(y) = \cos y$, we obtain

$\cos y = -1 + (y-\pi)^2/2! - (y-\pi)^4/4! + \cdots$

We also express x in powers of x - 1, and then the D.E. takes the form

$$\frac{dy}{dx} = [(x-1)+1]+[-1+(y-\pi)^2/2! - (y-\pi)^4/4! + \cdots]$$

or

$$\frac{dy}{dx} = x - 1 + (y-\pi)^2/2 - (y-\pi)^4/24 + \cdots.$$

Now substituting (6) and (7) into this, we obtain

$c_1 + 2c_2(x-1) + 3c_3(x-1)^2 + 4c_4(x-1)^3 + 5c_5(x-1)^4 + \cdots$

$= x-1 + [c_1(x-1) + c_2(x-1)^2 + \cdots]^2/2$

$- [c_1(x-1) + c_2(x-1)^2 + \cdots]^4/24 + \cdots$

$= (x-1) + (c_1^2/2)(x-1)^2 + c_1 c_2 (x-1)^3$

$+ (c_1 c_3 + c_2^2/2 + c_1^4/24)(x-1)^4 + \cdots.$

From this, $c_1 = 0$, $2c_2 = 1$, $3c_3 = c_1^2/2$, $4c_4 = c_1 c_2$, $5c_5 = c_1 c_3 + c_2^2/2 + c_1^2/24$; and from these $c_1 = 0$, $c_2 = 1/2$, $c_3 = 0$, $c_4 = 0$, $c_5 = (0 + 1/8 + 0)/5 = 1/40$. Substituting these into (6), we again obtain

$$y = \pi + (x-1)^2/2 + (x-1)^5/40 + \cdots.$$

Section 8.3, Page 409

4. Since the initial y value is 0, we choose the zeroth approximation ϕ_0 to be the function defined by

$\phi_0(x) = 0$ for all x. The nth approximation ϕ_n for $n \geq 1$ is given by

$$\phi_n(x) = y_0 + \int_{x_0}^{x} f[t, \phi_{n-1}(t)]dt$$

$$= 0 + \int_{0}^{x} \left\{ 1 + t[\phi_{n-1}(t)]^2 \right\} dt, \quad n \geq 1.$$

Using this formula for n = 1, 2, 3, $\cdots$, we obtain successively

$$\phi_1(x) = \int_{0}^{x} \left\{ 1 + t[\phi_0(t)]^2 \right\} dt = \int_{0}^{x}(1+0)dt = x,$$

$$\phi_2(x) = \int_{0}^{x} \left\{ 1 + t[\phi_1(t)]^2 \right\} dt = \int_{0}^{x}(1+t^3)dt$$

$$= x + x^4/4,$$

$$\phi_3(x) = \int_{0}^{x} \left\{ 1 + t[\phi_2(t)]^2 \right\} dt$$

$$= \int_{0}^{x} [1 + t(t + t^4/4)^2]dt$$

$$= \int_{0}^{x} (1 + t^3 + t^6/2 + t^9/16)dt$$

$$= x + x^4/4 + x^7/14 + x^{10}/160.$$

5. Since the initial y value is 0, we choose the zeroth approximation ϕ_0 to be the function defined by $\phi_0(x) = 0$ for all x. The nth approximation ϕ_n for $n \geq 1$ is

given by

$$\phi_n(x) = y_0 + \int_{x_0}^{x} f[t, \phi_{n-1}(t)]dt$$

$$= 0 + \int_0^x \left\{ e^t + [\phi_{n-1}(t)]^2 \right\} dt, \qquad n \geq 1.$$

Using this formula for $n = 1, 2, 3, \cdots$, we obtain successively

$$\phi_1(x) = \int_0^x (e^t + 0)dt = e^t \Big|_0^x = e^x - 1,$$

$$\phi_2(x) = \int_0^x [e^t + (e^t - 1)^2]dt$$

$$= \int_0^x (e^{2t} - e^t + 1)dt$$

$$= e^{2t}/2 - e^t + t \Big|_0^x = e^{2x}/2 - e^x + x + 1/2,$$

$$\phi_3(x) = \int_0^x [e^t + (e^{2t}/2 - e^t + t + 1/2)^2]dt$$

$$= \int_0^x (e^{4t}/4 - e^{3t} + te^{2t} + 3e^{2t}/2$$

$$\qquad - 2te^t + t^2 + t + 1/4)dt$$

$$= [e^{4t}/16 - e^{3t}/3 + e^{2t}(2t - 1)/4 + 3e^{2t}/4$$

$$\qquad -2e^t(t-1) + t^3/3 + t^2/2 + t/4]_0^x$$

$$= e^{4x}/16 - e^{3x}/3 + xe^{2x}/2 + e^{2x}/2$$

$$- 2xe^x + 2e^x + x^3/3 + x^2/2 + x/4 - 107/48.$$

Section 8.4B, Page 414

1.   a.   We use formula (8.49) with $f(x,y) = x - 2y$
and $h = 0.1$.   We find

(1) $x_1 = x_0 + h = 0.1$, $f(x_0,y_0) = f(0,1) = -2.000$,

$y_1 = y_0 + hf(x_0,y_0) = 1.000 + 0.1(-2.000) = 0.800$.

(2) $x_2 = x_1 + h = 0.2$, $f(x_1,y_1) = f(0.1,0.800) = -1.500$,

$y_2 = y_1 + hf(x_1,y_1) = 0.800 + 0.1(-1.500) = 0.650$.

(3) $x_3 = x_2 + h = 0.3$, $f(x_2,y_2) = f(0.2,0.650) = -1.100$,

$y_3 = y_2 + hf(x_2,y_2) = 0.650 + 0.1(-1.100) = 0.540$.

(4) $x_4 = x_3 + h = 0.4$, $f(x_3,y_3) = f(0.3, 0.540) = -0.780$,

$y_4 = y_3 + hf(x_3,y_3) = 0.540 + 0.1(-0.780) = 0.462$.

The desired approximations are $y_1 = 0.800$, $y_2 = 0.650$,
$y_3 = 0.540$, and $y_4 = 0.462$, respectively.

   b.   We use formula (8.49) with
$f(x,y) = x - 2y$ and $h = 0.05$.   We find

(1) $x_1 = x_0 + h = 0.05$, $f(x_0,y_0) = f(0,1) = -2.000$,

$y_1 = y_0 + hf(x_0,y_0) = 1.000 + 0.05(-2.000) = 0.900$.

(2) $x_2 = x_1 + h = 0.1$, $f(x_1,y_1) = f(0.05,0.900) = -1.750$,

$y_2 = y_1 + hf(x_1,y_1) = 0.900 + 0.05(-1.750) = 0.812$,

(3) $x_3 = x_2 + h = 0.15$, $f(x_2, y_2) = f(0.1, 0.812) = -1.524$,

$\qquad y_3 = y_2 + hf(x_2, y_2) = 0.812 + 0.05(-1.524) = 0.736$.

(4) $x_4 = x_3 + h = 0.2$, $f(x_3, y_3) = f(0.15, 0.736) = -1.322$,

$\qquad y_4 = y_3 + hf(x_3, y_3) = 0.736 + 0.05(-1.322) = 0.670$.

(5) $x_5 = x_4 + h = 0.25$, $f(x_4, y_4) = f(0.2, 0.670) = -1.140$,

$\qquad y_5 = y_4 + hf(x_4, y_4) = 0.670 + 0.05(-1.140) = 0.613$,

(6) $x_6 = x_5 + h = 0.3$, $f(x_5, y_5) = f(0.25, 0.613) = -0.976$

$\qquad y_6 = y_5 + hf(x_5, y_5) = 0.613 + 0.05(-0.976) = 0.564$.

(7) $x_7 = x_6 + h = 0.35$, $f(x_6, y_6) = f(0.3, 0.564) = -0.828$,

$\qquad y_7 = y_6 + hf(x_6, y_6) = 0.564 + 0.05(-0.828) = 0.523$,

(8) $x_8 = x_7 + h = 0.4$, $f(x_7, y_7) = f(0.35, 0.523) = -0.696$,

$\qquad y_8 = y_7 + hf(x_7, y_7) = 0.523 + 0.05(-0.696) = 0.488$.

The desired approximations are $y_2 = 0.812$, $y_4 = 0.670$, $y_6 = 0.564$, and $y_8 = 0.488$, respectively.

     c. The given D.E. is linear. We write it in the standard form as $dy/dx + 2y = x$. An I.F. is $e^{\int 2dx} = e^{2x}$. Multiplying through by this, we obtain $e^{2x} dy/dx + 2e^{2x}y = xe^{2x}$ or $\frac{d}{dx}[e^{2x} y] = xe^{2x}$. Integrating and simplifying, we find $y = x/2 - 1/4 + ce^{-2x}$. Applying the I.C., we find $c = 5/4$. Thus we obtain the exact solution $y = x/2 - 1/4 + 5e^{-2x}/4$. The values of

this at x = 0.1, 0.2, 0.3, and 0.4 are readily found to
be 0.823, 0.688, 0.586, and 0.512.

Section 8.4C, Page 418

1. a. Here $f(x,y) = 3x + 2y$, $x_0 = 0$, $y_0 = 1$, and
$h = 0.1$. We shall first approximate the value of y at
$x_1 = x_0 + h = 0.1$. We use formula (8.50) to find the
first approximation to this. We have

$$y_1^{(1)} = y_0 + hf(x_0,y_0) = 1.000 + (0.1)(2.000) = 1.200.$$

We now use (8.52) to find the second approximation
$y_1^{(2)}$. Since $f(x_1,y_1^{(1)}) = f(0.1, 1.200) = 2.700$, we have

$$y_1^{(2)} = y_0 + \frac{f(x_0,y_0) + f(x_1,y_1^{(1)})}{2} \quad h$$

$$= 1.000 + \left(\frac{2.000 + 2.700}{2}\right)(0.1) = 1.235.$$

We next use (8.53) to find the third approximation $y_1^{(3)}$.
Since $f(x_1,y_1^{(2)}) = f(0.1, 1.235) = 2.770$, we have

$$y_1^{(3)} = y_0 + \frac{f(x_0,y_0) + f(x_1,y_1^{(2)})}{2} \quad h$$

$$= 1.000 + \frac{2.000 + 2.770}{2}(0.1) = 1.239.$$

In like manner, we proceed to find the fourth approximation $y_1^{(4)}$. Since $f(x_1,y_1^{(3)}) = f(0.1, 1.239) = 2.778$,
we find

$$y_1^{(4)} = y_0 + \frac{f(x_0,y_0) + f(x_1,y_1^{(3)})}{2} h$$

$$= 1.000 + \frac{2.000 + 2.778}{2} (0.1) = 1.239.$$

Since $y_1^{(3)}$ and $y_1^{(4)}$ are the same to the required number of decimal places, we take their common value as the approximation y_1 to the value of the solution y at $x_1 = 0.1$. Thus we have

$$y_1 = 1.239 \quad .$$

We now proceed to approximate the value of y at $x_2 = x_1 + h = 0.2$. We use formulas (8.54) with $y_1 = 1.239$. We find successively

$$y_2^{(1)} = y_1 + hf(x_1,y_1) = 1.239 + 0.1(2.778) = 1.517.$$

$$y_2^{(2)} = y_1 + \frac{f(x_1,y_1) + f(x_2,y_2^{(1)})}{2} h$$

$$= 1.239 + \frac{2.778 + 3.634}{2}(0.1) = 1.560 \quad .$$

$$y_2^{(3)} = y_1 + \frac{f(x_1,y_1) + f(x_2,y_2^{(2)})}{2} h$$

$$= 1.239 + \frac{2.778 + 3.720}{2} (0.1) = 1.564 \quad .$$

$$y_2^{(4)} = y_1 + \frac{f(x_1,y_1) + f(x_2,y_2^{(3)})}{2} h$$

$$= 1.239 + \frac{2.778 + 3.728}{2} = 1.564 \quad .$$

Since $y_2^{(3)}$ and $y_2^{(4)}$ are the same to the required number of decimal places, we take their common value as the approximation y_2 to the value of the solution y at $x_2 = 0.2$. Thus we have

$$y_2 = 1.564 \ .$$

We now proceed to approximate the value of y at $x_3 = x_2 + h = 0.3$. We use formulas analogous to (8.54), with each subscript 1 replaced by 2, each subscript 2 replaced by 3, and $y_2 = 1.564$. We find successively

$$y_3^{(1)} = y_2 + hf(x_2,y_2) = 1.564 + 0.1(3.728) = 1.937.$$

$$y_3^{(2)} = y_2 + \frac{f(x_2,y_2) + f(x_3,y_3^{(1)})}{2} \ h$$

$$= 1.564 + \frac{3.728 + 4.174}{2} \ (0.1) = 1.959.$$

$$y_3^{(3)} = y_2 + \frac{f(x_2,y_2) + f(x_3,y_3^{(2)})}{2} \ h$$

$$= 1.564 + \frac{3.728 + 4.818}{2} \ (0.1) = 1.991.$$

$$y_3^{(4)} = y_2 + \frac{f(x_2,y_2) + f(x_3,y_3^{(3)})}{2} \ h$$

$$= 1.564 + \frac{3.728 + 4.882}{2} \ (0.1) = 1.995.$$

$$y_3^{(5)} = y_2 + \frac{f(x_2,y_2) + f(x_3,y_3^{(4)})}{2} \ h$$

$$= 1.564 + \frac{3.728 + 4.889}{2} (0.1) = 1.995.$$

Since $y_3^{(4)}$ and $y_3^{(5)}$ are the same to the required
number of decimal places, we take their common value as
the approximation y_3 to the value of the solution y at
$x_3 = 0.3$. We thus find

$$y_3 = 1.995 .$$

   b.   Here $f(x,y) = 3x + 2y$, $x_0 = 0$, $y_0 = 1$, and
$h = 0.05$. We shall first approximate the value of y at
$x_1 = x_0 + h = 0.05$. We use formula (8.50) to find the
first approximation to this. We have

$$y_1^{(1)} = y_0 + hf(x_0,y_0) = 1.000 + (0.05)(2.000) = 1.100.$$

We now use (8.52) to find the second approximation
$y_1^{(2)}$.

$$y_1^{(2)} = y_0 + \frac{f(x_0,y_0) + f(x_1,y_1^{(1)})}{2} h$$

$$= 1.000 + \frac{2.000 + 2.350}{2} (0.05) = 1.109.$$

We next use (8.53) to find the third approximation
$y_1^{(3)}$.

$$y_1^{(3)} = y_0 + \frac{f(x_0,y_0) + f(x_1,y_1^{(2)})}{2} h$$

$$= 1.000 + \frac{2.000 + 2.368}{2} (0.05) = 1.109 .$$

Since $y_1^{(2)}$ and $y_1^{(3)}$ are the same to the required
number of decimal places, we take their common value as
the approximation y_1 to the value of the solution at
$x_1 = 0.05$. Thus we have

$$y_1 = 1.109 \ .$$

We now proceed to approximate the value of y at
$x_2 = x_1 + h = 0.1$. We use formula (8.54) with
$y_1 = 1.109$. We find successively $y_2^{(1)} = y_1 + hf(x_1,y_1)$
$= 1.109 + 0.05(2.368) = 1.227.$

$$y_2^{(2)} = y_1 + \frac{f(x_1,y_1) + f(x_2,y_2^{(1)})}{2} h$$

$$= 1.109 + \frac{2.368 + 2.754}{2} (0.05) = 1.237.$$

$$y_2^{(3)} = y_1 + \frac{f(x_1,y_1) + f(x_2,y_2^{(2)})}{2} h$$

$$= 1.109 + \frac{2.368 + 2.774}{2} (0.05) = 1.238.$$

$$y_2^{(4)} = y_1 + \frac{f(x_1,y_1) + f(x_2,y_2^{(3)})}{2} h$$

$$= 1.109 + \frac{2.368 + 2.775}{2} (0.05) = 1.238.$$

Since $y_2^{(3)}$ and $y_2^{(4)}$ are the same to the required
number of decimal places, we take their common value as
the approximation y_2 to the value of the solution y at

$x_2 = 0.1.$ Thus we obtain

$$y_2 = 1.238.$$

This is the first result required in part (b).

To find the next result required in part (b), first
proceed to find the approximation y_3 to the value of y
at $x_3 = x_2 + h = 0.15.$ Use formulas analogous to
(8.54), with each subscript 1 replaced by 2, each
subscript 2 replaced by 3, and $y_2 = 1.238.$ Having
found y_3 in this way, then proceed to find the
approximation y_4 to the value of y at $x_4 = x_3 + h = 0.2.$
Use formulas analogous to (8.54), with each subscript 1
replaced by 3, each subscript 2 replaced by 4, and the
approximation y_3 first found. The approximation y_4
obtained in this way is the second result required in
part (b). The third result is then obtained analo-
gously.

c. The D.E. is linear. In standard form it is
$dy/dx - 2y = 3x;$ and an I.F. is $e^{-2x}.$ Multiplying
through by this, we have $d[e^{-2x}y]/dx = 3xe^{-2x}.$
Integrating and simplifying, we find $y = -3x/2 - 3/4$
$+ ce^{2x}.$ Applying the I.C., we find $c = 7/4.$ Hence we
obtain the exact solution $y = -3x/2 - 3/4 + 7e^{2x}/4.$
Using this, the values of the exact solution (to

three decimal places) at x = 0.1, 0.2, and 0.3 are
readily found to be 1.237, 1.561, and 1.989, respectively.

Section 8.4D, Page 423

1. Here $f(x,y) = 3x + 2y$, $x_0 = 0$, $y_0 = 1$, and
$h = 0.1$. Using these, we calculate successively k_1,
k_2, k_3, k_4, and k_0 defined by (8.57). We find
$k_1 = hf(x_0,y_0) = 0.1f(0,1) = 0.1(2) = 0.20000$. Since
$x_0 + h/2 = 0.05$ and $y_0 + k_1/2 = 1 + (0.20000)/2 = 1.10000$,
we find

$k_2 = hf(x_0 + h/2, y_0 + k_1/2) = 0.1 \, f(0.05, 1.10000)$

$\quad = 0.1(2.35000) = 0.23500.$

Since $y_0 + k_2/2 = 1 + (0.23500)/2 = 1.11750$, we find

$k_3 = hf(x_0 + h/2, y_0 + k_2/2) = 0.1f(0.05, 1.11750)$

$\quad = 0.1(2.38500) = 0.23850.$

Since $x_0 + h = 0.1$ and $y_0 + k_3 = 1.23850$, we find

$k_4 = hf(x_0 + h, y_0 + k_3) = 0.1f(0.1, 1.23850)$

$= 0.1(2.777) = 0.27770.$

Then $k_0 = (k_1 + 2k_2 + 2k_3 + k_4)/6$

$\quad = (0.20000 + 0.47000 + 0.47700 + 0.27770)/6$

$\quad = 0.23745.$

We round off k_0 by the rule of rounding off so as to leave the last retained digit even, and thus we take $k_0 = 0.2374$. Then by (8.58), the approximate value of the solution at $x_1 = 0.1$ is

$$y_1 = 1 + 0.2374 = 1.2374.$$

We now proceed to approximate the value of the solution at $x_2 = x_1 + h = 0.2$. We have $x_1 = 0.1$ and $y_1 = 1.2374$. Using these, we calculate successively k_1, k_2, k_3, k_4, and k_1 defined by (8.59). We find

$$k_1 = hf(x_1,y_1) = 0.1f(0.1, 1.2374) = 0.1(2.77480)$$

$= 0.27748$. Since $x_1 + h/2 = 0.15$ and $y_1 + k_1/2 = 1.2374 + 0.27748/2 = 1.37614$, we find

$$k_2 = hf(x_1 + h/2, y_1 + k_1/2) = 0.1f(0.15, 1.37614)$$

$$= 0.1(3.20228) = 0.32023.$$

Since $y_1 + k_2/2 = 1.2374 + 0.16011 = 1.39751$, we find

$$k_3 = hf(x_1 + h/2, y_1 + k_2/2) = 0.1f(0.15, 1.39751)$$

$$= 0.1(3.24502) = 0.32450.$$

Since $x_1 + h = 0.2$ and $y_1 + k_3 = 1.56190$, we find

$$k_4 = hf(x_1 + h, y_1 + k_3) = 0.1f(0.2, 1.56190)$$

$$= 0.1(3.72380) = 0.37238.$$

Then $K_1 = (k_1 + 2k_2 + 2k_3 + k_4)/6$

$= (0.27748 + 0.64046 + 0.64900 + 0.37238)/6$

$= 0.32322.$

We round off K_1, obtaining $K_1 = 0.3232$. Then by (8.60), the approximate value of the solution at $x_2 = 0.2$ is

$$y_2 = y_1 + K_1 = 1.2374 + 0.3232 = 1.5606.$$

Finally, we proceed to approximate the value of the solution at $x_3 = x_2 + h = 0.3$. We have $x_2 = 0.2$ and $y_2 = 1.5606$. Using these we calculate successively k_1, k_2, k_3, k_4, and $K_n = K_2$ using the formulas in the middle of page 420 of the text with $n = 2$. We find

$k_1 = hf(x_2,y_2) = 0.1f(0.2,1.5606)$

$= 0.1(3.72120) = 0.37212.$

Since $x_2 + h/2 = 0.25$ and $y_2 + k_1/2 = 1.5606 + 0.37212/2$ $= 1.74666$, we find $k_2 = hf(x_2 + h/2, y_2 + k_1/2) =$ $0.1f(0.25, 1.74666) = 0.1(4.24332) = 0.42433.$

Since $y_2 + k_2/2 = 1.5606 + 0.42433/2 = 1.77277$, we find $k_3 = hf(x_2 + h/2, y_2 + k_2/2) = 0.1f(0.25, 1.77277)$

$0.1(4.29553) = 0.42955.$

Since $x_2 + h = 0.3$ and $y_2 + k_3 = 1.99015$, we find $k_4 = hf(x_2 + h, y_2 + k_3) = 0.1f(0.3, 1.99015)$

$= 0.1(4.88030) = 0.48803.$

Then $K_2 = (k_1 + 2k_2 + 2k_3 + k_6)/6$

$= (0.37217 + 0.84866 + 0.85910 + 0.48803)/6$

$= 0.42799.$

We round off K_2, obtaining $K_2 = 0.4280$. Then the
approximate value of the solution at $x_3 = 0.3$ is
$y_3 = y_2 + K_2 = 1.5606 + 0.4280 = 1.9886$.

b. The exact solution of the problem was found in
Exercise 1 of Section 8.4C. It is $y = -3x/2 - 3/4$
$+ 7e^{2x}/4$. Using this, the values of the exact solution
(to four decimal places) at $x = 0.1$, 0.2, and 0.3 are
readily found to be 1.2374, 1.5607, and 1.9887,
respectively.

Section 8.4E, Page 425

1. We apply the formulas (8.63), (8.64), and (8,65)
with $n = 3$. We set
$x_0 = 0$, $x_1 = 0.1$, $x_2 = 0.2$, $x_3 = 0.3$,
$y_0 = 1.000$, $y_1 = 1.2375$, $y_2 = 1.5607$, $y_3 = 1.9887$, and
$x_4 = 0.4$. Then using $f(x,y) = 3x + 2y$, we find
$y_1' = f(x_1,y_1) = f(0.1, 1.2375) = 2.7750$,
$y_2' = f(x_2,y_2) = f(0.2, 1.5607) = 3.7214$,
$y_3' = f(x_3,y_3) = f(0.3, 1.9887) = 4.8774$.

We now use (8.63) with $n = 3$ and $h = 0.1$ to determine
$y_4^{(1)}$. We have
$$y_4^{(1)} = y_0 + 4h(2y_3' - y_2' + 2y_1')/3$$
$$= 1.000 + 0.4(9.7548 - 3.7214 + 5.5500)/3$$

$$= 2.5445.$$

We now use (8.64) to determine $y_4{'}^{(1)}$. We find

$$y_4{'}^{(1)} = f(x_4, y_4^{(1)}) = f(0.4, 2.5445) = 6.2889.$$

We now use (8.65) with n = 3 and h = 0.1 to determine $y_4^{(2)}$. We obtain

$$y_4^{(2)} = y_2 + h(y_4{'}^{(1)} + 4y_3{'} + y_2{'})/3$$

$$= 1.5607 + 0.1(6.2889 + 19.5096 + 3.7214)/3$$

$$= 2.5447.$$

Since $y_4^{(1)}$ and $y_4^{(2)}$ thus determined do not agree to four decimal places, we calculate $E = (y_4^{(2)} - y_4^{(1)})/29$. We find E = 0.0000, which is negligible to four decimal places. Thus we take the number $y_4^{(2)} = 2.5447$ as the approximate value of the solution at $x_4 = 0.4$ and denote it by y_4. Thus $y_4 = 2.5447$ is the desired result.

CHAPTER 9

Section 9.1A, Page 434

4. We have

$$\mathcal{L}\left\{f(t)\right\} = \int_0^\infty e^{-st} f(t)dt = \int_0^3 4e^{-st}dt + \int_3^\infty 2e^{-st} \, dt$$

$$= -4e^{-st}/s \, \Big|_0^3 + \lim_{R\to\infty}[-2e^{-st}/s]\Big|_3^R$$

$$= -4e^{-3s}/s + 4/s + \lim_{R\to\infty}[-2e^{-sR}/s + 2e^{-3s}/s]$$

$$= -4e^{-3s}/s + 4/s + 0 + 2e^{-3s}/s$$

$$= 2(2 - e^{-3s})/s, \text{ where } s > 0.$$

5. We have

$$\mathcal{L}\left\{f(t)\right\} = \int_0^\infty e^{-st} f(t)dt = \int_0^2 te^{-st} \, dt + \int_2^\infty 3e^{-st} \, dt$$

$$= -e^{-st}(st+1)/s^2 \, \Big|_0^2 + \lim_{R\to\infty} [-3e^{-st}/s]\Big|_2^R$$

$$= -e^{-2s}(2s+1)s^2 + 1/s^2 + \lim_{R\to\infty}[-3e^{-sR}/s + 3e^{-2s}/s]$$

$$= -e^{-2s}(2s+1)/s^2 + 1/s^2 + 0 + 3e^{-2s}/s$$

358

$$= 1/s^2 + e^{-2s}/s - e^{-2s}/s^2, \text{ where } s > 0.$$

Section 9.1B, Page 439

3. We use the identity $\sin^3 at = (3 \sin at - \sin 3at)/4$. Applying Theorem 9.2, we have

$\mathcal{L}\{\sin^3 at\} = 3 \mathcal{L}\{\sin at\}/4 - \mathcal{L}\{\sin 3at\}/4$. By (9.5), $\mathcal{L}\{\sin at\} = a/(s^2 + a^2)$ and $\mathcal{L}\{\sin 3at\} = 3a/(s^2 + 9a^2)$. Thus

$$\mathcal{L}\{\sin^3 at\} = \frac{3a}{4(s^2+a^2)} - \frac{3a}{4(s^2+9a^2)} = \frac{6a^3}{(s^2+a^2)(s^2+9a^2)}.$$

Now note that $(\sin^3 at)' = 3a \sin^2 at \cos at$. Then by Theorem 9.3, $\mathcal{L}\{f'(t)\} = s \mathcal{L}\{f(t)\} - f(0)$, so with $f(t) = \sin^3 at$ we have, $\mathcal{L}\{3a \sin^2 at \cos at\} = s \mathcal{L}\{\sin^3 at\} - 0$. Thus $\mathcal{L}\{\sin^2 at \cos at\} = s \mathcal{L}\{\sin^3 at\}/3a = \dfrac{2a^2 s}{(s^2 + a^2)(s^2 + 9a^2)}$.

6. We apply Theorem 9.4 with $n = 2$. We have $\mathcal{L}\{f''(t)\} = s^2 \mathcal{L}\{f(t)\} - sf(0) - f'(0)$. Letting $f(t) = t^4$, $f'(t) = 4t^3$, $f''(t) = 12t^2$, this becomes $\mathcal{L}\{12t^2\} = s^2 \mathcal{L}\{t^4\} - 0 - 0$, so $\mathcal{L}\{t^4\} = \mathcal{L}\{12t^2\}/s^2 = 24/s^5$.

7. We apply Theorem 9.2 to obtain $\mathcal{L}\{f''(t)\} + 3 \mathcal{L}\{f'(t)\} + 2 \mathcal{L}\{f(t)\} = \mathcal{L}\{0\}$. Then, using (9.11) and (9.13), we have

$[s^2 \mathcal{L}\{f(t)\} - sf(0) - f'(0)] + 3[s \mathcal{L}\{f(t)\} - f(0)]$
$+ 2 \mathcal{L}\{f(t)\} = 0$. Now applying the conditions $f(0) = 1$,
$f'(0) = 2$, this becomes
$s^2 \mathcal{L}\{f(t)\} - s - 2 + 3s \mathcal{L}\{f(t)\} - 3 + 2 \mathcal{L}\{f(t)\} = 0$,
or $(s^2 + 3s + 2) \mathcal{L}\{f(t)\} = s + 5$. Hence
$\mathcal{L}\{f(t)\} = (s + 5)/(s^2 + 3s + 2)$.

9. We let $f(t) = t^2$. Then, using information pro-
vided in Exercises 5 or 6, $F(s) = \mathcal{L}\{f(t)\} = \mathcal{L}\{t^2\} = 2/s^3$.
Then by Theorem 9.5, we have

$$\mathcal{L}\{e^{at} t^2\} = \mathcal{L}\{e^{at} f(t)\} = F(s-a) = 2/(s-a)^3.$$

12. We let $f(t) = \sin bt$. Then $F(s) = \mathcal{L}\{f(t)\}$
$= \mathcal{L}\{\sin bt\} = b/(s^2 + b^2)$. Then by Theorem 9.6,

$$\mathcal{L}\{t^3 \sin bt\} = (-1)^3 \frac{d^3}{ds^3} [F(s)]. \qquad (1)$$

As in Example 9.16, $\frac{d^2}{ds^2} [F(s)] = \frac{6bs^2 - 2b^3}{(s^2 + b^2)^3}$.

From this, $\frac{d^3}{ds^3} [F(s)] = \frac{24bs(b^2 - s^2)}{(s^2 + b^2)^4}$. Hence by (1),

$$\mathcal{L}\{t^3 \sin bt\} = \frac{24bs(s^2 - b^2)}{(s^2 + b^2)^4} .$$

Section 9.1C, Page 447

1. The given function is $5u_6(t)$. Using formula

(9.21), we have $\mathcal{L}\left\{5u_6(t)\right\} = 5e^{-6s}/s$.

4. We may express the values of f in the form

$$f(t) = \begin{cases} 2 - 0, & 0 < t < 5, \\ 2 - 2, & t > 5. \end{cases}$$

Thus $f(t)$ can be expressed as $2 - 2u_5(t)$. Then using formulas (9.2) and (9.21), we find $\mathcal{L}\left\{f(t)\right\} = 2\,\mathcal{L}\left\{1\right\}$ $- 2\,\mathcal{L}\left\{u_5(t)\right\} = 2/s - 2e^{-5s}/s = 2(1 - e^{-5s})/s$.

7. We may express the values of f in the form

$$f(t) = \begin{cases} 1 + 0 + 0 - 0, & 0 < t < 2, \\ 1 + 1 + 0 - 0, & 2 < t < 4, \\ 1 + 1 + 1 - 0, & 4 < t < 6, \\ 1 + 1 + 1 - 3, & t > 6. \end{cases}$$

Thus $f(t)$ can be expressed as

$$1 + u_2(t) + u_4(t) - 3u_6(t).$$

Then using formulas (9.2) and (9.21), we find

$$\mathcal{L}\left\{f(t)\right\} = \mathcal{L}\left\{1\right\} + \mathcal{L}\left\{u_2(t)\right\} + \mathcal{L}\left\{u_4(t)\right\}$$

$$- 3\,\mathcal{L}\left\{u_6(t)\right\} = 1/s + e^{-2s}/s + e^{-4s}/s - 3e^{-6s}/s$$

$$= (1 + e^{-2s} + e^{-4s} - 3e^{-6s})/s.$$

11. We must first express $f(t)$ for $t > 2$ in terms of $t - 2$. That is, we express t as $(t - 2) + 2$ and write

$$f(t) = \begin{cases} 0, & 0 < t < 2, \\ (t - 2) + 2, & t > 2 \quad. \end{cases}$$

This is the translated function defined by

$$u_2(t) \, \phi(t-2) \;=\; \begin{cases} 0, \; 0 < t < 2, \\ \phi(t-2), \; t > 2, \end{cases}$$

where $\phi(t) = t + 2$. By Theorem 9.7,

$$\mathcal{L}\{u_2(t)\phi(t-2)\} = e^{-2s} \mathcal{L}\{\phi(t)\} = e^{-2s} \mathcal{L}\{t+2\}$$
$$= e^{-2s}(1/s^2 + 2/s), \text{ where we have used (9.3) and (9.2).}$$

   14.   We express the values of f in the form

$$f(t) = \begin{cases} 2t - 0, \; 0 < t < 5 \\ 2t - 2(t-5), \; t > 5. \end{cases}$$

Thus $f(t)$ can be expressed as $f(t) = 2t - u_5(t)\phi(t-5)$,
where $\phi(t) = 2t$. By Formula (9.3), $\mathcal{L}\{2t\} = 2/s^2$. By
Theorem 9.7, $\mathcal{L}\{u_5(t)\phi(t-5)\} = e^{-5s} \mathcal{L}\{\phi(t)\}$
$= e^{-5s} \mathcal{L}\{2t\} = 2e^{-5s}/s^2$. Thus $\mathcal{L}\{f(t)\} = \mathcal{L}\{2t\}$
$- \mathcal{L}\{u_5(t)\phi(t-5)\} = 2/s^2 - 2e^{-5s}/s^2 = 2(1-e^{-5s})/s^2$.

   17.   We express the values of f in the form

$$f(t) = \begin{cases} 0 - 0, \; 0 < t < 4, \\ (t-4) - 0, \; 4 < t < 7, \\ (t-4) - (t-7), \; t > 7. \end{cases}$$

Thus $f(t)$ can be expressed as

$$f(t) = u_4(t) \, \phi(t-4) - u_7(t) \, \psi(t-7),$$

where $\phi(t) = t$ and $\psi(t) = t$. By formula (9.3),
$\mathcal{L}\{\phi(t)\} = \mathcal{L}\{\psi(t)\} = \mathcal{L}\{t\} = 1/s^2$. Then by
Theorem 9.7

$$\mathcal{L}\{f(t)\} = \mathcal{L}\{u_4(t)\phi(t-4)\} - \mathcal{L}\{u_7(t)\psi(t-7)\}$$

$$= e^{-4s}\,\mathcal{L}\{\phi(t)\} - e^{-7s}\,\mathcal{L}\{\psi(t)\}$$

$$= e^{-4s}/s^2 - e^{-7s}/s^2$$

$$= (e^{-4s} - e^{-7s})/s^2.$$

Section 9.2A, Page 453

4. We express $F(s)$ as follows :

$$F(s) = \frac{5s}{s^2+4s+4} = \frac{5(s+2)-10}{(s+2)^2} = \frac{5}{s+2} - \frac{10}{(s+2)^2}\ .$$

Now, using Table 9.1, number 2 with a = -2, and number
8 with a = -2 and n = 1, we find

$$\mathcal{L}^{-1}\{F(s)\} = 5\,\mathcal{L}^{-1}\left\{\frac{1}{s+2}\right\} - 10\,\mathcal{L}^{-1}\left\{\frac{1}{(s+2)^2}\right\}$$

$$= 5e^{-2t} - 10te^{-2t} = 5e^{-2t}(1 - 2t)\ .$$

5. We write

$$F(s) = \frac{s+2}{s^2 + 4s + 7} = \frac{s + 2}{(s+2)^2 + (\sqrt{3})^2}\ .$$

Then using Table 9.1, number 12, with a = 2, b = $\sqrt{3}$, we
have

$$\mathcal{L}^{-1}\{F(s)\} = \mathcal{L}^{-1}\left\{\frac{s + 2}{(s+2)^2 + (\sqrt{3})^2}\right\} = e^{-2t}\cos\sqrt{3}t.$$

8. We first employ partial fractions. We have

$$\frac{s+1}{s^3+2s} = \frac{A}{s} + \frac{Bs+c}{s^2+2}$$ and hence $s+1 = (A+B)s^2 + Cs + 2A.$

Thus $A+B = 0$, $C = 1$, $2A = 1$. Hence $A = 1/2$, $B = -1/2$,

$C = 1$; and we have the partial fractions decomposition

$$\frac{s+1}{s^3+2s} = \frac{1}{2s} - \frac{s}{2(s^2+2)} + \frac{1}{s^2+2} \quad ;$$

so $\mathscr{L}^{-1}\left\{\frac{s+1}{s^3+2s}\right\} = \frac{1}{2}\mathscr{L}^{-1}\left\{\frac{1}{s}\right\} - \frac{1}{2}\mathscr{L}^{-1}\left\{\frac{s}{s^2+2}\right\}$

$$+ \frac{1}{\sqrt{2}}\mathscr{L}^{-1}\left\{\frac{\sqrt{2}}{s^2+2}\right\}.$$

Then using Table 9.1, numbers 1, 4, and 3, respectively,

we find

$$\mathscr{L}^{-1}\left\{\frac{s+1}{s^2+2s}\right\} = \frac{1}{2} - \frac{1}{2}\cos\sqrt{2}t + \frac{1}{\sqrt{2}}\sin\sqrt{2}\,t.$$

10. We first employ partial fractions. We have

$$\frac{s+5}{s^4+3s^3+2s^2} = \frac{s+5}{s^2(s+1)(s+2)} = \frac{A}{s} + \frac{B}{s^2} + \frac{C}{s+1} + \frac{D}{s+2} \quad ,$$

and hence

$s+5 = As(s+1)(s+2) + B(s+1)(s+2) + Cs^2(s+2) + Ds^2(s+1)$.

Letting $s = 0$, we find $2B = 5$, so $B = 5/2$; letting

$s = -1$, we find $C = 4$; letting $s = -2$, we find $-4D = 3$,

so D = -3/4; and letting s = 1, we have 6A + 6B + 3C
+2D = 6 so A = 1 - B - C/2 - D/3 = -13/4. Thus we have
the partial fractions decomposition

$$\frac{s+5}{s^4+3s^3+2s^2} = -\frac{13}{4}(\frac{1}{s}) + \frac{5}{2}(\frac{1}{s^2}) + 4(\frac{1}{s+1}) - \frac{3}{4}(\frac{1}{s+2}) \ .$$

Then using Table 9.1, numbers 1, 7, 2, and 2
respectively, we find

$$\mathcal{L}^{-1}\left\{F(s)\right\} = -13/4 + 5t/2 + 4e^{-t} - 3e^{-2t}/4 \ .$$

13. We write

$$F(s) = \frac{2s + 7}{(s+3)^4} = \frac{2(s+3) + 1}{(s+3)^4} = \frac{2}{(s+3)^3} + \frac{1}{(s+3)^4} \quad .$$

Then

$$\mathcal{L}^{-1}\left\{F(s)\right\} = \mathcal{L}^{-1}\left\{\frac{2!}{(s+3)^3}\right\} + \frac{1}{6}\mathcal{L}^{-1}\left\{\frac{3!}{(s+3)^4}\right\} \ .$$

Using Table 9.1, number 8, with a = -3 and n = 2 and 3,
respectively, we find

$$\mathcal{L}^{-1}\left\{F(s)\right\} = t^2 e^{-3t} + t^3 e^{-3t}/6 = t^2 e^{-3t}(1 + t/6).$$

16. We employ partial fractions. We have

$$\frac{2s + 6}{8s^2 - 2s - 3} = \frac{A}{4s - 3} + \frac{B}{2s + 1}$$

and hence 2s + 6 = A(2s+1) + B(4s-3). Letting s = -1/2,
we find B = -1; and letting s = 3/4, we find A = 3.

Thus we have the partial fractions decomposition

$$\frac{2s + 6}{8s^2 - 2s - 3} = \frac{3}{4s - 3} - \frac{1}{2s + 1} = \frac{3}{4(s - 3/4)} - \frac{1}{2(s + 1/2)} \ .$$

Then, using Table 9.1, number 2, we find

$$\mathcal{L}^{-1} \left\{ F(s) \right\} = 3e^{3t/4}/4 - e^{-t/2}/2.$$

   19. F(s) is of the form $e^{-as} \Phi(s)$, where $a = \pi$
and $\Phi(s) = \dfrac{5s + 6}{s^2 + 9}$. By number 16 of Table 9.1,

$$\mathcal{L}^{-1} \left\{ e^{-as} \Phi(s) \right\} = u_a(t)\phi(t - a), \text{ where } u_a \text{ is}$$

defined by (9.32) and $\phi(t) = \mathcal{L}^{-1} \left\{ \Phi(s) \right\}$. [See
Theorem 9.7] We must find $\phi(t)$. Using numbers 4 and 3,
respectively, we find $\phi(t) = \mathcal{L}^{-1} \left\{ \Phi(s) \right\} =$

$$\mathcal{L}^{-1} \left\{ \frac{5s + 6}{s^2 + 9} \right\} = 5 \mathcal{L}^{-1} \left\{ \frac{s}{s^2 + 9} \right\} + 2 \mathcal{L}^{-1} \left\{ \frac{3}{s^2 + 9} \right\}$$

$= 5 \cos 3t + 2 \sin 3t.$ Thus

$$\phi(t - a) = \phi(t - \pi) = 5 \cos 3(t - \pi) + 2 \sin 3(t - \pi)$$

$$= -5 \cos 3t - 2 \sin 3t.$$

Then by number 16,

$$\mathcal{L}^{-1} \left\{ e^{-\pi s} \Phi(s) \right\} = u_\pi(t) \phi(t - \pi), \text{ that is,}$$

$$\mathcal{L}^{-1} \left\{ F(s) \right\} = u_\pi(t)(-5 \cos 3t - 2 \sin 3t)$$

$$= \begin{cases} 0, & 0 < t < \pi \ , \\ -5 \cos 3t - 2 \sin 3t, & t > \pi \ . \end{cases}$$

22. $F(s)$ is of the form $e^{-as} \Phi(s)$, where $a = 3$ and $\Phi(s) = \dfrac{2s + 9}{s^2 + 4s + 13}$. By number 16 of Table 9.1, $\mathcal{L}^{-1} \left\{ e^{-as} \Phi(s) \right\} = u_a(t)\phi(t-a)$, where u_a is defined by (9.32) and $\phi(t) = \mathcal{L}^{-1} \left\{ \Phi(s) \right\}$. We must find $\phi(t)$. We have

$$\frac{2s + 9}{s^2 + 4s + 13} = \frac{2(s + 2)}{(s+2)^2 + (3)^2} + \frac{5}{(s+2)^2 + (3)^2} .$$

Using numbers 12 and 11, respectively, we find

$$\phi(t) = \mathcal{L}^{-1} \left\{ \Phi(s) \right\} = \mathcal{L}^{-1} \left\{ \frac{2s + 9}{s^2 + 4s + 13} \right\}$$

$$= 2 \mathcal{L}^{-1} \left\{ \frac{s + 2}{(s+2)^2 + (3)^2} \right\} + \frac{5}{3} \mathcal{L}^{-1} \left\{ \frac{3}{(s+2)^2 + (3)^2} \right\} .$$

$$= 2e^{-2t} \cos 3t + (5/3)e^{-2t} \sin 3t. \quad \text{Thus}$$

$$\phi(t-a) = \phi(t-3) = e^{-2(t-3)}[2 \cos 3(t-3)$$

$$+ (5/3) \sin 3(t - 3)].$$

Then by number 16,

$$\mathcal{L}^{-1} \left\{ e^{-3s} \Phi(s) \right\} = u_3(t)\phi(t-3), \text{ that is,}$$

$$\mathcal{L}^{-1} \left\{ F(s) \right\} = u_3(t)\, e^{-2(t-3)}[2 \cos 3(t-3)$$

$$+ (5/3) \sin 3(t-3)]$$

$$= \begin{cases} 0, \; 0 < t < 3, \\ e^{-2(t-3)}[2 \cos 3(t-3) + (5/3) \sin 3(t-3)], \; t > 3. \end{cases}$$

27. We write $F(s) = \dfrac{2}{s^2-2s+5} + \dfrac{2e^{-(\pi s)/2}}{s^2-2s+5}$,   (1)

and first determine $\mathcal{L}^{-1}\left\{\dfrac{2}{s^2-2s+5}\right\}$. By Table 9.1,
number 11, we have

$$\mathcal{L}^{-1}\left\{\dfrac{2}{s^2-2s+5}\right\} = \mathcal{L}^{-1}\left\{\dfrac{2}{(s-1)^2+4}\right\} \qquad (2)$$

$$= e^t \sin 2t.$$

Now letting $\Phi(s) = 2/(s^2-2s+5)$ and
$\phi(t) = e^t \sin 2t$, we see that (2) is
$\mathcal{L}^{-1}\left\{\Phi(s)\right\} = \phi(t)$. Then by Table 9.1, number 16,

$$\mathcal{L}^{-1}\left\{\dfrac{2e^{-(\pi s)/2}}{s^2-2s+5}\right\} = \mathcal{L}^{-1}\left\{e^{-(\pi s)/2}\,\Phi(s)\right\}$$

$$= u_{\pi/2}(t)\phi(t-\pi/2) = \begin{cases} 0, & 0 < t < \pi/2, \\ \phi(t-\pi/2), & t > \pi/2, \end{cases}$$

$$= \begin{cases} 0, & 0 < t < \pi/2, \\ e^{t-\pi/2}\sin 2(t-\pi/2), & t > \pi/2. \end{cases}$$ That is,

$$\mathcal{L}^{-1}\left\{\dfrac{2e^{-(\pi s)/2}}{s^2-2s+5}\right\} = \begin{cases} 0, & 0 < t < \pi/2, \\ -e^{t-\pi/2}\sin 2t, & t > \pi/2. \end{cases} \qquad (3)$$

Then, using (1), (2), and (3), we find

$$\mathcal{L}^{-1}\left\{F(s)\right\}= \begin{cases} e^t \sin 2t - 0,\ 0 < t < \pi/2, \\ e^t \sin 2t - e^{t-\pi/2} \sin 2t,\ t > \pi/2, \end{cases}$$

$$= \begin{cases} e^t \sin 2t,\ 0 < t < \pi/2, \\ (1 - e^{-\pi/2})e^t \sin 2t,\ t > \pi/2. \end{cases}$$

Section 9.2B, Page 457

4. We write $1/s(s^2 + 4s + 13)$ as the product $F(s)G(s)$, where $F(s) = 1/s$ and $G(s) = 1/(s^2+4s+13)$. By Table 9.1, number 1, $f(t) = \mathcal{L}^{-1}\left\{F(s)\right\}$ $= \mathcal{L}^{-1}\left\{1/s\right\} = 1$, and by number 11,

$g(t) = \mathcal{L}^{-1}\left\{G(s)\right\} = \mathcal{L}^{-1}\left\{1/(s^2+4s+13)\right\}$

$= \frac{1}{3}\mathcal{L}^{-1}\left\{\dfrac{3}{(s+2)^2+9}\right\} = \frac{1}{3}e^{-2t}\sin 3t.$ Then

$$\mathcal{L}^{-1}\left\{H(s)\right\} = \mathcal{L}^{-1}\left\{F(s)G(s)\right\} = f(t) * g(t)$$

$$= \int_0^t f(\tau)g(t-\tau)d\tau = \int_0^t [1]\left[\frac{1}{3}e^{-2(t-\tau)}\sin 3(t-\tau)\right]d\tau$$

or

$$\mathcal{L}^{-1}\left\{H(s)\right\} = \mathcal{L}^{-1}\left\{G(s)F(s)\right\} = g(t) * f(t)$$

$$= \int_0^t g(\tau)f(t-\tau)d\tau = \int_0^t \left[\frac{1}{3}e^{-2\tau}\sin 3\tau\right][1]\,d\tau.$$

We evaluate the latter of these two integral
expressions.

$$\mathcal{L}^{-1}\left\{H(s)\right\} = \frac{1}{3}\frac{e^{-2\tau}}{13}(-2\sin 3\tau - 3\cos 3\tau)\Big|_0^t$$

$$= \frac{1}{39}[3 - e^{-2t}(2\sin 3t + 3\cos 3t)].$$

5. We write $H(s) = 1/s^2(s+3)$ as the product
$F(s)G(s)$, where $F(s) = 1/s^2$ and $G(s) = 1/(s+3)$. By
Table 9.1, number 7, $f(t) = \mathcal{L}^{-1}\left\{F(s)\right\} = \mathcal{L}^{-1}\left\{1/s^2\right\}$
$= t$, and by number 2, $g(t) = \mathcal{L}^{-1}\left\{G(s)\right\}$
$= \mathcal{L}^{-1}\left\{1/(s+3)\right\} = e^{-3t}$. Then

$$\mathcal{L}^{-1}\left\{H(s)\right\} = \mathcal{L}^{-1}\left\{F(s)G(s)\right\} = f(t) * g(t)$$

$$= \int_0^t f(\tau)g(t-\tau)d\tau = \int_0^t \tau e^{-3(t-\tau)}\, d\tau.$$

or

$$\mathcal{L}^{-1}\left\{H(s)\right\} = \mathcal{L}^{-1}\left\{G(s)F(s)\right\} = g(t) * f(t)$$

$$= \int_0^t g(\tau)f(t-\tau)d\tau = \int_0^t e^{-3\tau}(t-\tau)d\tau.$$

We evaluate the former of these two integral
expressions. We find

$$\mathcal{L}^{-1}\left\{H(s)\right\} = e^{-3t}[e^{3\tau}(3\tau-1)/9]_0^t = (-1+3t+e^{-3t})/9.$$

Section 9.3, Page 469

3. <u>Step 1.</u> Taking the Laplace Transform of both sides of the D.E., we have

$$\mathcal{L}\left\{\frac{d^2y}{dt^2}\right\} - 5\,\mathcal{L}\left\{\frac{dy}{dt}\right\} + 6\,\mathcal{L}\left\{y(t)\right\} = \mathcal{L}\left\{0\right\} \quad .(1)$$

Denoting $\mathcal{L}\left\{y(t)\right\}$ by Y(s) and applying Theorem 9.4, we have the following expressions for $\mathcal{L}\left\{\frac{d^2y}{dt^2}\right\}$ and $\mathcal{L}\left\{\frac{dy}{dt}\right\}$:

$$\mathcal{L}\left\{\frac{d^2y}{dt^2}\right\} = s^2Y(s) - sy(0) - y'(0), \quad \mathcal{L}\left\{\frac{dy}{dt}\right\} = sY(s) - y(0).$$

Applying the I.C.'s to these, they become

$$\mathcal{L}\left\{\frac{d^2y}{dt^2}\right\} = s^2Y(s) - s - 2, \quad \mathcal{L}\left\{\frac{dy}{dt}\right\} = sY(s) - 1.$$

Substituting these expressions into the left member of (1) and using $\mathcal{L}\left\{0\right\} = 0$, (1) becomes

$[s^2Y(s) - s - 2] - 5[sY(s) - 1] + 6Y(s) = 0$ or

$(s^2 - 5s + 6)Y(s) - s + 3 = 0.$

Step 2. Solving the preceeding for Y(s), we have

$$Y(s) = \frac{s - 3}{s^2 - 5s + 6} = \frac{1}{s-2} \quad .$$

Step 3. We must now determine

$$y(t) = \mathcal{L}^{-1}\left\{\frac{1}{s-2}\right\} \quad .$$

By Table 9.1, number 2, we immediately find

$$y = e^{2t} \quad .$$

7. <u>Step 1.</u> Taking the Laplace Theorem of both
sides of the D.E., we have

$$\mathcal{L}\left\{\frac{d^2y}{dt^2}\right\} - \mathcal{L}\left\{\frac{dy}{dt}\right\} - 2\,\mathcal{L}\left\{y(t)\right\} = \mathcal{L}\left\{18e^{-t}\sin 3t\right\} . \tag{1}$$

Denoting $\mathcal{L}\left\{y(t)\right\}$ by $Y(s)$ and applying Theorem 9.4,
we have the following expressions for $\mathcal{L}\left\{\dfrac{d^2y}{dt^2}\right\}$ and
$\mathcal{L}\left\{\dfrac{dy}{dt}\right\}$:

$$\mathcal{L}\left\{\frac{d^2y}{dt^2}\right\} = s^2Y(s) - sy(0) - y'(0),$$

$$\mathcal{L}\left\{\frac{dy}{dt}\right\} = sY(s) - y(0). \quad \text{Applying the I.C.'s to}$$

these, they become

$$\mathcal{L}\left\{\frac{d^2y}{dt^2}\right\} = s^2Y(s) - 3, \quad \mathcal{L}\left\{\frac{dy}{dt}\right\} = sY(s).$$

Substituting these expressions into the left member of
(1), this left member becomes
$[s^2Y(s) - 3] - sY(s) - 2Y(s)$ or $(s^2 - s - 2)Y(s) - 3$.
By Table 9.1, number 11, the right member of (1)
becomes,

$$\frac{54}{(s+1)^2 + 9} \quad .$$

Thus (1) reduces to

$$(s^2 - s - 2)Y(s) - 3 = \frac{54}{(s+1)^2 + 9} \quad .$$

Step 2. Solving the preceeding for $Y(s)$, we have

$$Y(s) = \frac{3s^2 + 6s + 84}{(s+1)(s-2)[(s+1)^2 + 9]} \quad .$$

Step 3. We must now determine

$$y(t) = \mathcal{L}^{-1} \left\{ \frac{3s^2 + 6s + 84}{(s+1)(s-2)(s^2+2s+10)} \right\} \quad .$$

We employ partial fractions. We have

$$\frac{3s^2 + 6s + 84}{(s+1)(s-2)(s^2+2s+10)} = \frac{A}{s+1} + \frac{B}{s-2} + \frac{Cs + D}{s^2+2s+10} \qquad \text{or}$$

$$3s^2 + 6s + 84 = A(s - 2)(s^2 + 2s + 10) \qquad (2)$$
$$+ B(s + 1)(s^2 + 2s + 10) + (Cs + D)(s+1)(s - 2)$$

or

$$3s^2 + 6s + 84 = (A + B + C)s^3 + (3B - C + D)s^2$$
$$+ (6A + 12B - 2C - D)s + (-20A + 10B - 2D).$$

From this, we obtain

$$\begin{cases} A + B + C = 0, \quad 3B - C + D = 3, \\ 6A + 12B - 2C - D = 6, \quad -20A + 10B - 2D = 84. \end{cases} \qquad (3)$$

Letting $s = -1$ in (2), we find $-27A = 81$, so $A = -3$; and letting $s = 2$ in (2), we find $54B = 108$, so $B = 2$.

Using the values for A and B, we find from (3), that
C = 1, D = -2. Thus we have

$$\mathcal{L}^{-1}\left\{\frac{3s^2 + 6s + 84}{(s+1)(s-2)(s^2+2s+10)}\right\} = -3\,\mathcal{L}^{-1}\left\{\frac{1}{s+1}\right\}$$

$$+ 2\,\mathcal{L}^{-1}\left\{\frac{1}{s-2}\right\} + \mathcal{L}^{-1}\left\{\frac{s-2}{s^2+2s+10}\right\}$$

$$= -3\,\mathcal{L}^{-1}\left\{\frac{1}{s+1}\right\} + 2\,\mathcal{L}^{-1}\left\{\frac{1}{s-2}\right\}$$

$$+ \mathcal{L}^{-1}\left\{\frac{s+1}{(s+1)^2 + 9}\right\} - \mathcal{L}^{-1}\left\{\frac{3}{(s+1)^2 + 9}\right\}\,.$$

Then by Table 9.1, numbers 2, 2, 12, and 11,
respectively, we find
$$y = -3e^{-t} + 2e^{2t} + e^{-t}\cos 3t - e^{-t}\sin 3t\,.$$

11. <u>Step 1.</u> Taking the Laplace Transform of both
sides of the D.E., we have

$$\mathcal{L}\left\{\frac{d^3y}{dt^3}\right\} - 5\,\mathcal{L}\left\{\frac{d^2y}{dt^2}\right\} + 7\,\mathcal{L}\left\{\frac{dy}{dt}\right\} - 3\,\mathcal{L}\{y(t)\}$$

$$\hspace{8cm}(1)$$

$$= \mathcal{L}\{20\sin t\}\,.$$

Denoting $\mathcal{L}\{y(t)\}$ by $Y(s)$ and applying Theorem 9.4,
we have the following expressions for $\mathcal{L}\left\{\frac{d^3y}{dt^3}\right\}$,

$$\mathcal{L}\left\{\frac{d^2y}{dt^2}\right\}\,,\text{ and }\mathcal{L}\left\{\frac{dy}{dt}\right\}:$$

$$\begin{cases} \mathcal{L}\left\{\dfrac{d^3y}{dt^3}\right\} = s^3\, Y(s) - s^2 y(0) - sy'(0) - y''(0), \\[4mm] \mathcal{L}\left\{\dfrac{d^2y}{dt^2}\right\} = s^2\, Y(s) - sy(0) - y'(0), \\[4mm] \mathcal{L}\left\{\dfrac{dy}{dt}\right\} = sY(s) - y(0). \end{cases}$$

Applying the I.C.'s to these, they become

$$\mathcal{L}\left\{\dfrac{d^3y}{dt^3}\right\} = s^3 Y(s)+2, \quad \mathcal{L}\left\{\dfrac{d^2y}{dt^2}\right\} = s^2 Y(s),$$

$$\mathcal{L}\left\{\dfrac{dy}{dt}\right\} = sY(s).$$

Substituting these expressions into the left member of (1), this left member becomes

$$s^3 Y(s) + 2 - 5s^2 Y(s) + 7sY(s) - 3Y(s)$$

or $\qquad (s^3 - 5s^2 + 7s - 3)\, Y(s) + 2.$

By Table 9.1, number 3, the right member of (1) becomes $20/(s^2 + 1)$. Thus (1) reduces to

$$(s^3 - 5s^2 + 7s - 3)\, Y(s) + 2 = 20/(s^2 + 1).$$

Step 2. Solving the preceeding for $Y(s)$, we have

$$Y(s) = \frac{18 - 2s^2}{(s^3 - 5s^2 + 7s - 3)(s^2 + 1)}$$

or

$$Y(s) = \frac{-2(s^2 - 9)}{(s - 1)^2(s - 3)(s^2 + 1)}$$

or finally

$$Y(s) = \frac{-2(s + 3)}{(s - 1)^2(s^2 + 1)} \quad .$$

Step 3. We must now determine

$$y(t) = \mathcal{L}^{-1}\left\{\frac{-2s - 6}{(s-1)^2(s^2+1)}\right\} \quad .$$

We employ partial fractions. We have

$$\frac{-2s - 6}{(s-1)^2(s^2+1)} = \frac{A}{s-1} + \frac{B}{(s-1)^2} + \frac{Cs+D}{s^2+1}$$

or $-2s-6 = A(s-1)(s^2+1) + B(s^2+1) + (Cs+D)(s-1)^2$

or $-2s-6 = (A+C)s^3 + (-A+B-2C+D)s^2$
$$+ (A+C-2D)s + (-A+B+D) \quad .$$

From this, we obtain

$$\begin{cases} A + C = 0, & -A + B - 2C + D = 0, \\ A + C - 2D = -2, & -A + B + D = -6. \end{cases}$$

The first and third of these given $C = -A$, $D = 1$. Then
the second and fourth reduce to $A + B = -1$, $-A + B = -7$,
respectively, from which $A = 3$, $B = -4$. Hence $A = 3$,
$B = -4$, $C = -3$, $D = 1$; and we have

$$\mathcal{L}^{-1}\left\{\frac{-2s-6}{(s-1)^2(s^2+1)}\right\} = 3\,\mathcal{L}^{-1}\left\{\frac{1}{s-1}\right\} - 4\,\mathcal{L}^{-1}\left\{\frac{1}{(s-1)^2}\right\}$$

$$- 3\,\mathcal{L}^{-1}\left\{\frac{s}{s^2+1}\right\} + \mathcal{L}^{-1}\left\{\frac{1}{s^2+1}\right\} \ .$$

Then by Table 9.1, numbers 2, 8, 4, and 3, respectively, we find

$$y = 3e^t - 4te^t - 3\cos t + \sin t \ .$$

13. <u>Step 1</u>. Taking the Laplace Transform of both sides of the D.E., we have

$$\mathcal{L}\left\{\frac{d^2y}{dt^2}\right\} - 3\,\mathcal{L}\left\{\frac{dy}{dt}\right\} + 2\,\mathcal{L}\left\{y(t)\right\} = \mathcal{L}\left\{h(t)\right\}, \tag{1}$$

Denoting $\mathcal{L}\left\{y(t)\right\}$ by $Y(s)$, applying Theorem 9.4 as in the previous exercises, and then applying the I.C.'s we find that

$$\mathcal{L}\left\{\frac{d^2y}{dt^2}\right\} = s^2 Y(s) \quad \text{and} \quad \mathcal{L}\left\{\frac{dy}{dt}\right\} = sY(s).$$

Substituting these expressions into the left member of (1), this left member becomes

$$s^2 Y(s) - 3sY(s) + 2Y(s)$$

or $(s^2 - 3s + 2)Y(s)$. By the definition of the Laplace Transform,

$$\mathcal{L}\left\{h(t)\right\} = \int_0^\infty e^{-st}h(t)dt = \int_0^4 2e^{-st}dt = \frac{2}{s}(1 - e^{-4s}) \ .$$

Thus (1) reduces to $(s^2 - 3s + 2)Y(s) = \frac{2}{s}(1 - e^{-4s})$.

     Step 2. Solving the preceeding for Y(s), we obtain

$$Y(s) = \frac{2(1 - e^{-4s})}{s(s - 1)(s - 2)} \qquad .$$

     Step 3. We must now determine

$$y(t) = \mathcal{L}^{-1}\left\{\frac{2}{s(s-1)(s-2)}\right\} - \mathcal{L}^{-1}\left\{\frac{2e^{-4s}}{s(s-1)(s-2)}\right\}. \tag{2}$$

We first apply partial fractions to the first term on the right of (2). We have

$$\frac{2}{s(s-1)(s-2)} = \frac{A}{s} + \frac{B}{s-1} + \frac{C}{s-2} \qquad ,$$

and from this we readily find A = 1, B = -2, C = 1. Thus

$$\mathcal{L}^{-1}\left\{\frac{2}{s(s-1)(s-2)}\right\} = \mathcal{L}^{-1}\left\{\frac{1}{s}\right\} - 2\mathcal{L}^{-1}\left\{\frac{1}{s-1}\right\}$$

$$+ \mathcal{L}^{-1}\left\{\frac{1}{s-2}\right\} \qquad .$$

By Table 9.1, numbers 1, 2, and 2 respectively, we find

$$\mathcal{L}^{-1}\left\{\frac{2}{s(s-1)(s-2)}\right\} = 1 - 2e^t + e^{2t} \qquad .$$

Letting $F(s) = \dfrac{2}{s(s-1)(s-2)}$ and $f(t) = 1 - 2e^t + e^{2t}$,

we thus have $\mathcal{L}^{-1}\left\{F(s)\right\} = f(t)$. We now consider

$$\mathcal{L}^{-1}\left\{\frac{2e^{-4s}}{s(s-1)(s-2)}\right\} = \mathcal{L}^{-1}\left\{F(s)e^{-4s}\right\} \ . \ \text{By}$$

Theorem 9.7, $\mathcal{L}^{-1}\left\{F(s)e^{-4s}\right\} = u_4(t)f(t-4)$,

$$= \begin{cases} 0, & 0 < t < 4, \\ f(t-4), & t > 4, \end{cases} \quad = \begin{cases} 0, & 0 < t < 4, \\ 1 - 2e^{t-4} + 2e^{2(t-4)}, & t > 4. \end{cases}$$

Thus from (2), we find $y = f(t) - u_4(t)f(t-4) =$

$$\begin{cases} 1 - 2e^t + e^{2t}, & 0 < t < 4, \\ 1 - 2e^t + e^{2t} - [1 - 2e^{t-4} + e^{2(t-4)}], & t > 4. \end{cases}$$

Hence,

$$y = \begin{cases} 1 - 2e^t + e^{2t}, & 0 < t < 4, \\ 2(e^{-4} - 1)e^t + (1 - e^{-8})e^{2t}, & t > 4. \end{cases}$$

16. <u>Step 1.</u> Taking the Laplace Transform of both
sides of the D.E., we have

$$\mathcal{L}\left\{\frac{d^2y}{dt^2}\right\} + 6\,\mathcal{L}\left\{\frac{dy}{dt}\right\} + 8\,\mathcal{L}\left\{y(t)\right\} = \mathcal{L}\left\{h(t)\right\} \ . \quad (1)$$

Denoting $\mathcal{L}\left\{y(t)\right\}$ by $Y(s)$, applying Theorem 9.4 as
in previous exercises, and then applying the I.C.'s,
we find that

$$\mathcal{L}\left\{\frac{d^2y}{dt^2}\right\} = s^2Y(s) - s + 1, \quad \mathcal{L}\left\{\frac{dy}{dt}\right\} = sY(s) - 1.$$

Substituting these expressions into the left member of (1), this left member becomes

$[s^2Y(s)-s+1] + 6[sY(s)-1] + 8Y(s)$ or

$(s^2 + 6s + 8)Y(s) - s - 5$. By the definition of the Laplace Transform,

$$\mathcal{L}\left\{h(t)\right\} = \int_0^\infty e^{-st}h(t)dt = \int_0^{2\pi} 3e^{-st}dt$$

$$= \frac{3}{s}(1 - e^{-2\pi s}).$$

Thus (1) reduces to $(s^2+6s+8)Y(s) - s - 5 = \frac{3}{s}(1-e^{-2\pi s})$.

     <u>Step 2.</u> Solving the preceeding for $Y(s)$, we obtain

$$Y(s) = \frac{s + 5}{s^2 + 6s + 8} + \frac{3(1 - e^{-2\pi s})}{s(s^2 + 6s + 8)} \quad .$$

     <u>Step 3.</u> We must now determine

$$y = \mathcal{L}^{-1}\left\{\frac{s + 5}{(s+2)(s+4)}\right\} + \mathcal{L}^{-1}\left\{\frac{3(1 - e^{-2\pi s})}{s(s+2)(s+4)}\right\}. \quad (4)$$

We first apply partial fractions to the first term on the right of (2). We have

$$\frac{s+5}{(s+2)(s+4)} = \frac{A}{s+2} + \frac{B}{s+4} \quad ,$$

and from this we readily find A = 3/2, B = -1/2. Thus

$$\mathcal{L}^{-1}\left\{\frac{s+5}{(s+2)(s+4)}\right\} = \frac{3}{2}\mathcal{L}^{-1}\left\{\frac{1}{s+2}\right\} - \frac{1}{2}\mathcal{L}^{-1}\left\{\frac{1}{s+4}\right\}.$$

By Table 9.1, number 2, we find

$$\mathcal{L}^{-1}\left\{\frac{s+5}{(s+2)(s+4)}\right\} = \frac{3}{2}e^{-2t} - \frac{1}{2}e^{-4t}. \qquad (3)$$

We now consider $\mathcal{L}^{-1}\left\{\frac{3(1-e^{-2\pi s})}{s(s+2)(s+4)}\right\}$.

We first apply partial fractions to $3/s(s+2)(s+4)$. We have

$$\frac{3}{s(s+2)(s+4)} = \frac{A}{s} + \frac{B}{s+2} + \frac{C}{s+4} \quad ,$$

and from this we readily find A = 3/8, B = -3/4, C = 3/8. Thus

$$\mathcal{L}^{-1}\left\{\frac{3}{s(s+2)(s+4)}\right\} = \frac{3}{8}\mathcal{L}^{-1}\left\{\frac{1}{s}\right\} - \frac{3}{4}\mathcal{L}^{-1}\left\{\frac{1}{s+2}\right\} + \frac{3}{8}\mathcal{L}^{-1}\left\{\frac{1}{s+4}\right\}.$$

By Table 9.1, numbers 1, 2, and 2, respectively, we find

$$\mathcal{L}^{-1}\left\{\frac{3}{s(s+2)(s+4)}\right\} = \frac{3}{8} - \frac{3}{4}e^{-2t} + \frac{3}{8}e^{-4t}.$$

Letting $F(s) = \dfrac{3}{s(s+2)(s+4)}$ and $f(t) = \dfrac{3}{8} - \dfrac{3}{4}e^{-2t} + \dfrac{3}{8}e^{-4t}$, we thus have $\mathcal{L}^{-1}\left\{F(s)\right\} = f(t)$. We now consider $\mathcal{L}^{-1}\left\{\dfrac{3e^{-2\pi s}}{s(s+2)(s+4)}\right\} = \mathcal{L}^{-1}\left\{F(s)e^{-2\pi s}\right\}$.

By Theorem 9.7,

$$\mathcal{L}^{-1}\left\{F(s)e^{-2\pi s}\right\} = u_{2\pi}(t)f(t-2\pi) = \begin{cases} 0, \ 0 < t < 2\pi, \\ f(t-2\pi), \ t > 2\pi, \end{cases}$$

$$= \begin{cases} 0, \ 0 < t < 2\pi, \\ \frac{3}{8} - \frac{3}{4}e^{-2(t-2\pi)} + \frac{3}{8}e^{-4(t-2\pi)}, \ t > 2\pi. \end{cases}$$

Thus

$$\mathcal{L}^{-1}\left\{\frac{3(1-e^{-2\pi s})}{s(s+2)(s+4)}\right\}$$

$$= \begin{cases} \frac{3}{8} - \frac{3}{4}e^{-2t} + \frac{3}{8}e^{-4t}, \ 0 < t < 2\pi, \\ \frac{3}{8} - \frac{3}{4}e^{-2t} + \frac{3}{8}e^{-4t} - \frac{3}{8} + \frac{3}{4}e^{-2(t-2\pi)} - \frac{3}{8}e^{-4(t-2\pi)}, \end{cases}$$

$$t > 2\pi,$$

$$= \begin{cases} \frac{3}{8} - \frac{3}{4}e^{-2t} + \frac{3}{8}e^{-4t}, \ 0 < t < 2\pi, \\ \frac{3}{4}(e^{4\pi}-1)e^{-2t} + \frac{3}{8}(1-e^{8\pi})e^{-4t}, \ t > 2\pi. \end{cases} \qquad (4)$$

Hence, using (3) and (4), (2) becomes

$$y = \begin{cases} \frac{3}{8} + \frac{3}{4}e^{-2t} - \frac{1}{8}e^{-4t}, \ 0 < t < 2\pi, \\ \frac{3}{4}(e^{4\pi}+1)e^{-2t} - \frac{1}{8}(1+3e^{8\pi})e^{-4t}, \ t > 2\pi. \end{cases}$$

18. <u>Step 1.</u> Taking the Laplace Transform of both sides of the D.E., we have

$$\mathcal{L}\left\{\frac{d^2y}{dt^2}\right\} + \mathcal{L}\left\{y(t)\right\} = \mathcal{L}\left\{h(t)\right\} . \qquad (1)$$

Denoting $\mathcal{L}\left\{y(t)\right\}$ by $Y(s)$, applying Theorem 9.4 as in previous exercises, and then applying the I.C.'s, we find that

$$\mathcal{L}\left\{\frac{d^2y}{dt^2}\right\} = s^2Y(s) - 2s - 3 .$$

Substituting this expression into the left member of (1), this left member becomes $(s^2+1)Y(s) - 2s - 3$. We now find $\mathcal{L}\left\{h(t)\right\}$. We have

$$h(t) = \begin{cases} t, & 0 < t < \pi, \\ \pi, & t > \pi, \end{cases} = \begin{cases} t - 0, & 0 < t < \pi, \\ t - (t-\pi), & t > \pi, \end{cases}$$

and hence $h(t) = t - u_\pi(t)f(t-\pi)$, where $f(t) = t$. Thus, using Theorem 9.7 and Table 9.1, number 7, we find $\mathcal{L}\left\{h(t)\right\} = \mathcal{L}\left\{t\right\} - \mathcal{L}\left\{u_\pi(t)f(t-\pi)\right\}$

$$= \frac{1}{s^2} - e^{-\pi s}\left(\frac{1}{s^2}\right) = \frac{1 - e^{-\pi s}}{s^2} \qquad . \text{ Thus (1) reduces to}$$

$$(s^2+1)Y(s) - 2s - 3 = \frac{1-e^{-\pi s}}{s^2} .$$

      Step 2. Solving the preceeding for $Y(s)$, we obtain

$$Y(s) = \frac{2s+3}{s^2+1} + \frac{1 - e^{-\pi s}}{s^2(s^2+1)} .$$

      Step 3. We must now determine

$$y = \mathcal{L}^{-1}\left\{\frac{2s+3}{s^2+1}\right\} + \mathcal{L}^{-1}\left\{\frac{1}{s^2(s^2+1)}\right\}$$

$$- \mathcal{L}^{-1}\left\{\frac{e^{-\pi s}}{s^2(s^2+1)}\right\} . \qquad (2)$$

Using Table 9.1, numbers 4 and 3, we at once find

$$\mathcal{L}^{-1}\left\{\frac{2s+3}{s^2+1}\right\} = 2\cos t + 3\sin t. \qquad (3)$$

We now apply partial fractions to the second term in
the right member of (2). We have

$$\frac{1}{s^2(s^2+1)} = \frac{A}{s} + \frac{B}{s^2} + \frac{Cs+D}{s^2+1}$$

and from this we readily find A = 0, B = 1, C = 0,
D = -1. Thus

$$\mathcal{L}^{-1}\left\{\frac{1}{s^2(s^2+1)}\right\} = \mathcal{L}^{-1}\left\{\frac{1}{s^2}\right\} - \mathcal{L}^{-1}\left\{\frac{1}{s^2+1}\right\}.$$

Then using Table 9.1, numbers 7 and 3, we find

$$\mathcal{L}^{-1}\left\{\frac{1}{s^2(s^2+1)}\right\} = t - \sin t. \qquad (4).$$

Now letting $F(s) = \dfrac{1}{s^2(s^2+1)}$ and $f(t) = t - \sin t$,

(4) states that $\mathcal{L}^{-1}\left\{F(s)\right\} = f(t)$. Using Theorem
9.7, we now find

$$\mathcal{L}^{-1}\left\{\frac{e^{-\pi s}}{s^2(s^2+1)}\right\} = \mathcal{L}^{-1}\left\{F(s)e^{-\pi s}\right\} = u_\pi(t)f(t-\pi)$$

$$= \begin{cases} 0, & 0 < t < \pi, \\ f(t-\pi), & t > \pi, \end{cases} = \begin{cases} 0, & 0 < t < \pi, \\ t - \pi - \sin(t-\pi), & t > \pi, \end{cases}$$

That is

$$\mathcal{L}^{-1}\left\{\frac{e^{-\pi s}}{s^2(s^2+1)}\right\} = \begin{cases} 0, & 0 < t < \pi, \\ t - \pi + \sin t, & t > \pi, \end{cases} \qquad (5)$$

Thus, using (3), (4), and (5), (2) becomes

$$y = \begin{cases} (2\cos t + 3\sin t)+(t-\sin t)-0, \ 0 < t < \pi, \\ (2\cos t + 3\sin t)+(t-\sin t)-(t-\pi+\sin t), t > \pi \end{cases}$$

$$= \begin{cases} 2\sin t + 2\cos t + t, \ 0 < t < \pi, \\ \sin t + 2\cos t + \pi, \ t > \pi. \end{cases}$$

Section 9.4, Page 473

3. **Step 1.** Taking the Laplace Transform of both sides of each D.E. of the system, we have

$$\begin{cases} \mathcal{L}\left\{\frac{dx}{dt}\right\} - 5\,\mathcal{L}\left\{x(t)\right\} + 2\,\mathcal{L}\left\{y(t)\right\} = 3\,\mathcal{L}\left\{e^{4t}\right\}, \\ \mathcal{L}\left\{\frac{dy}{dt}\right\} - 4\,\mathcal{L}\left\{x(t)\right\} + \mathcal{L}\left\{y(t)\right\} = \mathcal{L}\left\{0\right\}. \end{cases} \quad (1)$$

We denote $\mathcal{L}\left\{x(t)\right\}$ by $X(s)$ and $\mathcal{L}\left\{y(t)\right\}$ by $Y(s)$. Applying Theorem 9.3 and the given I.C.'s, we express $\mathcal{L}\left\{\frac{dx}{dt}\right\}$ and $\mathcal{L}\left\{\frac{dy}{dt}\right\}$ in terms of $X(s)$ and $Y(s)$, respectively, as follows:

$$\begin{cases} \mathcal{L}\left\{\frac{dx}{dt}\right\} = s\,X(s) - x(0) = sX(s) - 3, \\ \mathcal{L}\left\{\frac{dy}{dt}\right\} = s\,Y(s) - y(0) = sY(s). \end{cases} \quad (2)$$

By Table 9.1, number 2, $\mathcal{L}\left\{e^{4t}\right\} = \frac{1}{s-4}$. Thus, from this and (2), we see that (1) becomes

$$\begin{cases} s\,X(s) - 3 - 5\,X(s) + 2Y(s) = \frac{3}{s-4}, \\ s\,Y(s) - 4X(s) + Y(s) = 0, \end{cases}$$

or
$$\begin{cases} (s - 5) \ X(s) + 2Y(s) = \dfrac{3s - 9}{s - 4} & , \\[2mm] -4 \ X(s) + (s+1) \ Y(s) = 0. \end{cases}$$

Step 2. We solve this system for the two
unknowns $X(s)$ and $Y(s)$. We have

$$\begin{cases} (s+1)(s-5) \ X(s) + 2(s+1) \ Y(s) = \dfrac{(3s-9)(s+1)}{s - 4} & , \\[2mm] -8 \ X(s) + 2(s+1) \ Y(s) = 0. \end{cases}$$

Subtracting, we obtain

$$(s^2 - 4s + 3) \ X(s) = \frac{(3s - 9)(s + 1)}{s - 4} \quad ,$$

from which we find

$$X(s) = \frac{3(s-3)(s+1)}{(s-1)(s-3)(s-4)} = \frac{3(s + 1)}{(s-1)(s-4)} \quad .$$

In like manner, we find

$$Y(s) = \frac{12}{(s-1)(s-4)} \quad .$$

Step 3. We must now determine

$$x(t) = \mathcal{L}^{-1}\left\{ X(s) \right\} = \mathcal{L}^{-1}\left\{ \frac{3(s+1)}{(s-1)(s-4)} \right\} \quad .$$

and

$$y(t) = \mathcal{L}^{-1}\left\{ Y(s) \right\} = \mathcal{L}^{-1}\left\{ \frac{12}{(s-1)(s-4)} \right\} \quad .$$

We first find $x(t)$. We employ partial fractions.
We have

$$\frac{3(s+1)}{(s-1)(s-4)} = \frac{A}{s-1} + \frac{B}{s-4} \quad .$$

from which we readily find A = -2, B = 5. Thus

$$x(t) = -2 \mathcal{L}^{-1}\left\{\frac{1}{s-1}\right\} + 5 \mathcal{L}^{-1}\left\{\frac{1}{s-4}\right\} \quad ,$$

and using Table 9.1, number 2, we obtain

$$x(t) = -2e^t + 5e^{4t} \quad .$$

Similarly, we find y(t). Again employing partial fractions, we obtain

$$y(t) = -4 \mathcal{L}^{-1}\left\{\frac{1}{s-1}\right\} + 4 \mathcal{L}^{-1}\left\{\frac{1}{s-4}\right\} \quad .$$

Again using Table 9.1, number 2, we obtain

$$y(t) = -4e^t + 4e^{4t} \quad .$$

7. <u>Step 1</u>. Taking the Laplace Transform of both sides of each D.E. of the system, we have

$$\begin{cases} 2\mathcal{L}\left\{\frac{dx}{dt}\right\} + \mathcal{L}\left\{\frac{dy}{dt}\right\} - \mathcal{L}\left\{x(t)\right\} - \mathcal{L}\left\{y(t)\right\} = \mathcal{L}\left\{e^{-t}\right\} \\ \qquad\qquad\qquad\qquad\qquad\qquad\qquad\qquad\qquad (1) \\ \mathcal{L}\left\{\frac{dx}{dt}\right\} + \mathcal{L}\left\{\frac{dy}{dt}\right\} + 2\mathcal{L}\left\{x(t)\right\} + \mathcal{L}\left\{y(t)\right\} = \mathcal{L}\left\{e^t\right\}. \end{cases}$$

We denote $\mathcal{L}\left\{x(t)\right\}$ by X(s) and $\mathcal{L}\left\{y(t)\right\}$ by Y(s). Applying Theorem 9.3 and the given I.C.'s, we express $\mathcal{L}\left\{\frac{dx}{dt}\right\}$ and $\mathcal{L}\left\{\frac{dy}{dt}\right\}$ in terms of X(s) and Y(s), respectively, as follows:

$$\mathcal{L}\left\{\frac{dx}{dt}\right\} = s\,X(s) - x(0) = sX(s) - 2,$$

$$\mathcal{L}\left\{\frac{dy}{dt}\right\} = s\,Y(s) - y(0) = sY(s) - 1.$$

(2)

By Table 9.1, number 2, $\mathcal{L}\left\{e^{-t}\right\} = 1/(s+1)$,

$\mathcal{L}\left\{e^{t}\right\} = 1/(s-1)$. Thus, from this and (2), we see that (1) becomes

$$\begin{cases} 2sX(s) - 4 + sY(s) - 1 - X(s) - Y(s) = 1/(s+1), \\ sX(s) - 2 + sY(s) - 1 + 2X(s) + Y(s) = 1/(s-1), \end{cases}$$

or

$$\begin{cases} (2s-1)\,X(s) + (s-1)\,Y(s) = \dfrac{5s+6}{s+1}, \\ (s+2)\,X(s) + (s+1)\,Y(s) = \dfrac{3s-2}{s-1}. \end{cases}$$

Step 2. We solve this system for the two unknowns $X(s)$ and $Y(s)$. We have

$$\begin{cases} (s+1)(2s-1)X(s) + (s+1)(s-1)Y(s) = 5s + 6, \\ (s-1)(s+2)X(s) + (s+1)(s-1)Y(s) = 3s - 2. \end{cases}$$

Subtracting, we obtain $(s^2 + 1)\,X(s) = 2s + 8$, from which we find

$$X(s) = \frac{2s + 8}{s^2 + 1} .$$

In like manner, we find

$$Y(s) = \frac{s^3 - 12s^2 - s + 14}{(s+1)(s-1)(s^2+1)} .$$

Step 3. We must now determine

$$x(t) = \mathcal{L}^{-1}\left\{X(s)\right\} = \mathcal{L}^{-1}\left\{\frac{2s + 8}{s^2 + 1}\right\}$$

and

$$y(t) = \mathcal{L}^{-1}\left\{Y(s)\right\} = \mathcal{L}^{-1}\left\{\frac{s^3 - 12s^2 - s + 14}{(s+1)(s-1)(s^2+1)}\right\} .$$

We first obtain x(t). Using Table 9.1, numbers 4 and 3, we find

$$x(t) = 2\mathcal{L}^{-1}\left\{\frac{s}{s^2+1}\right\} + 8\mathcal{L}^{-1}\left\{\frac{1}{s^2+1}\right\}$$

$$= 2 \cos t + 8 \sin t.$$

We proceed to find y(t). We first employ partial fractions. We have

$$\frac{s^3 - 12s^2 - s + 14}{(s+1)(s-1)(s^2+1)} = \frac{A}{s-1} + \frac{B}{s+1} + \frac{Cs+D}{s^2+1}$$

or

$$s^3 - 12s^2 - s + 14 = A(s+1)(s^2+1) + B(s-1)(s^2+1)$$

$$+ (Cs+D)(s-1)(s+1). \tag{3}$$

or

$$s^3 - 12s^2 - s + 14 = (A+B+C)s^3 + (A-B+D)s^2$$

$$+ (A+B-C)s + (A-B-D).$$

From this, we have

$$\begin{cases} A + B + C = 1, & A - B + D = -12, \\ A + B - C = -1, & A - B - D = 14. \end{cases} \tag{4}$$

Letting s = 1 in (3), we find A = 1/2; and letting

$s = -1$ in (3), we find $B = -1/2$. Using these values and (4), we find $C = 1$, $D = -13$. Thus we have

$$y(t) = \frac{1}{2} \mathcal{L}^{-1} \left\{ \frac{1}{s-1} \right\} - \frac{1}{2} \mathcal{L}^{-1} \left\{ \frac{1}{s+1} \right\}$$

$$+ \mathcal{L}^{-1} \left\{ \frac{s}{s^2+1} \right\} - 13 \mathcal{L}^{-1} \left\{ \frac{1}{s^2+1} \right\} .$$

Using Table 9.1, numbers 2, 2, 4, and 3, respectively, we find

$$y(t) = \frac{1}{2} e^t - \frac{1}{2} e^{-t} + \cos t - 13 \sin t.$$

10. **Step 1.** Taking the Laplace Transform of both sides of each D.E. of the system, we have

$$\begin{cases} \mathcal{L}\left\{\frac{d^2x}{dt^2}\right\} - 3 \mathcal{L}\left\{\frac{dx}{dt}\right\} + \mathcal{L}\left\{\frac{dy}{dt}\right\} + 2 \mathcal{L}\left\{x(t)\right\} - \mathcal{L}\left\{y(t)\right\} = 0, \\[2mm] \mathcal{L}\left\{\frac{dx}{dt}\right\} + \mathcal{L}\left\{\frac{dy}{dt}\right\} - 2 \mathcal{L}\left\{x(t)\right\} + \mathcal{L}\left\{y(t)\right\} = 0. \end{cases} \quad (1)$$

We denote $\mathcal{L}\left\{x(t)\right\}$ by $X(s)$ and $\mathcal{L}\left\{y(t)\right\}$ by $Y(s)$. Applying Theorem 9.4 and the given I.C.'s, we express $\mathcal{L}\left\{\frac{d^2x}{dt^2}\right\}$ and $\mathcal{L}\left\{\frac{dx}{dt}\right\}$ in terms of $X(s)$ and $\mathcal{L}\left\{\frac{dy}{dt}\right\}$ in terms of $Y(s)$ as follows:

$$\mathcal{L}\left\{\frac{d^2x}{dt^2}\right\} = s^2 X(s), \quad \mathcal{L}\left\{\frac{dx}{dt}\right\} = s X(s), \quad \mathcal{L}\left\{\frac{dy}{dt}\right\} = s Y(s) + 1.$$

Substituting these expressions into (1), we see that (1) becomes

$$\begin{cases} s^2X(s) - 3sX(s) + sY(s) + 1 + 2X(s) - Y(s) = 0, \\ sX(s) + sY(s) + 1 - 2X(s) + Y(s) = 0, \end{cases}$$

or

$$\begin{cases} (s^2 - 3s + 2)X(s) + (s - 1)Y(s) = -1, \\ (s-2)X(s) + (s + 1)Y(s) = -1. \end{cases}$$

Step 2. We solve this system for the two unknowns $X(s)$ and $Y(s)$. We have

$$\begin{cases} (s^2 - 3s + 2)(s+1)X(s) + (s-1)(s+1)Y(s) = -(s+1), \\ (s-2)(s-1)X(s) + (s+1)(s-1)Y(s) = -(s-1). \end{cases}$$

Subtracting, we obtain $(s^3 - 3s^2 + 2s)X(s) = -2$ from which we find

$$X(s) = -\frac{2}{s(s-1)(s-2)} \quad .$$

In like manner, we find

$$Y(s) = -\frac{s-2}{s(s-1)} \quad .$$

Step 3. We must now determine

$$x(t) = \mathcal{L}^{-1}\left\{X(s)\right\} = \mathcal{L}^{-1}\left\{\frac{-2}{s(s-1)(s-2)}\right\} \quad , \text{ and}$$

$$y(t) = \mathcal{L}^{-1}\left\{Y(s)\right\} = \mathcal{L}^{-1}\left\{\frac{-s+2}{s(s-1)}\right\} \quad .$$

We first find $x(t)$, using partial fractions. We have

$$\frac{-2}{s(s-1)(s-2)} = \frac{A}{s} + \frac{B}{s-1} + \frac{C}{s-2}$$

or $-2 = A(s-1)(s-2) + Bs(s-2) + Cs(s-1)$. Letting
$s = 0$, we find $A = -1$; letting $s = 1$, we find $B = 2$;
and letting $s = 2$, we find $C = -1$. Thus we have

$$x(t) = -\mathcal{L}^{-1}\left\{\tfrac{1}{s}\right\} + 2\,\mathcal{L}^{-1}\left\{\tfrac{1}{s-1}\right\} - \mathcal{L}^{-1}\left\{\tfrac{1}{s-2}\right\}$$

Now using Table 9.1, numbers 1, 2, and 2, respectively,
we find $x(t) = -1 + 2e^{t} - e^{2t}$.

  We now find $y(t)$, again using partial fractions.
We have

$$\frac{-s+2}{s(s-1)} = \frac{A}{s} + \frac{B}{s-1} \qquad .$$

We readily find $A = -2$, $B = 1$; and hence

$$y(t) = -2\,\mathcal{L}^{-1}\left\{\tfrac{1}{s}\right\} + \mathcal{L}^{-1}\left\{\tfrac{1}{s-1}\right\} \qquad .$$

Then from Table 9.1, numbers 1 and 2, we find

$$y(t) = -2 + e^{t} \qquad .$$